T0174206

Ecology, Uncertainty and Policy

Ecology, Uncertainty and Policy

MANAGING ECOSYSTEMS FOR SUSTAINABILITY

EDITED BY

J.W. HANDMER ■ T.W. NORTON ■ S.R. DOVERS

Routledge
Taylor & Francis Group

LONDON AND NEW YORK

First published 2001 by Pearson Education Limited

Published 2013 by Routledge

2 Park Square, Milton Park, Abingdon, Oxon OX14 4RN

711 Third Avenue, New York, NY 10017, USA

Routledge is an imprint of the Taylor & Francis Group, an informa business

ISBN-13: 978-0-13-016121-5 (pbk)

British Library Cataloguing-in-Publication Data
A catalogue record for this book is available from the British Library

Library of Congress Cataloging-in-Publication Data
Ecology, uncertainty and policy : managing ecosystems for sustainability /
    edited by J.W. Handmer, T.W. Norton, S.R. Dovers.
        p.   cm.
    Includes bibliographical references (p.   ).
    ISBN 0–13–016121–7 (alk. paper)
    1. Ecosystem management.   2. Environmental policy.   3. Sustainable
development.   I. Handmer, John W.   II. Norton Tony.   III. Dovers,
Stephen.

QH75.E28   2001
333.95—dc21   2001021977

Contents

Contributors   xi
Foreword   xiv

1. | Ignorance, uncertainty and ecology: key themes   1
S.R. Dovers, T.W. Norton and J.W. Handmer

Introduction   1
Risk, uncertainty, ignorance and policy   3
Problem definition   4
Ignorance: precaution and other approaches   7
Ecology and policy   10
Challenges for ecology   14
Surprise and contingencies   17
Conclusion   19
References   19

2. | Challenges to ecosystem management and some
implications for science and policy   26
T.W. Norton

Introduction   26
Ecosystem complexity   27
Ecology and ecosystem management   28
Contemporary approaches to ecosystem assessment   30
Improving ecosystem management   33
Implications for institutional arrangements   37
Conclusions   39
Acknowledgements   39
References   40

3. | Bioregional approaches to conservation: local strategies to deal with uncertainty   43

N. Johnson, K. Miller and M. Miranda

Introduction   43
Uncertainty, bioregional management and the importance of local capacity   44
Obstacles to effective local participation in bioregional management   49
Strategies to engage local communities in bioregional management   56
References   63

4. | Sentimental ecology, science and sustainable ecosystem management   66

R. Mason and S. Michaels

The Adirondack Park   66
Park history   70
The blowdown   73
Species reintroduction   74
Conclusion   79
References   80

5. | Limitless lands and limited knowledge: coping with uncertainty and ignorance in northern Australia   83

J. Woinarski and F. Dawson

Introduction   83
Case studies   89
Conclusions   104
Acknowledgements   107
References   107

6. | Global warming: science as a legitimator of politics and trade?   116

S. Boehmer-Christiansen

Introduction   116
The developing law of climate change: Kyoto and the sustainability of bureaucracy   117

An alliance against 'unsustainable' coal use 124
Earth systems research is Big Science and
underpins IPCC science advice 127
The functioning of the IPCC 129
Policy advice from science and politicization
of the IPCC 132
Conclusions 134
Note 135
References 135

7. | Sustainability, uncertainty and global fisheries 138
R. Rayfuse and M. Wilder

Introduction 138
The nature of fisheries 140
The concept of sustainability in the fisheries
context 141
International fisheries management 141
The difficulties of international fisheries
management: lack of certainty 142
The difficulties of international fisheries
management: lack of enforcement 145
The responses of international fisheries
management to uncertainty 146
Responses of international fisheries to
enforcement problems 151
Conclusion 163
References 164

8. | Sustainability, uncertainty and environmental
policy: lessons from New Zealand's pastoral
high country 167
M. Harte and J. Gough

Introduction 167
The South Island pastoral high country: an evolving
social, institutional and ecological mosaic 168
Uncertainty, sustainable management and policy
development 176
Improving the policy development process 180

Conclusion 187
References 188

9. | Fire and biodiversity: understanding and
managing the impacts of fire on forest
biodiversity in south eastern Australia 191
J.E. Williams

Introduction 191
Fire and the structure, function and biodiversity of
forests in south eastern Australia 192
Modelling approaches 195
Fire protection versus fire management 197
Some implications for resource managers and policy 198
Conclusions 202
Acknowledgements 203
References 203

10. | Wetlands: policy ahead of knowledge? 209
P. Adam

Introduction 209
What is a wetland? 209
Wetland boundaries 213
The value of wetlands 215
Assessing wetland functions 220
The urgency of action 224
Conflicting agendas 225
Mitigation 227
The way ahead 228
References 231

11. | Acid rain and critical loads: science policy
processes in the United Kingdom 236
A. Tickle

Introduction 236
The evolution of UK acid rain science 238
Issues for uncertainty, sustainability and policy 252
Acknowledgements 256
References 257

12. | Uncertainty, epistemic communities and
    public policy   262
    K.J. Walker

    Introduction   262
    Policy and policy processes   266
    Policy distortions   271
    Policy communities and networks   272
    Science and public policy   279
    Conclusion: authoritative knowledge and public policy 285
    References   286

13. | Managing ecosystems for sustainability:
    challenges and opportunities   291
    J.W. Handmer, S.R. Dovers and T.W. Norton

    Ecological science may be a threat to our lifestyles 291
    Ignorance and uncertainty in science and all
    knowledge   293
    The urgency of mitigation policies for ecological
    degradation   294
    Implementing policy   296
    Monitoring: its function and dimensions   296
    Scale: space and time   298
    Democracy: can it deliver ecological sustainability? 299
    Moving ahead while coping with new challenges   300
    References   302

    Index   305

Contributors

Associate Professor Paul Adam is Director of First Year Biology at the University of New South Wales. He has written extensively on wetland ecology and management policy, and also on the ecology of Australian rainforests.

Dr Sonja Boehmer-Christiansen is a Reader in Geography at the University of Hull. She is currently researching international efforts to translate environmental protection into trade, aid and development policies, such as funding the 'sustainable use' of genetic resources and the 'transfer' of clean energy technology. Until 1995 she worked as Senior Research Fellow at the Science Policy Research Unit (SPRU, University of Sussex). She has given evidence to the Australian Government against the Kyoto Protocol.

Dr Stephen Dovers is a Senior Fellow at the Centre for Resource and Environmental Studies, Australian National University. Within the field of sustainable development he concentrates on policy development and analysis, and on environmental history.

Dr John Handmer is Professor of Environmental Geography at Middlesex University, London. He also holds professorial positions at the Australian National University and RMIT University in Melbourne, Australia. His research deals with hazards, contingency planning and sustainable development.

Dr Michael Harte is the New Zealand Seafood Industry Council's General Manager, Policy and Science. Dr Harte has 15 years experience as an economist and resource manager in the private, government and academic

sectors in New Zealand and Canada. Janet Gough is a mathematician/
economist, and Senior Policy Analyst with ERMA New Zealand, the
agency that evaluates all applications for new organisms (including
GMOs) and hazardous substances in New Zealand, and private re-
searcher/consultant. She has more than 20 years experience in resource
management including risk perceptions and risk communication.

Dr Nels Johnson is Deputy Director for Biological Resources, Dr Kenton
Miller is Vice President for International Conservation and Development,
and Marta Miranda is a research associate at the World Resources Institute, Washington, DC. They are international leaders in bioregional planning for biodiversity planning.

Dr Robert Mason is Director of Environmental Studies and Associate Professor of Geography and Urban Studies at Temple University, Philadelphia.
His principal research interests are in environmental policy making, urban
growth management, and protected areas planning in the USA and Japan.
Recent publications include *Contested Lands: Conflict and Compromise
in New Jersey's Pine Barrens* and the *Atlas of US Environmental Issues*.

Dr Sarah Michaels is on the faculty at the University of Colorado. Her
research interests are in comparative environmental policy, institutional
arrangements for environmental decision making, knowledge transfer
for managing natural resources, and policy research methodologies. Recently, she has published articles in *Environmental Management*, *Policy
Studies Journal*, *Environmental Conservation*, *Land Use Policy*, and *Sustainable Development*. Before joining the University of Colorado she
was at Tufts and Auckland.

Dr Tony Norton is Professor of Environmental Sustainability and Head,
Department of Geospatial Science, RMIT University, Melbourne, Australia. He has published over 200 papers and 7 books on a variety of
topics concerned with biodiversity conservation and environmental policy.

Dr Rosemary Rayfuse is a Senior Lecturer in International Law at the
University of New South Wales, Sydney, Australia. Martijn Wilder is a
Partner at Baker & McKenzie, Sydney, Australia, advising on general
public international law and specifically both domestic and international
environmental law. He is Adjunct Lecturer in International Environmental Law at the University of New South Wales, Sydney, Secretary of the
Australian Branch of the International Law Association, and President

of TRAFFIC (Oceania) and a Trustee of and adviser to the World Wide Fund for Nature (Australia). He has published widely in a range of areas relating to international law, including Antarctica, wildlife, fisheries and climate change.

Dr Andrew Tickle lectures at Birkbeck College, University of London. His research addresses environmental politics in Europe and currently focuses on the role of environmental groups in post-socialist countries. Previously he worked for Greenpeace.

Dr Ken Walker is an independent researcher. According to a recent reviewer, he "pioneered the study of environmental politics in Australia". He edited the collection *Australian Environmental Policy*, and is author of the textbook *The Political Ecology of Environmental Policy: An Australian Introduction*, published in 1992 and 1994 respectively by NSW University Press. From 1976 to 1995 he developed and taught groundbreaking interdisciplinary courses in environmental policy and politics in Australia, Africa and the United States. He writes mainly in the fields of environmental policy and environmental political theory, but has subsidiary interests in technology, technology transfer, and technology history in its social context.

Dr Jann Williams is a Senior Fellow in the Department of Geospatial Science, RMIT University. She is one of Australia's leading experts on fire ecology and the management of eucalypt-dominated ecosystems and is currently the President of the Ecological Society of Australia.

Dr John Woinarski is one of Australia's foremost ecologists. He is Principal Research Scientist with the Parks and Wildlife Commission of the Northern Territory of Australia and involved in biodiversity conservation research across northern Australia. Freya Dawson lectures in the Law Faculty, University of Wollongong, and is a doctoral candidate at the University of Sydney working on legal aspects of biodiversity conservation. She is a recognized expert on this topic in Australia.

Foreword

Paradigm shifts occur infrequently, but when they do, such shifts create discomfort for almost all involved. A paradigm shift is occurring in environmental management and is driven by the vision of sustainability. Policy objectives and management outcomes are increasingly driven by the need to meet sustainability goals, and this is reflected across a diverse range of sectors and institutions within society. Public policies on the energy, transport and mining sectors in North America, the management of fisheries in the Southern Oceans, and the use of scarce water resources in the Middle East, for example, now strongly reflect the vision of sustainability.

Sustainability has added a wealth of new dimensions and issues for policy makers and managers to consider. As well as economic rationalism and the emerging issue of internationalization, governments, industry and society must address issues of social justice and ecological sustainability. In terms of ecological sustainability, new environmental laws, new biodiversity conservation protocols and the use of indigenous ecological knowledge now feature in the environmental policy of many countries in South East Asia and Latin America. Such changes place enormous demands on institutions responsible for policy development and implementation. These changes also require the sciences and the policy 'arena' to reconsider how they interact, communicate, learn and adapt.

Ecology, Uncertainty and Policy: Managing Ecosystems for Sustainability explores these issues using a range of international case studies and theoretical contributions. A key focus is the interplay between science and public policy and how this can be promoted to secure policy outcomes more in keeping with sustainable development. Topics such as imperfect knowledge, contested ideas, risk and uncertainty, and precaution emerge as common threads across regions, sectors and policy areas, resulting in major policy or institutional failures and bad environmental outcomes.

Effective environmental policy requires effective implementation. Meeting this need increasingly requires institutional capacity building and negotiated co-operative arrangements between a range of groups. Different contributors to this book indicate the need to expand the concept of policy implementation away from sole reliance on government bodies, to encompass all interest groups or stakeholders. It also means acknowledgement of other constraints on the policy process, such as the dominance in government of certain ideologies.

Environmental managers cannot wait for certainty before they make decisions. Rather, decisions have to be taken in the face of uncertainty, but recognizing ignorance and providing for the ability to learn and adapt as new knowledge becomes available. Necessarily, ignorance and uncertainty are pervasive themes of this book. All groups seeking to influence policy may exploit, encourage and hide uncertainty or use it to undermine policy proposals or discredit scientists and others with different views. Environmental policy and environmental management strategies are often framed and implemented according to the degree and type of known or perceived uncertainty. The limitations of adopting a probabilistic approach to uncertainty can be easily appreciated by considering the failures of management agencies to deal adequately with (biologically or ecologically) threatening processes or by examining the strategies used in fields such as politics, law or medicine to deal with the concern. Such limitations illustrate the need for novel and precautionary approaches to management and also for contingency planning. This book rightly identifies contingency planning as a critical issue that has yet to be satisfactorily dealt with internationally.

Ecology, Uncertainty and Policy: Managing Ecosystems for Sustainability also presents important discourses on the limitations of ecology and the environmental sciences themselves. Factors such as the high degree of uncertainty in most aspects of ecology, and the highly contested nature of many ecological concepts are explored to illustrate the nature of the impact of 'sustainability' on environmental policy formulation and environmental management. These explorations are strengths of the text and foreshadow important emerging fields of study over the next decade.

Professor Kim Lowell
Centre de recherche en géomatique,
Université Laval,
Québec, Canada

1 Ignorance, uncertainty and ecology: key themes

Stephen Dovers, Tony Norton and
John Handmer*

Introduction

This book explores the relationship between science and policy. The particular emphasis is on the ecological sciences, and on the policy imperatives associated with sustainability. After the World Commission on Environment and Development (WCED 1987) and 1992 UN Conference on Environment and Development (UNCED) and its outcomes (United Nations 1992), the world has struggled to implement the research and policy agenda of sustainability. This agenda integrates the traditional environmental issues of resource degradation and depletion, biodiversity, pollution and wastes with issues of security, economic development, poverty, and fairness to future generations.

Since then, many countries have developed strategies, many organizational and statutory changes have been made, and much research and analysis undertaken. Ten years on from UNCED, the year 2002 will see major reassessment of progress, and it is unlikely that the judgement will involve other than recognition of some advances but acknowledgement that much more has yet to be achieved. The situation in Australia, often promoted as a leader in sustainable development, summarizes the situation in many countries well enough. A review of the implementation of Australia's 1992 National Strategy for Ecologically Sustainable Development found many commitments, policies and programmes, but a lack of 'good policy practice', inadequate implementation, and poorly co-ordinated institutional arrangements (Productivity Commission 1999). The ecological

* Some of the arguments in this chapter first appeared in an article in the journal *Biodiversity and Conservation* (Dovers *et al.* 1996).

rationality underpinning the sustainability idea, and central to the themes of this book, has not had great purchase in public policy and institutions.

At the time of UNCED, Harrison (1992: 315) described sustainable development, not too grandly, as universally accepted as the prime goal of human progress. Sustainability demands integration of three dimensions:

- *ecological considerations*: the maintenance of essential ecological processes and life support systems, and the protection of biodiversity;
- *social considerations*: human health and well being, equity, social justice, public participation; and
- *economic considerations*: economic growth, efficiency and diversification, international competitiveness, cost-effective policies.

The challenge of this integration is just as pressing, although the heady rhetoric and hopes of a decade ago have given way to a more sober realization of the immensity of the task. Adequate integration of these considerations suggests that they should not be too unequal in their information base, political status or accreditation in policy processes. Ecology – or rather an understanding of ecological systems – is crucial to sustainability. This chapter examines the status of the ecological dimension of sustainability, using the following propositions:

- Ecology does not enjoy the influence on policy it deserves.
- Ecology is not a well-recognized player in policy debates.
- Most ecologists would wish otherwise.
- The causes of this are mixed, with implications for both ecologists and decision makers.

It might be argued that the social dimension suffers in a similar way to the economic, but that is not the topic here. This chapter seeks to expose some general themes that are pursued in later chapters.

For sustainability to be pursued seriously, there are a number of policy imperatives and principles now stated as basic, including inter- and intra-generational equity, maintaining ecological integrity, and the precautionary principle (WCED 1987; Australia, The Commonwealth 1992a; United Nations 1992). Two key imperatives are increasing the knowledge base, and rethinking institutional arrangements and policy processes to cope with uncertainty. If we knew the range of outcomes of our current choices the issues would be clear, but we do not; uncertainty pervades all issues in sustainability (Funtowicz and Ravetz 1990; Common and Perrings 1992; Dovers and Handmer 1992; Faber *et al*. 1992). Waiting for scientific 'proof' is no longer viable, if it ever was. It is the rule with environmental issues that the situation is apparently serious, but highly uncertain, and we cannot await unequivocal evidence of cause and effect. The costs may be too great if we delay; ecosystems may be further degraded, species and genetic diversity lost forever. Policies must be formulated, decisions made and actions initiated in the face of uncertainty and ignorance. This poses fundamental questions about the role of science in policy. Are our existing policy- and decision-making processes and institutional and political arrangements adequate? Are ecologists adequately informed of the information needs of decision makers? Are decision makers cognizant of the nature of science and its limitations? What can ecologists do to improve their penetration of policy processes?

Risk, uncertainty, ignorance and policy

Sustainability problems have many attributes that define them both as substantive biophysical issues and as problems for policy (Table 1.1). For each of these attributes there are risks to be assessed, uncertainties to be pondered, and ignorance to be overcome. The attribute of mensurability underpins all others. Even day-to-day decisions tax our knowledge and predictive capacities; whether to release effluent into a stream, log an area of forest, or approve a new chemical compound. For more complex issues the uncertainties are often profound. With human population–environment linkages, the sheer magnitude, complexity, and political and moral implications of decisions may create policy paralysis (Harrison 1992; Myers 1993; Dovers and Norton 1994). With the impacts of climate change at regional scales, the magnitude and even direction of change is unclear, yet realistic energy reform to cut greenhouse gas emissions may require a purposeful reshaping of patterns of production and consumption unprecedented in human history (Patterson 1990; Houghton *et al*. 1992; Dovers 1995a). Conserving the biodiversity is a similarly complex problem (Common and Norton 1992; Wilson 1992; Barbier *et al*. 1994). The information richness and sensitivity of an ecologically sustainable society suggests a profound departure from the present patchy state of research and monitoring of natural and human systems and their interactions. As Peet (1992: 211) put it:

> In its proactive mode, sustainable development policy is built on research to provide good baseline information. We need to know how ecosystems behave and what makes them vulnerable to breakdown. We need to know about thresholds for irreversible damage, about critical limiting factors and indicators of environmental health, and about the state of particular resources within these systems.

At present we do not know these things, and redoubled efforts to improve knowledge and understanding of the relevant biophysical, ecological and human systems are demanded. Yet fully adequate information to support decisions will rarely, if ever, be achieved: uncertainty will always remain. Ecology has a central role to play in the process of balancing the imperative of policy action in the near term with the temptation to delay to seek further evidence.

Problem definition

Addressing issues of sustainability requires a means to characterize and define the nature and scale of different problems. Table 1.1 presents a set of attributes constructed for problem framing and scaling to inform policy choice (Dovers 1995b, 1997). A simple classification of policy problems in sustainability can be constructed using these attributes:

1. *Micro-problems*: the majority of problem attributes at the lower end of each scale. These are spatially and temporally discrete, not overly complex or uncertain, nor requiring large resource commitments or the development of new technologies or policy processes and, if particularly topical, then only on a local or sectoral scale. These problems may be resolved through existing institutional arrangements and policy processes such as environmental impact assessment, development approval, and effluent licensing. An example in ecosystem management would be the development of a management or recovery plan for a single reserve or species.

2. *Meso-problems*: the majority of attributes in the middle range. These are significant problems and may be prominent on the public agenda, but do not pose systemic threats to the present pattern of production and consumption, or overwhelming challenges to existing policy processes. A process or decision affecting a large number of micro-problems would fall into this category (e.g. impact assessment

Table 1.1 Attributes and scale descriptors for framing policy problems in sustainability (from Dovers 1995b; 1997).

Problem-framing attributes:
   1. Spatial scale of cause or effect:
     Local – national – regional – international – global
   2. Magnitude of possible impacts:
 2a: Impacts on natural systems:
     Minor – moderate – severe – catastrophic
 2b: Impacts on human systems:
     Minor – moderate – severe – catastrophic
   3. Temporal scale of possible impacts:
 3a: Timing:
     Near-term (months, years) – medium-term (years, decades) – long-term (decades, centuries)
 3b: Longevity of possible impacts:
     Short-term (months, years) – medium-term (years, decades) – long-term (decades, centuries)
   4. Reversibility:
     Easily/quickly reversed – difficult/expensive to reverse – irreversible
   5. Mensurability of factors and processes:
     Well-known – risk – uncertainty – ignorance
   6. Degree of complexity and connectivity:
     Discrete, linear – complex, involving multiple feedbacks and linkages

Response-framing attributes:
   7. Nature of cause/s:
     Discrete, simple – fundamental, systemic
   8. Relevance to the polity:
     Irrelevant/beyond jurisdiction – primary responsibility
   9. Tractability:
 9a: Availability of means:
     Fully sufficient – available instruments/arrangements/technologies – totally insufficient
 9b: Acceptability of means:
     Negligible opposition – moral/social/political/economic barriers – insurmountable opposition
 10. Public concern:
10a: Level of public concern:
     Low – moderate – high
10b: Basis of public concern:
     Widely shared – moderate variance in understanding – disparate perceptions
 11. Existence of goals:
     Clearly stated – generally stated – absent

---

procedures), as would problems that can be fully addressed within one country. An example would be the management of an important vegetation association across a broad geographical range.

3. *Macro-problems*: the majority of attributes in the upper range. These are multifaceted, complex, fraught with uncertainty, spatially and temporally diffuse, highly connected to other issues, and threatening major disruption of human and natural systems. They are often amalgams of lower-order problems, classified together on scientific and policy agendas due to cause and effect linkages that demand integrated research and response approaches. Examples include biodiversity, climate change, and growth in human populations and rates of resource consumption. Biodiversity loss is a global macro-problem.

The above classification provides a general perspective of scale. To match this to different types of uncertainty we can draw on Funtowicz and Ravetz (1990; 1991), who offer a heuristic model for approaching problems in science for policy, using two axes: decision stakes and system uncertainty. They define three 'problem-solving strategies':

1. '*Applied science*' is used where both uncertainty and decision stakes are low, in the face of 'technical uncertainty'. These are micro-problems, where we would use the term *quantifiable risk* (probability distributions can be assigned to possible outcomes). The available decision support techniques (see below, and Norton, Chapter 2) are generally operational at this level.

2. '*Professional consultancy*' involving qualitative judgement and skill, is used where either uncertainty or decision stakes are moderate, in the face of 'methodological uncertainty'. These are meso-problems, and we use the term *uncertainty* (direction of change is known, but probability distributions cannot be assigned). Available techniques are of questionable validity for these problems.

3. '*Post-normal science*' is used where uncertainty and decision stakes are high, where, in Funtowicz and Ravetz's description, facts are uncertain, values in dispute, stakes high and decisions urgently needed. Here, we face what they term 'epistemological uncertainty'. These are macro-problems of sustainability, and we would suggest the term *ignorance*. Available techniques are not useful in any practical policy support sense here, and may even mislead.

This is one view of the diversity of types of uncertainty. In policy terms, this can be equated to Dunn's (1981) categorization of well-, moderately and ill-structured policy problems. However, ignorance and uncertainty represent such critically important factors that more detail is warranted.

Ignorance: precaution and other approaches

A number of researchers have constructed taxonomies of risk, uncertainty and ignorance. The most detailed, and we believe most useful at a general level, is that of Smithson (1989), shown in Figure 1.1. Smithson uses ignorance as the first-order descriptor, differentiated into error (to be ignorant of) and irrelevance (to ignore). Risk (probability) and (scientific) uncertainty are relegated to minor levels in this taxonomy.

From Smithson's taxonomy, and drawing also on discussions by Suter *et al.* (1987), Faber *et al.* (1992) and Wynne (1992), we can identify a number of key points:

- ignorance is not unitary, but comprises different types and aspects;
- ignorance does not exist simply 'out there' in the environment, but is socially constructed;
- while some ignorance can be reduced through research and monitoring, there will always be ignorance that is to all intents and purposes irreducible;

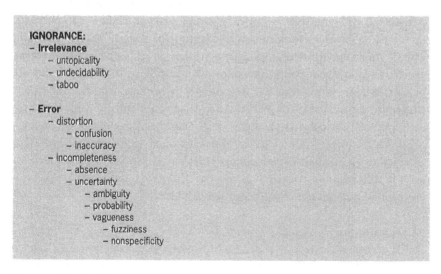

```
IGNORANCE:
– Irrelevance
      – untopicality
      – undecidability
      – taboo

– Error
      – distortion
            – confusion
            – inaccuracy
      – incompleteness
            – absence
            – uncertainty
                  – ambiguity
                  – probability
                  – vagueness
                        – fuzziness
                        – nonspecificity
```

Figure 1.1 Smithson's (1989) taxonomy of ignorance

- in the face of ignorance, we may face ecological surprise or threshold effects; and
- ignorance will not be uniform between individuals and across society.

A crucial point is that ignorance increases when we act on the basis of what we think we know: 'scientific uncertainty can be seen to be important not in itself, supposedly measurable on some objective scale, but as a function of (in relation to) the extent of technological or policy commitment riding on the body of knowledge concerned' (Wynne 1992: 123). These considerations should influence our approach to decision making. Issues such as these inform the framework for 'ignorance auditing' developed by Dòvers and Handmer (1995). This comprises a detailed checklist of types, gradients, sources and causes, thus providing a mechanism for accounting for risk, uncertainty and ignorance in policy. One message for natural scientists emerging from this sketch is that traditional scientific uncertainty – 'out there' in the environment – is only one part of the picture, and perhaps not the most important. Ignorance generated within human systems of governance, policy and law may be, in principle, easier to address. It also serves to provide a background against which to view various techniques and approaches.

The major recent development in dealing with uncertainty in environmental policy has been adoption of the 'precautionary principle' (PP) in policy and law. Internationally, the PP is enunciated in, for example, the Rio Declaration on Environment and Development (1992) at Principle 15, and the UN Framework Convention on Climate Change (1992) at Article 3.3. In Australia, it appears, along with other ESD principles, in the National Strategy for Ecologically Sustainable Development (Australia, The Commonwealth 1992a: 8), the National Greenhouse Response Strategy (Australia, The Commonwealth 1992b: 68), and literally hundreds of subsequent national and state policies. More importantly, ESD principles are expressed in over 120 Australian laws and the precautionary principle in particular is the basis of a small but growing body of jurisprudence (Gullett 1998; Dovers and Handmer 1999; Stein 2000). The Council of Australian Governments' 1992 Intergovernmental Agreement on the Environment (IGAE) gives the following definition (Schedule, *National Environment Protection Act* 1994):

> Precautionary principle –
> Where there are threats of serious or irreversible environmental damage, lack of full scientific certainty should not be used as a reason for postponing measures to prevent environmental degradation. In the application

of the precautionary principle, public and private decisions should be guided by:

(i) careful evaluation to avoid, wherever practicable, serious or irreversible damage to the environment; and

(ii) an assessment of the risk-weighted consequences of various options.

The implications of adopting the precautionary principle are far from clear (see Gundling 1990; Bodansky 1991; Cameron and Abouchar 1991; Wynne 1992; Dovers and Handmer 1995; O'Riordan and Cameron 1994; Raffensperger and Tickner 1999). It is generally proposed that the PP implies that: uncertainty should not delay implementing protection measures; 'anticipatory' or 'preventative' approaches to environmental protection should replace reaction; and the burden of proof should shift towards those proposing development. Clearer interpretations will most likely emerge, over time, through the courts. One Australian legal interpretation, albeit concerning a micro-problem, emerged through the Land and Environment Court of New South Wales. Overturning permission for a road development impacting on threatened species, Stein (1993) stated that the principle is a 'statement of common sense' and that 'where uncertainty or ignorance exists concerning the nature or scope of environmental harm (whether this follows from policies, decisions or activities), decision makers should be cautious'. In a contrasting judgement in the same court, Talbot (1994) did not accord the PP such status.

Whatever the final interpretations, there are problems with the principle. The elements of the principle ('serious or irreversible environmental damage', 'full scientific certainty', 'careful evaluation', 'assessment of risk-weighted consequences') are value-laden and qualitative descriptors (Dovers and Handmer 1995). The principle is a moral injunction, open to interpretation. O'Riordan (1992) described the precautionary principle as equally political and scientific. As currently stated and interpreted, we suggest that it is, rather, overwhelmingly political. The PP, however, is not the only available decision-support mechanism; others exist and can be used to inform policy. These techniques and approaches include (the list is selective):

- quantitative risk assessment (Warner 1992);
- maximax, minimax and minimax regret criterion, used in cost-benefit analysis (Pearce and Nash 1981: 83–88; Common 1995);
- safe minimum standards (Bishop 1978; 1993);
- performance assurance bonds (Costanza and Perrings 1990);
- the NUSAP (numeral–unit–spread–assessment–pedigree) notation system proposed by Funtowicz and Ravetz (1990);

- ecological or environmental risk assessment (Suter *et al.* 1987; Hunsaker *et al.* 1990; Graham *et al.* 1991; Beer and Ziolkowski 1995);
- population viability analysis (Boyce 1992; Burgman *et al.* 1993; Norton and Possingham 1993);
- adaptive management (Holling 1978; Gunderson *et al.* 1995; Dovers and Mobbs 1997);
- recourse to courts of law; and
- various approaches based on negotiation, mediation and conflict resolution.

Public participation is viewed by some as a means of coping with uncertainty, as is the pursuit of so-called 'no regrets' options. More a policy optimization strategy than a methodological approach, the latter involves seeking areas where social or economic benefits also exist, justifying reform despite uncertainties over environmental costs and benefits. All these techniques have limitations, are subject to criticism, and many require a political judgement to precede their application. While they offer means of operationalizing the precautionary principle in some situations, they are in general suited only to micro-problems. The imperative to account for ignorance is becoming stronger. For example, in Australia there is likely to be an increasing expectation of public and corporate policy accord with the 1995 Risk Management Standard (AS/NZS 4360), with detailed procedures already being discussed (Cross 1996; Norton *et al.* 1996). Without near-perfect information or widely accepted decision-support techniques to render insufficient information into unequivocal answers, decisions will remain firmly political. That is, although perhaps informed by 'hard' information, decisions will depend on judgements, values, morals, ideology, guesses, pragmatism, convenience and expedience. Scientists often call a decision 'political' in a derogatory sense; policy analysts know that this is simply the way the world works. Rational, value-free policy processes do not and cannot exist (Rees 1990; Davis *et al.* 1993; Considine 1994; Doyle and Kellow 1995). Accepting that policy is not the product of objective, rational processes reminds us that science is not the only rationality. It recasts the question of the frustrated scientist – Why don't they listen to scientific advice? – to ask what ecology needs to do in order to become a stronger player in the policy game.

Ecology and policy

Without ecology, there would hardly be a sustainability debate. Ecology provides much of – or points to the lack of – the information that defines these issues. The science–policy relationship is never easy, and we would suggest that the situation for ecology is especially hard. Meso- and macro-policy problems in sustainability are different in both kind and degree from those in other policy fields, for the following reasons (Dovers 1997; see also Beck 1992; Walker 1994; Adam 1995; Doyle and Kellow 1995):

- broadened and variable spatial scales (compared to those determining political and economic systems);
- deepened and variable temporal scales (compared to those determining political and economic systems);
- the possibility of absolute ecological limits to human activity;
- irreversible impacts, and related policy urgency;
- complexity within and connectivity between problems;
- pervasive risk, uncertainty and ignorance;
- typically cumulative rather than discrete impacts;
- new moral dimensions (e.g. concerning other species or future generations);
- 'systemic' problem causes, embedded deeply in patterns of production, consumption, settlement and governance;
- lack of available, uncontested research methods, policy instruments and management approaches;
- lack of defined policy, management and property rights, roles and responsibilities;
- demands (and justification) for increased community participation in both policy formulation and actual management;
- the presence of major assets, services and processes that are not traded and thus cannot be satisfactorily represented by economic value; and
- sheer novelty as a suite of policy problems.

Ecology has not influenced policy development in Australia or internationally as much as many ecologists think it should (e.g. Haber 1992; Noble 1994; Dovers et al. 1996). It can be argued that the ecological profession is not an accepted part of the policy community. The difficulties of translating the uncertainties inherent in the findings of ecological science into standards useful for policy were demonstrated by Hohl and Tisdell (1993), and Worster (1993) questions the status and ability of ecology in the global sustainability debate. Problems with interpreting the oft-used term 'ecological integrity' are discussed by Noss (1995). Underdal (1989: 259) identified conditions which will weaken the impact of scientific inputs to policy: tentative hypotheses; unclear 'cure' to the problem; remote effects, affecting periphery or not clearly visible to the public; problems developing slowly; political conflict; cross-issue 'contamination'; and ad hoc decision-making processes. All these apply to ecology. Four interrelated factors seem crucial in causing the weakness of the ecological dimension of sustainability: ecological theory is incomplete; natural systems are complex and difficult to study; many natural systems are subject to unprecedented rates of change; and communication across sectors and disciplines is limited.

The knowledge base of ecology is grossly incomplete. The theory, principles and other intellectual and technical apparatus required for developing a process understanding of ecosystems and their response to human-induced disturbance is underdeveloped and limited. Unlike, say, physics and chemistry, much ecological theory is highly contested. Debates over evolution and co-evolution, competition and species coexistence, species diversity and ecosystem stability/disturbance, and ecosystem resilience recur in ecological fora (e.g. Peters 1991; Polis 1994; Walter and Paterson 1994). Cherrett's (1988) survey of British ecologists revealed the lack of consensus over the explanatory power and acceptability of most ecological concepts. More practically, basic data on the distribution and abundance of most species and the way they interact have not been collected. Estimates of global species richness vary among authorities by an order of magnitude, controversy remains as to what constitutes a species, and few natural systems are sufficiently understood to permit reliable characterizations of energy flows and dynamics.

Natural systems are often complex and difficult to study; indeed, ecological science increasingly emphasizes complexity, heterogeneity and subtlety. As Fairweather (1993: 4–5) put it, 'one aspect of our science that irks managers is the uncertainty inherent in scientific knowledge, due both to its probabilistic mode of discourse and the fact that we probably don't know enough about many ecosystems'. Ecologists constantly unearth complexity: unexpected nutrient cycles, in-species variation, reliance on highly seasonal food sources or rare refuges, or the ever-widening parameters of variation in hydrological and climatic factors. While this builds a truer picture of ecological systems, it complicates decision making by emphasizing vulnerability under non-average conditions. The overwhelming bulk of species and ecosystems have yet to receive detailed analysis to reveal such interdependencies and, given current patterns of research and its resourcing, are unlikely to do so. The inherent complexity of ecological systems may mean that further study will not improve understanding of their dynamics within the timeframes that policy and decision makers have to obey. Nor might such studies lead to greater certainty about the response of a system to human interference. Additional research may illuminate previously unknown properties that indicate even greater complexity, compounding prediction and management. From the point of view of decision makers, funding more ecological research and monitoring may not make their task easier in the short term.

A further factor complicating the use of ecological understanding in policy processes is that many systems are now subject to unprecedented rates of change as a result of human activities (Botkin 1990). Such change complicates attempts to improve understanding of ecosystems: they become moving targets. Studies may be compromised by unexpected impacts arising during their implementation. Long-term studies – where they exist – may be difficult to interpret or be incorrectly interpreted if changes in baseline conditions occur and are not detected or are inadequately accommodated in initial study design. Long-term ecological monitoring is crucial, yet insufficient attention is paid to it and it is poorly resourced, in part due to the lack of both political and scientific kudos attached to it.

Existing mechanisms for using scientific information, or accounting for the lack of it, in policy processes are often inadequate. Uncertainties may be deemed too hard to deal with or simply inconvenient or irrelevant. This results from poor understanding of science in the community generally and by policy makers, which is compounded by poor understanding by most ecologists of the policy process and of the information requirements of governments. Given the focus here on ecology, we can speculate as to why ecologists might not concern themselves with the study of policy processes. The first reason is simply time: in a complex science there are often other things to do. Second, traditional ecological training rarely includes 'non-scientific' subjects. Third, the political and policy urgency of ecological issues is relatively recent, and the relevant policy processes immature. Fourth, there are problems with crossing disciplinary boundaries (e.g. Heberlein 1988; Machlis 1992). Finally, accessible literature on environmental policy processes has been hard to find, although this is improving (e.g. in Australia, Walker 1994; Doyle and Kellow 1995). Ecology is not unique in this regard. For example, economics and the law – two other disciplines centrally relevant to sustainability – connect insufficiently to other disciplines (Common 1995; Dovers 1999).

The issue of communication is crucial. The status of ecological knowledge is unlikely to change radically in the short to medium term, even given increased resources, and rates of ecosystem change are more likely to increase than stabilize or decrease. Hence, for the next few decades a fundamental challenge will be to manage in the face of considerable ignorance and uncertainty. Both policy makers and ecologists need to improve the basis of communication.

Ecological information is time-consuming and expensive to acquire, and supply rarely meets demand. How can we more effectively use what information we possess? How can we improve existing policy-development processes and better inform decision makers? To explore this, we have elsewhere considered three recent Australian policy processes: the ecologically sustainable development (ESD) process; the Resource Assessment Commission forest inquiry; and the Industry Commission greenhouse inquiry (Dovers *et al.* 1996). These cases expose some major, generic issues, including: scientific representation in the policy process; the importance of the manner in which policy problems are initially constructed and thus pursued within a process; the crucial role of lasting institutional bases for processes; unease over scientific uncertainty; and the irrevocably political nature of policy decisions.

Two key themes emerged from the analysis of Dovers *et al.* (1996). First, inadequacies in current processes aside, reconciling the demand for decisions in the near term with incomplete knowledge is an enormous and complex task; there are no precedents for tackling the macro-problems of sustainability. The second is the place of 'political' versus 'scientific' decisions, and the fact that neither science nor policy seems comfortable with the present balance.

Challenges for ecology

The potential role of ecological information in the policy process is undoubtedly limited by institutional arrangements and the way issues of sustainability are conceptualized and addressed. These inadequacies are not necessarily always unrecognized and indeed are often enough exploited; some forms of ignorance (e.g. confusion, distortion) may benefit some policy actors (Smithson 1989). Open processes that allow informed debate are essential if options are to be identified and discussed. However, even when expert opinion is genuinely sought, many processes appear poorly equipped to accommodate it.

Ecology faces the same challenges as the sciences more generally, but the nature of sustainability problems makes the science–policy relationship even harder. Ecology can and should make a greater contribution to policy. Its failure to do so results from factors which are to a degree inescapable (the nature of the issues), but also to shortcomings on the part of both ecologists and policy makers. With respect to the latter, the first imperative is simply to acquaint themselves more closely with the practices and status of the relevant sciences. The second is to rethink the structure of institutions and policy processes to account for the need for better ecological information, definition of problems and goals, and improved policy monitoring.

Turning to ecology, we suggest a number of challenges relating to its own affairs and to its engagement with policy process. These may create opportunities for the enhancement of both the discipline of ecology and the quality of public policy.

1. The first challenge is to define clearly the bounds and manner of characterizing and communicating risk, uncertainty and ignorance in ecological research. It needs to be made clear what we know, how we know what we know and, beyond that, what might be. Ecologists should resist pretending to know what *is* at stake, and communicate more strongly what *might be* at stake. This would be more helpful for indicating approaches to ignorance and uncertainty that might better inform policy, and usefully invites consideration of decision-making processes that can work when key knowledge is highly contested.

2. There is a need to build greater in-discipline consensus over key issues and areas of policy concern. Fragmented and qualified answers are unavoidable, and ecologists must make clear the realities of complexity and uncertainty to decision makers. But fragmented and qualified answers are easily discounted or twisted in adversarial systems of government and law (see Walker, Chapter 12). Policy makers honestly seeking to construct sound policies may discern no clear direction from the ecological discourse. Where the debates so necessary to development of the discipline do occur, it needs to be made clear where these differences matter for policy and where they do not.

3. The limitations of quantitative risk-based approaches need to be made explicit. Consider ecological risk assessment (ERA) (e.g. Suter 1993) and population viability analysis (PVA) (e.g. Boyce 1992; Lindenmayer and Possingham 1994). These suffer the same limitations as any quantitative risk assessment – large information requirements, possible emphasis on what is measurable rather than what is important, and an inability to handle human factors. This is further explored by Norton (Chapter 2).

4. Curricula and fora should be fostered within ecology to enhance consideration of linkages with policy, and development of the necessary interdisciplinary approaches and frameworks.

5. Extending this, there is a need to engage policy processes more actively, communicating with policy makers as they interpret the precautionary principle. Insights from ecological research do not become

automatically apparent to people concerned with policy; they must be actively translated and communicated. New processes and institutions will continue to appear, and ecologists must keep a close eye on these if they wish to penetrate emerging policy debates.

5. Ecologists need to explore mechanisms of communication with broader audiences, to increase the currency of ecological insights in relevant public debates.

6. Greater emphasis is required on basic, long-term ecological research and monitoring.

With respect to monitoring, ecologists need to make clear the limits imposed on advice to policy makers by the lack of basic data (Underwood 1994). Long-term, integrated ecological monitoring must be emphasized, but is scientifically and politically unfashionable for various reasons. To avoid sceptical responses to pleas for more resources, the value of such basic work must be clearly matched to goals now enshrined in national and international sustainability policy. Poor information will in all likelihood mean poor policy, but more importantly it will mean that we cannot know whether policies are working. A telling comparison may be made with basic information systems supporting the social and economic dimensions of sustainability. If we are to integrate ecological, social and economic imperatives, where are the ecological equivalents of input–output tables, national accounts, censuses of population and housing, and labour market surveys? Social and economic policy abilities would be vastly compromised in the absence of these information inputs, yet policy concerning ecological aspects of sustainability proceeds in the absence of similar support. The change required to redress the imbalance would be profound: resourcing, effort and co-ordination orders of magnitude greater than at present. However, long-term monitoring will not help in the short term – in the language of sustainability, it is a morally demanded endowment of information to future generations. So it must always be balanced by better abilities to make decisions in the face of ignorance and uncertainty in the nearer term, guided by the precautionary principle and other approaches.

Where broader communication is concerned the complexity of ecological research findings needs to be built into 'big pictures', conveying the ecological condition and trend of ecosystems to a wider audience. This requires different forms of communication. Communication should not stop with specialized journals and conferences where ecologists talk only to other ecologists. The crucial communication paths are with other disciplinary audiences – economists, political scientists, lawyers – and with policy professionals. This does not mean popularizing, but it does require different styles of presentation and media of publishing. An outstanding international example is Wilson's (1992) *The Diversity of Life*. In Australia, examples include Kirkpatrick (1994), Lindenmayer (1994) and Morton (1994). Broader communication efforts must be recognized and rewarded, because at present they do not attract scientific kudos.

Surprise and contingencies

Even with better science–policy linkages and policy processes more sensitive to ignorance, there will be surprise (Holling 1986; Faber *et al.* 1992). Even in the fanciful eventuality of near-complete knowledge, we would still have incompetence, distortion, negligence, corruption and taboos. The question remains of reaction to the totally unexpected, or to the urgent. First, the likelihood of surprise reinforces the need for ignorance and uncertainties in ecological sciences to be clearly communicated to policy makers, to underline the possibility of the unexpected and thus the need for appropriate response mechanisms.

In ecosystem management, one avenue of urgent response is recovery plans for endangered species (e.g. Clark *et al.* 1994; Miller *et al.* 1994), but this implies crisis management and is highly particular in the broader context of biodiversity conservation at landscapes scales. Such broader scales are, it is increasingly stated, the most promising direction for ecology. Another approach is adaptive management, where management, research and policy are joined in an experimental or 'learning by doing' mode (Walters and Holling 1990). This promises to institutionalize linkages between science and policy and establish iterative learning mechanisms, towards the necessary 'civic science' (Lee 1993). But adaptive management needs a robust margin of ecosystem health to offer sufficient latitude. Crucially, it assumes institutional, policy and management arrangements of sufficient continuity and longevity to match the temporal scales of the natural systems being addressed, and this assumption is rarely met by existing arrangements (Dovers 1995c; Dovers and Mobbs 1997). The institutional conditions for adaptive management are very demanding (Lee 1993). Such qualifications aside, adaptive management should be further pursued and combined with the widely promoted but scarcely operationalized idea of 'policy learning' in public policy (Bennett and Howlett 1992; May 1992; Brewer and Clark 1994).

In general, there is a pro-sustainability case for maintaining spare capacity as a buffer against the unexpected. Adaptive management would seem to require this. While consistent with the precautionary principle, this runs counter to optimizing resource use (Dovers and Handmer 1993). Irreversibility means that if the prediction is wrong or the unexpected occurs, while policy learning may occur its value may be debatable if the population, species or ecosystem has gone. Precaution recommends buffers against catastrophe, or, if the environmental asset in question is very limited, then enhancement. Examples might include larger and/or more areas in conservation reserves, restoration of much-reduced habitats, or safety margins added to exploitation prescriptions (an argument against the notion of maximum sustainable yield).

There is potential in the idea of contingency planning, drawing on the field of emergency planning and management (Drabek and Hoetmer 1991; Handmer 1992). Contingency planning is basic to endangered species recovery, as epitomized in the US Endangered Species Act 1973, but is rare in ecosystem management other than in the guise of planning for oil spills. Emergency planning instructs that residual risk, uncertainty and ignorance should be embraced and explicitly provided for. This could be applied to conservation strategies or reserve management plans. Ecological emergencies differ from anthropogenic emergencies (e.g. floods, technological failures) in a number of important respects. Environmental effects may be slow to develop or become obvious, and the environment depends on others to articulate its problems. Comprehensive emergency planning for anthropogenic emergencies is a process involving all stakeholders (e.g. Emergency Management Australia 1990). A plan is only one outcome of the process, which should not stop at that point.

Relevant areas in ecosystem management would include preparedness for such eventualities as destruction of rare refuge habitats by wildfire, the conservation implications of severe droughts, pollution episodes, low-probability but possibly catastrophic impacts of exotic diseases or introduced species on flora and fauna, algal blooms in major aquatic systems, or oil spills. Contingency plans for algal blooms form part of the New South Wales (Australia) Algal Management Strategy, which is concerned with both ecological impacts and provision of safe drinking water supplies (Verhoeven 1995).

The initial stage for ecological contingency planning would, as in emergency management, involve identification of all conceivable hazard sources or impacts. This should cover both known hazards and those perceived to be unlikely. The process would then be defined by the specific context, but informed by the general rules that management plans and regimes should include reference to these potential hazards and allocation of related responsibilities, and that both proactive measures (reducing vulnerability) and reactive responses should be explored. Planning should also extend to the recovery phase to take advantage of opportunities arising from the emergency, using the 'policy windows' thus presented (Solecki and Michaels 1994). The idea of contingency planning could inform ecosystem and biodiversity management in the near term. It would have even greater usefulness in a possible future of better information flows and linkages where early warnings are more likely.

Conclusion

The directions for research, policy and management indicated here – maintaining 'spare capacity', contingency planning, greatly enhanced ecological monitoring – would require levels of resourcing far in excess of those current. Whether such commitments would be made is a political issue involving various competing social goals, including the maintenance of ecological integrity and preservation of biodiversity as stated in official policy. Even the rhetoric of official policy suggests that greater resources are required. The ability of ecology to argue for such increased attention might be strengthened by further development of the themes proposed here.

The ecological dimension of sustainability policy is relatively weak in its information and communication base and its political status. Communicating and managing uncertainty and ignorance are central to this. This introductory chapter has identified some general themes. Linkages between ecology and policy need to be strengthened, but this will take some effort. It will also require a close review of cases, events and processes from which lessons might be drawn. Some ecologists might see this as a distraction and diffusion of available skills, effort and resources. But it may be viewed proactively, as an opportunity for development of the discipline and for growth in resources, status and impact. The following chapters take up this challenge in a critical fashion, offering perspectives from both within and outside ecology.

References

Adam, B. (1995) Running out of time: global crisis and human engagement. In: Redclift, M. and Benton, T. (eds) *Social Theory and Global Environment*. New York: Routledge, 92–112.

Australia, The Commonwealth (1992a) *National Strategy for Ecologically Sustainable Development*, December 1992. Canberra: Australian Government Publishing Service.

Australia, The Commonwealth (1992b) *National Greenhouse Response Strategy*, December 1992. Canberra: Australian Government Publishing Service.

Barbier, E.B., Burgess, J.C. and Folke, C. (1994) *Paradise Lost? The Ecological Economics of Biodiversity*. London: Earthscan.

Beck, U. (1992) From industrial society to risk society: questions of survival, social structure and ecological enlightenment. *Theory, Culture and Society* 9: 97–123.

Beer, T. and Ziolkowski, F. (1995) *Environmental Risk Assessment: an Australian Perspective*. Canberra: Office of the Supervising Scientist.

Bennett, C.J. and Howlett, M. (1992) The lessons of learning: reconciling theories of policy learning and policy change. *Policy Sciences* 25: 275–94.

Bishop, R.C. (1978) Endangered species and uncertainty: the economics of a safe minimum standard. *American Journal of Agricultural Economics* 60: 10–18.

Bishop, R.C. (1993) Economic efficiency, sustainability, and biodiversity. *Ambio* 22: 69–73.

Bodansky, D. (1991) Scientific uncertainty and the precautionary principle. *Environment* 33(7): 4–5, 43–4.

Botkin, D.B. (1990) *Discordant Harmonies: a New Ecology for the Twenty-first Century*. Oxford: Oxford University Press.

Boyce, M.S. (1992) Population viability analysis. *Annual Review of Ecology and Systematics* 23: 481–506.

Brewer, G.D. and Clark, T.W. (1994) A policy perspective: improving implementation. In: Clark, T.W., Reading, R.P. and Clarke, A.L. (eds) *Endangered Species Recovery: Finding the Lessons, Improving the Process*. Washington DC: Island Press, 391–413.

Burgman, M.A., Ferson, S. and Akçakaya, H.R. (1993) *Risk Assessment in Conservation Biology*. New York: Chapman and Hall.

Cameron, J. and Abouchar, J. (1991) The precautionary principle: a fundamental principle of law and policy for the protection of the global environment. *Boston College International and Comparative Law Review* 14: 1–27.

Cherrett, J.M. (1988) Ecological concepts: a survey of the views of the members of the British Ecological Society. *Biologist* 35: 64–6.

Clark, T.W., Reading, R.P. and Clarke, A.L. (1994) (eds) *Endangered Species Recovery: Finding the Lessons, Improving the Process*. Washington DC: Island Press.

Common, M.S. (1995) *Sustainability and Policy: Limits to Economics*. Melbourne: Cambridge University Press.

Common, M.S. and Norton, T.W. (1992) Biodiversity: its conservation in Australia. *Ambio* 21: 258–65.

Common, M. and Perrings, C. (1992) Towards an ecological economics of sustainability. *Ecological Economics* 6: 7–34.

Considine, M. (1994) *Public Policy: a Critical Approach*. Melbourne: Macmillan.

Costanza, R. and Perrings, C. (1990) A flexible assurance bonding system for improved environmental management. *Ecological Economics* 2: 57–76.

Cross, J. (1996) The risk management standard. *Australian Journal of Emergency Management* 10(4): 4–7.

Davis, G., Wanna, J., Warhurst, J. and Weller, P. (1993) *Public Policy in Australia*. 2nd edn. Sydney: Allen and Unwin.

Dovers, S. (1995a) *Sustainable Energy Systems: Pathways for Australian Energy Reform*. Melbourne: Cambridge University Press.

Dovers, S.R. (1995b) A framework for scaling and framing policy problems in sustainability. *Ecological Economics* 12: 93–106.

Dovers, S.R. (1995c) Information, sustainability, and policy. *Australian Journal of Environmental Management* 2: 142–56.

Dovers, S.R. (1997) Sustainability: demands on policy. *Journal of Public Policy* 16: 303–318.

Dovers, S.R. (1999) Adaptive policy, institutions and management: challenges for lawyers and others. *Griffith Law Review* 8: 374–93.

Dovers, S.R. and Handmer, J.W. (1992) Uncertainty, sustainability and change. *Global Environmental Change* 2: 262–76.

Dovers, S.R. and Handmer, J.H. (1993) Contradictions in sustainability. *Environmental Conservation* 20: 217–22.

Dovers, S.R. and Handmer, J.W. (1995) Ignorance, the precautionary principle, and sustainability. *Ambio* 24: 92–7.

Dovers, S. and Handmer, J. (1999) Ignorance and the precautionary principle: toward an analytical framework. In: Harding, R. and Fisher, E. (eds) *Perspectives on the Precautionary Principle*. Sydney: Federation Press, 167–89.

Dovers, S.R. and Mobbs, C.D. (1997) An alluring prospect? Ecology and the requirements of adaptive management. In: Klomp, N. and Lunt, I. (eds) *Frontiers in Ecology: Building the Links*. London: Elsevier, 39–52.

Dovers, S.R. and Norton, T.W. (1994) Population, environment and sustainability: reconstructing the debate. *Sustainable Development* 2: 1–7.

Dovers, S.R., Norton, T.W. and Handmer, J.W. (1996) Uncertainty, ecology, sustainability, and policy. *Biodiversity and Conservation* 5: 1143–67.

Doyle, T. and Kellow, A. (1995) *Environmental Politics and Policy Making in Australia*. Melbourne: Macmillan.

Drabek, T.E. and Hoetmer, G.J. (1991) (eds) *Emergency Management: Principles and Practices for Local Government*. Washington DC: International City Management Association.

Dunn, W.N. (1981) *Public Policy Analysis: an Introduction*. Englewood Cliffs, NJ: Prentice-Hall.

Emergency Management Australia (1990) *Community Emergency Planning*. Canberra: Emergency Management Australia.

Faber, M., Manstetten, R. and Proops, J. (1992) Towards an open future: ignorance, novelty and evolution. In: Costanza, R., Norton, B.G. and Haskell, B.D.

(eds) *Ecosystem Health: New Goals for Environmental Management*. Washington DC: Island Press, 72–96.

Fairweather, P.G. (1993) Links between ecology and ecophilosophy, ethics and the requirements of environmental management. *Australian Journal of Ecology* 18: 3–19.

Funtowicz, S.O. and Ravetz, J.R. (1990) *Uncertainty and Quality in Science for Policy*. Dordrecht: Kluwer Academic.

Funtowicz, S.O. and Ravetz, J.R. (1991) A new scientific methodology for global environmental issues. In: Costanza, R. (ed.) *Ecological Economics: the Science and Management of Sustainability*. New York: Columbia University Press, 137–152.

Graham, R.L., Hunsaker, C.T. and O'Neill, R.V. (1991) Ecological risk assessment at the regional scale. *Ecological Applications* 1: 196–206.

Gullett, W. (1998) Environmental impact assessment and the precautionary principle: legislating caution in environmental protection. *Australian Journal of Environmental Management* 5: 146–58.

Gunderson, L.H., Holling, C.S. and Light, S.S. (1995) (eds) *Barriers and Bridges to the Renewal of Ecosystems and Institutions*. New York: Columbia University Press.

Gundling, L. (1990) The status in international law of the principle of precautionary action. In: Freestone, D. and Ijlstra, T. (eds) *The North Sea: Perspective on Regional Environmental Co-operation*. London: Graham and Trotman, 23–30.

Haber, W. (1992) On transfer of scientific information into political action: experiences from the German Council of Environmental Advisors. *Ecology International* 20: 3–13.

Handmer, J.W. (1992) Emergency management in Australia: concepts and characteristics. In: Parker, D.J. and Handmer, J.W. (eds) *Hazard Management and Emergency Planning*. London: James and James, 227–41.

Harrison, P. (1992) *The Third Revolution: Population, Environment, and a Sustainable World*. London: Penguin.

Heberlein, T.A. (1988) Improving interdisciplinary research: integrating the social and natural sciences. *Society and Natural Resources* 1: 5–16.

Hohl, A. and Tisdell, C.A. (1993) How useful are environmental safety standards in economics? – the example of safe minimum standards for protection of species. *Biodiversity and Conservation* 2: 168–81.

Holling, C.S. (1978) (ed.) *Adaptive Environmental Management and Assessment*. Chichester: Wiley.

Holling, C.S. (1986) The resilience of terrestrial ecosystems: local surprise and global change. In: Clark, W. and Munn, R. (eds) *Sustainable Development of the Biosphere*. Cambridge: Cambridge University Press, 292–317.

Houghton, J.T., Callander, B. and Varney, S.K. (1992) *Climate Change 1992: the Supplementary Report to the IPCC Scientific Assessment*. Cambridge: Cambridge University Press.

Hunsaker, C.T., Graham, R.L., Suter, G.W., O'Neill, R.V., Barnthouse, L.W. and Gardner, R.H. (1990) Assessing ecological risk on a regional scale. *Environmental Management* 14: 323–32.

Kirkpatrick, J. (1994) *A Continent Transformed: Human Impact on the Natural Vegetation of Australia*. Melbourne: Oxford University Press.

Lee, K. (1993) *Compass and Gyroscope: Integrating Science and Politics for the Environment*. Washington DC: Island Press.

Lindenmayer, D.B. (1994) Timber harvesting impacts on wildlife: implications for ecologically sustainable forest use. *Australian Journal of Environmental Management* 1: 56–68.

Lindenmayer, D. and Possingham, P. (1994) *The Risk of Extinction: Ranking Management Options for Leadbeater's Possum using Population Viability Analysis*. Canberra: Centre for Resource and Environmental Studies, Australian National University.

Machlis, G.E. (1992) On the contribution of sociology to biodiversity research and monitoring. *Biological Conservation* 62: 161–70.

May, P. (1992) Policy learning and policy failure. *Journal of Public Policy* 12: 331–54.

Miller, B., Reading, R., Conway, C., Jackson, J.A., Hutchins, M., Snyder, N., Forrest, S., Frazier, J. and Derrickson, S. (1994) A model for improving endangered species recovery programs. *Environmental Management* 18: 637–45.

Morton, S.R. (1994) European settlement and the mammals of arid Australia. In: Dovers, S. (ed.) *Australian Environmental History: Essays and Cases*. Melbourne: Oxford University Press, 141–66.

Murphy, D.D. and Noon, B.D. (1991) Coping with uncertainty in wildlife biology. *Journal of Wildlife Management* 55: 773–82.

Myers, N. (1993) Population, environment, and development. *Environmental Conservation* 20: 205–16.

Noble, I.R. (1994) Science, bureaucracy, politics and ecologically sustainable development. In: Norton, T.W. and Dovers, S.R. (eds) *Ecology and Sustainability of Southern Temperate Ecosystems*. Melbourne: CSIRO, 117–25.

Norton, T., Beer, T. and Dovers, S. (1996) (eds) *Risk and Uncertainty in Environmental Management: Proceedings of the 1995 Fenner Conference*. Canberra: Centre for Resource and Environmental Management, Australian National University.

Norton, T.W. and Possingham, H.P. (1993) Wildlife modelling for biodiversity conservation. In: Jakeman, A.J., Beck, B. and McAleer, M. (eds) *Modelling Change in Environmental Systems*. Sydney: Wiley and Sons, 243–65.

Noss, R.F. (1995) Ecological integrity and sustainability: buzzwords in conflict? In: Westra, L. and Lemons, J. (eds) *Perspectives on Ecological Integrity*. Dordrecht: Kluwer, 60–76.

O'Riordan, T. (1992) *The Precautionary Principle in Environmental Management*. Centre for Social and Economic Research on the Global Environment Working Paper 92–03. Norwich: CSERGE, University of East Anglia.

O'Riordan, T. and Cameron, J. (1994) (eds) *Interpreting the Precautionary Principle*. London: Earthscan.

Patterson, W.C. (1990) *The Energy Alternative: Changing the Way the World Works*. London: Optima.

Pearce, D.W. and Nash, C.A. (1981) *The Social Appraisal of Projects: a Text in Cost-Benefit Analysis*. London: Macmillan.

Peet, J. (1992) *Energy and the Ecological Economics of Sustainability*. Washington DC: Island Press.

Peters, R.H. (1991) *A Critique for Ecology*. Cambridge: Cambridge University Press.

Polis, G.A. (1994) Food webs, trophic cascades and community structure. *Australian Journal of Ecology* 19: 121–36.

Productivity Commission (1999) *Implementation of Ecologically Sustainable Development by Commonwealth Departments and Agencies*. Canberra: AusInfo.

Raffensperger, C. and Tickner, J. (1999) (eds) *Protecting Public Health and the Environment: Implementing the Precautionary Principle*. Washington DC: Island Press.

Rees, J. (1990) *Natural Resources: Allocation, Economics and Policy*. 2nd edn. London: Methuen.

Smithson, M. (1989) *Ignorance and Uncertainty: Emerging Paradigms*. New York: Springer-Verlag.

Solecki, W.D. and Michaels, S. (1994) Looking through the postdisaster policy window. *Environmental Management* 18: 587–95.

Stein, Justice P. (1993) Leatch v. Director-General National Parks and Wildlife Service and Shoalhaven City Council. *Local Government and Environment Reports of Australia* 81: 270–87.

Stein, P. (2000) Are decision-makers too cautious with the precautionary principle? *Environmental and Planning Law Journal* 17: 3–23.

Suter, G.W. (1993) *Ecological Risk Assessment*. Boca Raton, Fla: Lewis.

Suter, G.W., Barnthouse, L.W. and O'Neill, R.V. (1987) Treatment of risk in environmental impact assessment. *Environmental Management* 11: 295–303.

Talbot, Justice (1994) Nicholls v. Director National Parks and Wildlife Service, Forestry Commission of New South Wales, Minister for Planning. *Local Government and Environmental Reports of Australia* 84: 397–433.

Underdal, A. (1989) The politics of science in international resource management: a summary. In: Andresen, S. and Ostreng, W. (eds) *International Resource Management: the Role of Science and Politics*. London: Belhaven, 253–67.

Underwood, A.J. (1994) On beyond BACI: sampling designs that might reliably detect environmental disturbances. *Ecological Applications* 4: 3–15.

United Nations (1992) *Agenda 21: The United Nations Programme of Action from Rio*. New York: UN.

Verhoeven, T.J. (1995) Reducing the impacts of toxic blue green algae. In: Handmer, J.W. (ed.) *Community Preparedness for Disaster: World Disaster*

Reduction Day 1994. Canberra: Centre for Resource and Environmental Studies, Australian National University.

Walker, K.J. (1994) *The Political Economy of Environmental Policy: an Australian Introduction*. Sydney: University of New South Wales Press.

Walter, G.H. and Paterson, H.E.H. (1994) The implications of palaeontological evidence for theories of ecological communities and species richness. *Australian Journal of Ecology* 19: 241–50.

Walters, C.J. and Holling, C.S. (1990) Large-scale management experiments and learning by doing. *Ecology* 71: 2060–68.

Warner, F. (1992) (ed.) *Risk Assessment, Perception and Management: a Study Group Report*. London: The Royal Society.

Wilson, E.O. (1992) *The Diversity of Life*. New York: W.W. Norton & Company.

World Commission on Environment and Development (1987) *Our Common Future*. Oxford: Oxford University Press.

Worster, D. (1993) The shaky ground of sustainability. In: Sachs, W. (ed.) *Global Ecology: a New Arena of Political Conflict*. London: Zed Books, 132–45.

Wynne, B. (1992) Uncertainty and environmental learning: reconceiving science and policy in the preventative paradigm. *Global Environmental Change* 2: 111–27.

2 | Challenges to ecosystem management and some implications for science and policy

Tony Norton

Introduction

The ecosystems and biota we see today represent the evolution of life over a period of almost four billion years. Over the past several hundred million years the diversity of life has changed markedly as a result of evolutionary and other forces. Environmental fluctuations have led to changes in ecosystems and biota. Over the past 15 million years or so many regions of the planet have become more arid and this has led to changes in their constituent biota. As a result of changes in climate over the past two million years, glaciers have expanded and retreated a number of times. This has resulted in major expansions and contractions in the range of biota and ecosystems such as estuaries, forests and savannas. While these fluctuations have affected biological diversity, the rate of change has often been sufficiently gradual to allow organisms to adapt. They have, for example, been able to evolve, migrate and persist in refugia as a response to unfavourable conditions. While natural environmental changes and calamities have at times destroyed ecosystems and many organisms, populations of many species have survived and evolutionary processes have continued.

In stark contrast, the impacts of humans on the biosphere – particularly during the latter half of the twentieth century – have been and are profound. Today the loss of species may be in the order of 150 000 times the natural background rate, and the documented and inferred rate of loss of the planet's biological diversity over the past few decades represents the greatest episode of species loss since the end of the Cretaceous era (Wilson 1992). Given the sheer scale of the 'human enterprise' (Vitousek *et al.* 1986; Ehrlich and Ehrlich 1990), there seems little doubt that the conservation of the biosphere is intimately tied to the concept of sustainability and the progress that humanity makes towards it in the near future. The challenges that the conservation of the biosphere pose are diverse, dynamic and increasingly complex and uncertain (World Commission on Environment and Development 1987). Yet the need for wise ecosystem management would now seem more important than perhaps at any previous time in the evolution of humanity (United Nations Environment Programme 1995).

A move towards sustainability appears to require a more adequate integration of social, economic and ecological considerations. Ecological considerations may be focused around the need to maintain essential ecological processes and life-support systems, and protect biodiversity. In this contribution I discuss the difficulties that ecosystem management poses for the ecological sciences and the implications of this for policy makers and the policy process.

Ecosystem complexity

Most ecological systems are complex in nature, form and function. They often support a number of taxa that have similar functional roles and that help buffer the system against periodic disturbances. Such systems have acquired 'resilience' within the normal range of disturbances that they might experience over periods of hours to millenia. This is the nature of adaptation and evolution. Because of this, it appears uncommon for ecological systems to collapse suddenly in the absence of some dire, single event. Rather, the degradation of ecosystems and their capacity to supply human needs often results from cumulative changes to the system (Platt and Peet 1998).

These cumulative changes may take the form of various combinations of natural and human-induced impacts. Both forms of disturbance may interact to produce impacts on the system that are novel and irreversible (Pickett and White 1985; McCarthy and Burgman 1995). Depending on the state of the ecological system and the frequency of these events, changes may be sudden and extreme or they may be subtle and occur over periods of time that are difficult to detect without systematic measurement (Kendall 1998). The ecological responses these changes produce may be highly complex, synergistic and non-linear in form. They may even be chaotic (Turner and Dale 1991).

Unfortunately, the past responses of an ecological system may not provide a reliable basis upon which to set current management guidelines. Many ecological systems have been subject to cumulative impacts over the past few centuries that are likely to be well outside of their normal range (Proffitt and Devlin 1998). This means that many of the systems we attempt to manage today may already be susceptible to impacts and some may be prone to collapse. Significantly, it also means that the precise effect of the same impact on a system is likely to change over time. The modest change that a system has accommodated in the past could result in its collapse today. For example, depending on the state of a water system, the addition of a small amount of nutrients may, at one extreme, lead to the onset of an algal bloom or, at the other, produce no significant ecological difference in the system.

Ecology and ecosystem management

Ecology as a discipline is relatively young. Although partially based on ideas that were developed much earlier (e.g. by Plato and Aristotle in the fourth century BC, by Bacon in the 1600s, and by Malthus in the late 1700s), ecology as it is now understood emerged in the twentieth century and has only flourished in the past 50 years or so. It is difficult to estimate accurately the size of the ecological community since the professional societies do not represent all ecologists. Current membership of societies ranges in size from the Ecological Society of America with around 8000 members, to those with a few hundred. Probably fewer than 10 000 trained ecologists exist worldwide. Relatively few of these people are in positions of authority in government, industry or academia, although some leading ecologists play eminent roles in the policy arena and among community-based groups.

Because it is young, and the task is large, the knowledge base of ecology is very incomplete. Major disciplinary theory and concepts are few in number and include: the theory of evolution by natural selection, the concepts and mathematical analysis of population growth and demography, island biogeographic theory, niche theory, the concept of gradient analysis, small and declining population paradigms, and the concept of minimum viable population. Unlike, say, physics or chemistry, many ecological theories and concepts are still developing and are contested.

The 'core elements' of ecology are often challenged and debated. Recent examples of contested ideas include: What constitutes a species? What are the mechanisms of evolution, co-evolution, competition and species coexistence? What is the relationship between species diversity and ecosystem stability or resilience? Knowledge of most biota, ecosystems and their function is rudimentary. Less than 30 per cent of the total number of species present on the planet at this time are known to western science and basic ecological knowledge of most known species is exceptionally limited. Most natural systems vary differentially in space and time, thus making reliable prediction difficult or impossible. It has been demonstrated repeatedly that further research, while enhancing an understanding of ecosystems, may not necessarily lead to a greater ability to manage them.

Current dominant paradigms in ecology also have their limits. For example, the Popperian philosophy is dominant in ecological research whereby studies are typically formulated and directed at falsifying (null) hypotheses. This approach can work well in controlled environments such as glasshouse and other ex-situ studies but is limited for many in-situ situations due to many factors, not the least the rate of global change. It can be very difficult, if not impossible, to set up in the field experimental studies where all of the major driving variables are 'controlled'. While sophisticated experimental and statistical designs have been developed to redress some of these needs, interpretation of research data is not necessarily straightforward. This problem is exacerbated when sample sizes are unavoidably low (e.g. in the study of rare species and declining ecosystems) and the power of any statistical test (its ability to detect all but extreme changes) is low.

Another difficulty concerns the organization of the profession itself. At present, ecology is limited in its capacity to engage in and inform key policy processes that address issues of sustainability. Again, the reasons for this are often understandable. Like other sciences, ecology has been concerned with acquiring reliable knowledge of natural systems and their biota. Much of the infrastructure is related to this and to ensuring that research is empirically based and peer reviewed. The systems that are studied are typically highly complex and function over variable and often long time scales. It can frequently be extremely difficult to convey the complexity and non-linearity of natural systems to lay people and policy makers. The new agenda of sustainability further complicates discussion and analysis. Further, many regard the operating modes of ecologists and governments as incompatible. Ecologists might prefer to make assessments following comprehensive research of a problem, whereas government decisions are made on the basis of political needs; time lines are typically short and the most acceptable option or opinion may be pre-ferred among competing views and approaches to a messy problem.

The above points are made to illustrate some of the difficulties associ-ated with attempting to manage complex ecosystems in an informed manner, rather than to imply that *basic* ecological understanding is not sound or that ecologists are not actively contributing to sustainability. In fact, an impressive general understanding exists of the biosphere and its biogeochemical systems; the diversity of terrestrial and marine ecosystems that occur on the planet; broad centres of endemicity and high species richness; the geological record and its implications for evolution and the 'background rate' of species loss; the broad (ecological) functional roles played by a number of systems and species; the role that many ecosystems and species play in providing 'ecosystem services' essential to humans and their quality of life; the broad rate of environmental degradation occurring from the continental to the regional level; and many ecosystems and some species that are threatened by human activities.

The ecological profession is becoming better equipped to communicate with government and society at large. While tensions between 'science' and 'application', and issues of objectivity, values and advocacy in science remain, groups such as the Ecological Society of America (ESA-US) and the Society for Conservation Biologists have been pro-active in terms of interacting with government and policy makers. The ESA-US, for example, has established a permanent office with full-time staff in Washington DC to promote dialogue. This includes formulating various position statements (e.g. the Sustainable Biosphere Initiative; Lubchenko 1991) and respond-ing to bills, proposed amendments of Acts (e.g. US Endangered Species legislation) and many other key issues (Pitelka 1998; Platt and Peet 1998; Carpenter *et al.* 1998). An association of several societies for plant systematics has developed and promoted an international research agenda for systematics (Systematics Agenda 2000 1994). In Australia, the National Biodiversity Council (NBC) was formed in December 1994 to engage the community and government in issues concerning the conservation of biolo-gical diversity. The NBC is currently actively informing national debate over the use of several native ecosystems and the extraordinary high rate of clearance of native vegetation in many regions of the Australian continent.

Contemporary approaches to ecosystem assessment

Even so, much remains to be done and a fundamental challenge for society is how to manage ecological systems on a sustainable basis when it is difficult or simply not possible at this time to reliably identify many impacts and establish their ecological significance. This challenge has been expressed another way by ecological economics – that is, how to avoid impacts that may threaten the resilience of an ecosystem and lead to its collapse (Perrings *et al.* 1995).

Various approaches have been taken to this problem, ranging from simply ignoring it through to the use of relatively sophisticated 'decision-support' techniques. These may be qualitative, quantitative or a combination of the two and include: adaptive management (e.g. Holling 1973); safe minimum standards (e.g. Bishop 1978); precautionary principle (e.g. Dovers *et al.* 1996); performance assurance bonds (e.g. Costanza and Perrings 1990); quantitative risk assessment (e.g. Beer and Ziolkowski 1995); environmental impact assessment, and population viability analysis (e.g. Boyce 1992). All of these approaches have their limitations and some are discussed here as a reminder of the difficulties associated with the task. Difficulties associated with the use of the precautionary principle were covered in the introduction to this book.

Risk assessment has been widely employed in environmental management over the past two decades. It is now seen by some to be an integral part of high-quality environmental management systems (e.g. Beer and Ziolkowski 1995), although the best ways to use the technique remain to be resolved (Cocklin *et al.* 1992). Difficulties arise because, for example, technical, quantitative assessments of risk necessarily concentrate on aspects for which sufficient data are available. The extent to which these analyses are indicative of the likely response of the total system varies and can seriously hinder reliable interpretation. Further, the true quality and applicability of risk assessment can only be judged by the degree to which its predictions match observed impacts. This requires estimation rather than simply the detection of an impact, and necessitates that the effect be measured with minimal error and bias. If the size of an impact is relatively small compared to the long-term natural variability of the system, it may be difficult to detect with any degree of confidence. Rarely is the natural variability and disturbance regime (the nature, frequency and intensity of disturbances) reliably known for a system or its components.

In an attempt to deal with this, sophisticated environmental, ecological and biological modelling and simulation techniques have been developed to understand better the response of systems and biological populations to specified disturbances such as fire, grazing, logging, hunting/harvesting and controls on water flow (e.g. Burgman *et al.* 1993; Underwood 1994; Yodzis 1994; Naiman *et al.* 1998). The development and evaluation of simulation models and models based on dynamic programming require time series data on system dynamics (Norton and Possingham

1993). Acquiring such biophysical data is often difficult since few long-term data sets of the natural variability of systems, disturbance regimes and the response of systems to disturbances are available. Where data are available, it may be difficult to establish cause and effect relationships because the data were collected for other purposes and are not amenable to such analyses, or because the data collection protocol was inadequate. Underwood (1994), for example, discussed a number of the limitations of present approaches including the widely adopted Before–After–Control–Impact (BACI) experimental design. He reported that the conclusions drawn from such studies are often illogical.

Because of the inherent difficulties associated with assessing potential impacts, the findings of many environmental impact assessments have been of limited value. Buckley (1989), for example, concluded that the environmental impacts of development have rarely been predicted with accuracy in Australia. He compared the predictions made in environmental impact assessments (EIAs and synonymous activities) prior to development to the actual impacts determined by monitoring programmes undertaken after the development had commenced. Impact assessments for projects started prior to the late 1970s contained few testable predictions. Of some 1000 EIAs produced in Australia, Buckley (1989) reported that data sufficient to test predictions existed for only 19 projects. Of the 181 major predictions identified as testable from these projects, actual impacts proved as or less severe than predicted for 72 per cent of cases and more severe for 28 per cent. For the most critical predictions (n = 68), in terms of the significance of the potential environmental impact, 57 per cent of predictions proved as or less severe than predicted while 43 per cent were more severe. The precision of impact prediction (i.e. the ratio of actual and predicted magnitude of impact, with the smaller of the two as the numerator), ranged from 0.02–100 per cent in cases where the prediction was less severe and 0.16–96 per cent where predictions were more severe.

Significant difficulties in risk assessment also arise as a result of the use of various definitions of 'risk'. In terms of the institutions concerned with floods, hazards and emergencies, 'risk' has frequently been interchanged with the term 'hazard'. However, an increasingly common definition is that risk is the chance or probability of an event causing a potentially undesirable effect. Handmer (1995) indicated the tensions that may arise within and between institutions in terms of those approaching the issue of risk and hazards from a technical perspective compared to social and political perspectives. Environmental changes may have natural and/or human origins, but the impacts that result may not be understood outside of the human context. This means that any purely technical assessment of risk may fail to capture the true complexity of the issues at hand, and thus identify inadequate options for managing them. Dealing with risk is increasingly seen as a complex social as well as technical issue.

Improving ecosystem management

How can institutional capacity be enhanced to deal with cumulative environment change and the risks to ecosystems and the environment that result from human activities? Clearly, it is not easy – other than perhaps in the more extreme and obvious cases. At present, ecological understanding of most systems is insufficient to permit reliable, non-trivial predictions of their likely response to many perturbations. Well focused research and monitoring are important elements of any viable strategy to manage the environment on a sustainable basis. It is equally clear, however, that more research will not necessarily reduce the level of ignorance that exists with respect to many ecosystems, at least to permit the development of operationally effective management guidelines in the short to medium term (Norton 1996; Dovers *et al.* 1996). Although techniques are being developed and refined to address risk and uncertainty, many aspects are technically very difficult if not intractable. Certainly, many are intractable within the timeframes of current policy processes.

This suggests that a precautionary approach is required for undertaking environmental assessments. Such an approach will need to consider both the environmental and human dimensions of the issues at hand. Because these dimensions and the appropriate response to them will vary with situations, a flexible protocol for assessments is desirable. As a result, it is neither possible nor warranted to be too prescriptive about the conduct of such assessments. However, such assessments must be set in the appropriate biophysical, social and institutional context. Formal techniques are required for acquiring and analysing data and quantifying risks – where this is possible and appropriate (Norton *et al.* 1996). Synthesizing this information and dealing with uncertainty within a precautionary framework will involve value judgements and negotiation on a case-by-case basis. Such approaches will involve both legally binding and non-binding elements. The extent to which existing institutions and processes can maintain this flexibility, particularly when court challenges and litigation are involved, is not clear.

To help address and manage the risks and uncertainties posed by cumulative environmental change, the following areas are among those needing further consideration of how best to (1) develop and articulate a precautionary approach to environmental assessment and management; and (2) develop consistent guidelines for the acquisition and use of data, and the use of analytical methods for assessment. Reliable comparison of the nature of impacts is often difficult because few standard procedures exist for assessing environmental impacts and there are many competing assessment designs, each replete with their own inherent and often different assumptions. Hence consistent operating guidelines are required to assess cumulative changes and potential environmental impacts. These guidelines should specify the means by which environmental context is established, and detail the essential components of an ecosystem that should be evaluated – guidance that is lacking from many current processes.

Because of concerns over the ecological and environmental implications of employing inadequate methods and accepting incorrect hypotheses, the traditional technical means of identifying and characterizing significant impacts on a system are being challenged (e.g. Norton and Williams 1992). For much of the last century, the rejection of null hypotheses has been based on arbitrary levels of statistical significance that are very high; probabilities of 0.05 or 0.01 are the norm. Since few hypotheses about environmental impacts are amenable to controlled experimentation, and the ecological and environmental implications of accepting incorrect hypotheses may be serious and irreversible, the blanket application of such statistical standards appears inappropriate. It has been suggested that a cautious approach to environmental management should be most concerned about failing to detect significant impacts on a system, for instance in the case of anticipated global climate change (e.g. Orians 1993; Williams *et al.* 1994; Endter-Wada *et al.* 1998). A precautionary approach brings with it a new value system whereby new standards of impact and risk assessment are set to help minimize the likelihood of failing to detect impacts and/or underestimating the significance of an impact. These issues require further consideration.

The need to provide 'context' for ecological and environmental assessments and to allow for potential impacts resulting from cumulative change is increasingly recognized. Meeting this need is not straightforward, however. Typically, assessments are required for activities that are specific to a site or area, and time period. Rarely do these sites or areas encompass (in space and time) the total system and processes that may sustain the environmental attributes of concern. In the absence of this broader context, it is possible to fail to identify an impact or to miscalculate its potential ecological significance. Clearly, the potential for inadequate or flawed ecosystem assessments might be minimized if contextual information were routinely used during analyses. For example, from an ecological perspective information is required on: (1) the relationship of the site/ area of concern to the larger system that sustains it; (2) the natural spatio-temporal variability of the larger ecological system; (3) the state (condition) of the larger ecological system; and (4) the state of key attributes of the site or area in relation to the larger ecological system. Data are often available, or can be obtained within a reasonable timeframe, to address (1) and (3). Modern, computer-based environmental and biological data sets, remote-sensing technology and sophisticated analytical techniques typically permit the states of the sites, areas and systems of concern to be characterized at least at a broad scale (e.g. Woinarski 1992; United Nations Environment Programme 1995). A basic rationale for such approaches and case studies of the applications of techniques at the landscape and ecosystem scale can be found, for example, in Nix (1994). More difficulty (and controversy) arises in addressing (2) and (4), and technical aspects may only be addressed adequately through long-term monitoring.

Monitoring provides a basis for acquiring data on the natural variability of systems and their response to imposed changes. A broad range of environmental monitoring programmes are employed by government, research institutions and private industry. Examples include the monitoring of water quality in catchments supporting both urban areas and intensive agriculture in Europe, North America and eastern Australia. Research agencies monitor, among other things, physical and chemical changes in the atmosphere and Southern Oceans as a basis for understanding environmental change. In contrast, far fewer ecological monitoring programmes are extant and these are typically limited to relatively straightforward exercises such as tracking the recruitment and persistence of vascular plants on rehabilitated mine sites; assessing the effects of pollutants (e.g. sulphur dioxide, heavy metals) on native plants and animals (e.g. mangroves, wetlands, heathlands, rats, birds, shellfish) downstream from the point source; and recording fluctuations in the abundance of exotic plants and animals at disturbed sites (e.g. mine sites, forestry cut-overs).

To help evaluate the less obvious human impacts on natural systems, the monitoring of selected organisms has been suggested as a basis for tracking finer-scale fluctuations, say within and between populations of taxa. This general concept has been employed successfully to help monitor the environmental impact of some forms of waterborne and airborne pollution in the northern hemisphere. Three main categories of species which may prove useful for ecological monitoring were identified by Soulé and Kohm (1989) and discussed in the Global Biodiversity Assessment (United Nations Environment Programme 1995): 'indicators', 'keystones' and 'mobile links'. They suggested that indicator species be chosen for monitoring because they are representative of a particular use, ecosystem or management concern. A keystone species is one that, by being lost from a system, leads directly or indirectly to the virtual disappearance of several other species. Species which are important functional components of more than one food chain, plant–animal association, or ecosystem are termed mobile links. Examples of keystone taxa include: top carnivores; large herbivores and termites; species that maintain landscape features important to a variety of other species; pollinators and other mutualists; seed dispersers; and parasites and pathogenic microorganisms that may control population explosions of host species. However, the utility of the overall approach for informing sustainable management is yet to be adequately assessed and requires further research.

Of the ecological monitoring programmes aiming to collect long-term time series data, relatively few have been in progress for more than a few decades; most are much shorter (Common and Norton 1994). Most monitoring exercises are episodic and short-term because they are often driven by a particular research problem of relatively short-term interest, or are reliant on non-secure funding from government. There exists a need to systematize ecological monitoring so as to provide an enhanced predictive understanding of natural systems. Common and Norton (1994) argued that a prime objective of ecological monitoring should be to detect long-term and cumulative processes, changes and impacts.

The problem of detecting human disturbances in ecological systems is exacerbated when the ecological systems under study are uncommon or rare and sample sizes are unavoidably low. This limits the ability ('power') of any statistical test to detect all but extreme changes. The statistical power of methods used in quantitative risk assessment is based on the probability of rejecting the null hypothesis of 'no effect' when this is false. Calculation of the statistical power of an experimental test requires information on the number of replicates as well as the ratio between the size of an effect and the variability among the replicates. Matters are further complicated by the fact that developments are *not* randomly located in space nor replicated statistically, experimental techniques employed to assess potential impacts need explicitly to accommodate the lack of spatial replication and randomization. Improved means are required to provide biophysical context for such analyses. However, well-planned ecological monitoring could readily help redress many of the present limitations in this area.

Implications for institutional arrangements

The concept of sustainable development or ecologically sustainable development has been adopted by most nations and is reflected in much recent government policy and legislation. A number of institutional and legislative initiatives have been put in place to promote sound environmental management. However, many of these initiatives may ultimately fail in the absence of greater efforts to address the key issues raised above. In many situations, the available decision-support tools for assessing the merits of various resource uses and proposed uses are grossly inadequate. Hence, it can be difficult or simply not possible to provide reliable advice to government and other decision makers about the environmental impact of proposed or current human activities at the operational level. The reasons for these difficulties are reasonably clear and many were discussed earlier. The motivation for writing this chapter is that these difficulties should be more widely acknowledged and systematically addressed. Moreover, it is important to appreciate the fact that a number of key issues are very difficult to resolve given present scientific understanding and the available disciplinary apparatus. Many gaps in information, knowledge and understanding that underpin a move to sustainability cannot be filled in the near future. Some may never be filled. This information is significant because it provides another perspective when considering options for promoting sustainable development, particularly at the national and international level.

Communication and problem definition

Presently, the discipline of ecology is often not well placed to engage in and inform key policy processes that address issues of sustainability. The transfer of existing knowledge to the policy process is limited, and the development of research agendas and related activities may not be adequately informed by the current or prospective needs of decision makers.

Where the uncertainty is high, it is clear that normal scientific processes and professional consultancies are usually inadequate. Funtowicz and Ravetz (1990) proposed a simple model of the relationship between system uncertainty and the decision stakes. As the decision stakes and uncertainty increase, one moves from 'applied science' through 'professional consultancy' to 'post-normal science'. Presently, ecological research falls mostly within the area of 'applied science', whereas probably most issues awaiting resolution for sustainability fall within 'post-normal science'. In general, ecologists are currently poorly equipped to deal with issues beyond 'applied science' and which typically fall outside of their immediate peer group. However, this disciplinary limitation is by no means unique. A recent initiative to help broaden the range of disciplines considering issues of sustainability fundamental to the future of humanity has led to the creation of 'ecological economics' (e.g. Costanza 1989; Common 1995; Perrings et al. 1995; Endter-Wada 1998). An initial research agenda for ecological economics was proposed by Perrings et al. (1993) and some of the issues covered here build on their assessment.

Means to enhance the capacity of ecology to inform policy processes more usefully require immediate attention and much more debate. A number of changes to the way ecology is undertaken, taught and communicated are required if the discipline is to contribute significantly to sustainability policy. New educational curricula are required that expose people to the ecological and human dimensions of the spheres within which they operate and are accountable. Associated with this would be changes to the incentive structures and peer assessment employed in career development. This is essential because it takes time to appreciate adequately the complexity of the issues at hand, and to develop and exploit serious collaborations of a cross-disciplinary or multidisciplinary nature. Without the latter we may be poorly placed to tackle many major issues of sustainability in the twenty-first century. Existing institutions have an important role to play in promoting and supporting these changes and should be encouraged to do so.

Building capacity for long-term ecological monitoring

Long-term ecological monitoring is a central component of any move to sustainability because it provides a basis to track fluctuations in specified components of the environment and thereby evaluate the appropriateness of management regimes and options for achieving sustainability goals (e.g. Common and Norton 1994). Enhancing institutional capacity for monitoring requires a long-term commitment by society of adequate resources to support this activity. At the same time, further debate and agreement are essential as to which components of specified ecological systems will be monitored and how these data and information will be used. New international and national initiatives by governments and industry (e.g. UN Convention on Biological Diversity; state of the environment reporting; resource use strategies and agreements) provide opportunities to address the latter need in a more open, coherent and ecologically informed manner.

Ecosystem management and contingency planning

Contingency plans have been formulated to deal with major emergencies and threats to life and property. However, outside of the marine environment and contaminated sites, few formal plans have been developed to deal with major emergencies and disasters that threaten biological diversity and key ecosystems and biota providing essential ecosystem services. For example, extensive modification of landscapes and ecosystems in coastal, eastern Australia by Europeans now predisposes remnant ecosystems and their biota to collapse when exposed to phenomena such as catastrophic fire and extended drought. Costly mechanisms to ensure the conservation of remnant areas and populations of biota are likely to be of limited long-term value unless such contingencies are also in place. Clearly, contingency plans need to be established for a range of key ecosystems and biota. The nature of these plans and how they might cost-effectively exploit existing institutional arrangements and infrastructure needs much more attention.

Conclusions

Given the sheer scale of the human enterprise, it is no surprise that many institutions are struggling to address existing problems, let alone refocus their limited resources to tackle effectively the emerging issues of sustainability. Many professional disciplines have an important role to play in reshaping the institutions and policy processes required to manage the biosphere more wisely. The above arguments suggest that ecologists and policy makers need to become more familiar with issues of sustainability at all scales and become more aware of the human dimensions and policy frameworks within which such decisions are made. Similarly, policy makers and ecologists need to become more aware of the nature and limitations of ecology and science as means to support the goals of sustainability.

Acknowledgements

A draft of this contribution was prepared while the author was a visiting scientist in the Environment Department at the University of York, UK.

I especially thank the Head of Department, Professor Charles Perrings for his intellectual contributions and support during this period.

References

Beer, T. and Ziolkowski, F. (1995) *Environmental Risk Assessment: an Australian Perspective.* Canberra: Australian Government Publishing Servile.

Bishop, R. (1978) Endangered species and uncertainty: the economics of a safe minimum standard. *American Journal of Agricultural Economics* 60: 10–18.

Boyce, M. (1992) A critical review of population viability analysis. *Annual Review of Ecology and Systematics.*

Buckley, R.C. (1989) *Precision in Environmental Impact Prediction.* Resource and Environmental Studies 2. Canberra: Centre for Resource and Environmental Studies, The Australian National University.

Burgman, M.A., Ferson, S. and Akçakaya, H.R. (1993) *Risk Assessment in Conservation Biology.* New York: Chapman and Hall.

Carpenter, S.R., Caraco, N.F., Howarth, R.W., Sharpley, A.N. and Smith, V.H. (1998) Nonpoint pollution of surface waters with phosphorus and nitrogen. *Ecological Applications* 8: 559–68.

Cocklin, C., Parker, S. and Hay, J. (1992) Notes on cumulative environmental change I: concepts and issues. *Journal of Environmental Management* 35: 31–49.

Common, M.S. (1995) *Sustainability and Policy. Limits to Economics.* Cambridge: Cambridge University Press.

Common, M.S. and Norton, T.W. (1994) Biodiversity, natural resource accounting and ecological monitoring. *Environmental and Resource Economics* 4: 29–53.

Costanza, R. (1989) What is ecological economics? *Ecological Economics* 1: 1–7.

Costanza, R. and Perrings, C.A. (1990) A flexible assurance bonding system for improved environmental management. *Ecological Economics* 2: 57–76.

Dovers, S.R., Norton, T.W. and Handmer, J.W. (1996) Uncertainty, ecology, sustainability and policy. *Biodiversity and Conservation* 5: 1143–67.

Ehrlich, P.R. and Ehrlich, A.H. (1990) *The Population Explosion.* New York: Simon and Schuster.

Endter-Wada, J., Blahna, D., Krannich, R. and Brunson, M. (1998) A framework for understanding social science contributions to ecosystem management. *Ecological Applications* 8: 891–904.

Funtowicz, S.O. and Ravetz, J.R. (1990) *Uncertainty and Quality in Science for Policy.* Dordrecht: Kluwer.

Handmer, J.W. (1995) Issues emerging from natural hazard research and emergency management. In: Norton, T.W., Beer, T. and Dovers, S.R. (eds) *Risk and Uncertainty in Environmental Management,* 54–67. Proceedings of the 1995 Fenner Conference on the Environment, Australian Academy of Sciences and the Centre for Resource and Environmental Studies, The Australian National University, Canberra.

Holling, C.S. (1973) Resilience and stability of ecological systems. *Annual Review of Ecology and Systematics* 4: 1–23.

Kendall, B. (1998) Estimating the magnitude of environmental stochasticity in survivorship data. *Ecological Applications* 8: 184–193.

Lubchenko, J. (1991) The sustainable biosphere initiative: an ecological research agenda. *Ecology* 72: 371–412.

McCarthy, M. and Burgman, M. (1995) Coping with uncertainty in forest wildlife planning. *Forest Ecology and Management* 74: 23–36.

Naiman, R.J., Magnusson, J.J. and Firth, P.L. (1998) Integrating cultural, economic, and environmental requirements for fresh water. *Ecological Applications* 8: 569–70.

Nix, H.A. (1994) Assessment framework and information requirements. In: Jakeman, A.J. and Pittock, A.B. (eds) *Climate Impact Assessment Methods for Asia and the Pacific*. Canberra: Australian International Development Assistance Bureau, 3–8.

Norton, T.W. (1996) Conserving biological diversity in Australia's temperate eucalypt forests. *Forest Ecology and Management* 86: 128–42.

Norton, T.W. and Possingham, H.P. (1993) Wildlife modelling for biodiversity conservation. In: Jakeman, A.J., Beck, B. and McAleer, M. (eds) *Modelling Change in Environmental Systems*. Sydney: Wiley and Sons, 243–65.

Norton, T.W. and Williams, J.E. (1992) Habitat modelling and simulation for nature conservation: a need to deal systematically with uncertainty. *Mathematics and Computers in Simulation* 33: 379–84.

Norton, T.W., Beer, T. and Dovers, S.R. (1996) *Risk and Uncertainty in Environmental Management*. Proceedings of the 1995 Fenner Conference on the Environment, Australian Academy of Sciences and the Centre for Resource and Environmental Studies, The Australian National University, Canberra.

Orians, G.H. (1993) Policy implications for global climate change. In: Kareiva, P., Kingsolver, J.G. and Huey, R.B. (eds) *Biotic Interactions and Climate Change*. Sunderland, Mass: Sinauer Associates, 467–79.

Perrings, C.A., Folke, C. and Maler K.-G. (1993) The ecology and economics of biodiversity loss: the research agenda. *Ambio* 22: 201–11.

Perrings, C.A., Maler, K.-G., Folke, C., Holling, C.S. and Jansson, B.-O. (1995) *Biodiversity Conservation*. Dordrecht: Kluwer Academic Publishers.

Pickett, S.T.A. and White, P.S. (1985) *The Ecology of Natural Disturbance and Patch Dynamics*. Orlando, Fla: Academic Press.

Pitelka, L.F. (1998) Air pollution and terrestrial ecosystems. *Ecological Applications* 8: 627–8.

Platt, W.J. and Peet, R.K. (1998) Ecological concepts in conservation biology: lessons from Southeastern U.S. ecosystems. *Ecological Applications* 8: 907–8.

Proffitt, C.E. and Devlin, D.J. (1998) Are there cumulative effects in red mangroves from oil spills during seedling and sapling stages? *Ecological Applications* 8: 121–7.

Soulé, M.E. and Kohm, K.A. (1989) *Research Priorities for Conservation Biology*. Covelo, Ca: Island Press.

Systematics Agenda 2000 (1994) Produced by Systematics Agenda 2000 – a consortium of the American Society for Plant Taxonomists, the Society for Systematic Biologists, and the Willi Hennig Society in co-operation with the Association of Systematics Collections, New York.

Turner, M.G. and Dale, V.H. (1991) Modeling landscape disturbance. In: Turner, M.G. and Gardner, R.H. (eds) *Quantitative Methods in Landscape Ecology*. New York: Springer Verlag, 323–51.

Underwood, A.J. (1994) On beyond BACI: sampling designs that might reliably detect environmental disturbances. *Ecological Applications* 4: 3–15.

United Nations Environment Programme (1995) *Global Biodiversity Assessment*. Cambridge: Cambridge University Press.

Vitousek, P.M., Ehrlich, P.R., Ehrlich, A.H. and Matson, P. (1986) Human appropriation and the products of photosynthesis. *BioScience* 36: 368–73.

Williams, J.E., Norton, T.W. and Nix, H.A. (1994) *Climate Change and the Maintenance of Conservation Values in Terrestrial Ecosystems*. A report to the Climate Change and Marine Branch, Department of the Environment, Sport and Territories, Canberra.

Wilson, E.O. (1992) *The Diversity of Life*. London: Penguin.

Woinarski, J.C.Z. (1992) Biogeography and conservation status of mammals, birds and reptiles of north-western Australia: an inventory and base for planning an ecological reserve system. *Wildlife Research* 19: 665–705.

World Commission on Environment and Development (1987) *Our Common Future*. Oxford: Oxford University Press.

Yodzis, P. (1994) Predator–prey theory and management of multispecies fisheries. *Ecological Applications* 4: 51–8.

3 | Bioregional approaches to conservation: local strategies to deal with uncertainty

Nels Johnson, Kenton Miller and
Marta Miranda

Introduction

Managing for the conservation and sustainable use of biological resources – in developed and developing countries alike – involves making decisions in the face of considerable uncertainty. Traditional strategies for nature conservation focused principally upon national parks and reserves, and individually threatened species and populations. Science has shown, however, that these key measures need to be complemented with steps taken at geographic scales greater than those typically used to protect species or national parks. For example, successful restoration of an endangered species may require a land area several times larger than available protected areas. Moreover, critical ecological processes that shape habitat structures and species composition, including fire regimes, migration patterns, and nutrient cycles frequently operate at broad geographic scales. Thus, while managers must contend with uncertainty within the boundaries of protected areas, they will need to address greater levels of uncertainty at the bioregional scale. These areas include communal lands and private holdings, in an array of uses, from farm and harvested forests or fishing grounds to villages and infrastructure.

These bioregional management efforts are unlikely to succeed without local capacity to cope with scientific uncertainties, dynamic and unpredictable economic and social change, and the establishment of flexible and responsive policy and institutional arrangements. Local communities and decision makers everywhere, however, face obstacles that limit their ability to act as responsible stewards of ecosystems and their biological wealth. We believe greater investment to enable effective local community participation in bioregional management efforts is a key strategy for conserving and sustainably using biological resources in the face of uncertainty.

Uncertainty, bioregional management and the importance of local capacity

Uncertainty is a constant companion of natural resource managers and policy makers, and it comes in various forms. Managers and decision makers must not only contend with scientific uncertainty, but they also face considerable social and economic uncertainty, and capricious policy and institutional environments. These three categories of uncertainty are briefly reviewed below.

Scientific uncertainty is widely recognized as a problem in determining appropriate courses of action for the management of complex biological systems (UNEP 1995; Wright 1999). Such basic information as geographic range, reproductive rates, population size, and habitat requirements are known for only a few tens of thousands of the world's estimated 14 million species (e.g. UNEP 1995; Ceballos *et al.* 1998). When it comes to dynamic interactions between species and their environment, uncertainties are the rule rather than the exception. For example, debates over whether biodiversity contributes to ecosystem stability have raged for decades, and just when consensus seemed to have settled on there being no clear relationship, new research indicates species diversity does protect productivity in grassland ecosystems subjected to drought (Tilman *et al.* 1996). But does this relationship apply to other ecosystems? Is there a threshold effect in terms of species loss and ecosystem productivity decline? What indicators might help to determine whether such a threshold is being approached? While such uncertainties create problems for scientists asked to recommend appropriate policy and management responses, they create even greater confusion and challenges for managers on the ground.

Uncertainty in management decisions is also due to the complexities and dynamism of social and economic systems in which resource management decisions are made. A change in agricultural crop subsidies can lead to almost instantaneous changes in land use decisions by farmers, with uncertain impacts on ecosystems. A lack of institutional continuity also hampers the success of local conservation projects, as administrations change and agency directors sympathetic to community-based conservation are replaced by those who are not (Feldmann 1994; Dovers 1997). What can conservation strategies do to survive such rapid change? If the synergistic effects of local tenurial systems, regional demographic patterns, national economic development policies, and international trade and structural adjustment policies are the root causes of tropical deforestation (Repetto and Gillis 1988), which factors are most important in a specific place? What information and what access to the policy-making process at which levels are needed to counter destructive economic and social synergies?

Finally, these scientific, social and economic uncertainties are compounded by untested policy and institutional arrangements that are accompanying widespread *de jure* and *de facto* devolution and decentralization of natural resource management responsibilities. Local communities from Indonesia to Brazil, for example, have become more assertive in their efforts either to preserve customary local control of forest resources or to wrest back control from unresponsive and frequently ineffective centralized forest management institutions (Lynch and Talbott 1995). In the United States, frustration with centralized planning on public lands is leading both environmentalists and producer groups to demand greater local and regional involvement in the management of public lands (Riebsame 1996). Elsewhere, governments have deliberately moved to share management authority and responsibility over natural resources. For example, authority and responsibility for managing coastal resources in the Philippines has recently been transferred to local governments in recognition of the limited capacity the national government has to manage these resources in a country of 7000 islands with nearly 20 000 kilometres of coastline.

The prospect of new policy and institutional arrangements introduces more uncertainty into natural resource management decisions. What management goals and standards will be used? Who is involved in planning? What is the most appropriate scale of management that adequately addresses ecosystem functions, economic regions, and social dynamics? Where do local institutions get financial resources, technical assistance, and information that is not locally available? How are national and international interests in local biological resources to be accounted for?

Bioregional management approaches – that is, the co-operative planning and implementation of conservation efforts over a large geographic area or bioregion – have a number of characteristics that can help managers sustain biodiversity and natural resource production in the face of uncertainties in each of the categories described above. For example, in recognition of pervasive scientific uncertainties, bioregional management emphasizes research and monitoring, use of adaptive management techniques, and consideration of local and traditional knowledge that may address issues for which science and other formal disciplines can provide little guidance. Based on an analysis of conservation projects around the world, Miller (1996) identified 14 characteristics of a comprehensive bioregional approach to conservation (Table 3.1).

The study by Miller (1996) demonstrated that bioregional management approaches call for a new balance of power between central governmental and regional or local levels of public, communal and private management. Since, typically, bioregions are geographically larger than most established protected areas, management programmes need to be re-scaled to embrace whole ecosystem-wide spaces, such as upstream catchments, mountain slopes, and coastal zones. Thus, national policies and administrative mechanisms need to be shifted 'downward' to encourage and provide incentives to bioregional-level institutions, while conservation mechanisms need to scale 'upward' to step out of the boundaries of agency jurisdictions, working with other agencies and individuals throughout the bioregion, wherever biodiversity is found.

No longer can natural resource policies formulated in a capital city hundreds or thousands of kilometres away be thought of as a sufficient way to maintain biodiversity and ecological processes. Governments must be willing to recognize that centralized and overly-generalized 'recipes' for natural resource management are insufficient to address the array of activities at the regional and local levels. Thus, one fundamental strategy for the maintenance and restoration of biodiversity is to strike a new balance in the roles and responsibilities of central government agencies, and regional and local programmes and entities. Chief among the features of this shift is getting government to provide encouragement and incentives for local stewardship while offering solid scientific, technical and managerial orientation and guidance. At the other end of the scale, regional and local communities will need to develop the capacity to plan and implement the many tasks associated with mixed-use, mixed-ownership area management. Even though the inherently larger scale of bioregional management poses significant challenges for co-ordination and negotiation, involvement of local players can be a more efficient and cost-effective approach to management, a point that is not lost on cash-strapped central authorities now faced with shrinking budget-driven decentralization and devolution pressures.

To minimize social and economic uncertainties, bioregional management features broad and substantial involvement of all stakeholders, the use

Table 3.1   Key characteristics of bioregional management.

From worldwide experience of integrated conservation and development projects, ecosystem management efforts, and biosphere reserves, the following can be identified as common characteristics of bioregional management:

1. *Large regions* – Bioregional management programmes embrace regions large enough to include the habitats and ecosystem functions and processes needed to make biotic communities and populations ecologically viable over the long term. These regions must be able to accommodate migratory patterns, anticipate nature's time cycles, and absorb the impacts of climate change.
2. *Leadership and management* – The leadership to establish bioregional programmes may come from public agencies or from the community of residents and resource users. The tasks of convening stakeholders, negotiating vision statements, planning and implementing agreed activities can be shared co-operatively between public and private entities, or be fully community-based.
3. *Cores, corridors and matrices* – Core wild-land sites feature representative samples of the region's characteristic biodiversity. Ideally, such sites, which may already be designated as protected areas, are linked by corridors of natural or restored wild cover to permit migration and adaptation to global change. Both the core sites and corridors are nested within a matrix of mixed land uses and ownership patterns.
4. *Economic sustainability* – The livelihoods of people living and working within the bioregion, and especially in the matrix, are encouraged. Appropriate incentives to make optimal use of local resources, and apply sustainable technologies, are combined with a system for sharing the costs and benefits fairly.
5. *Full involvement of stakeholders* – All parties who can affect or benefit from the resources have the opportunity to be fully involved in planning and managing the bioregional programme. Key here is building the local capacity to participate, negotiate, and perform the various tasks involved.
6. *Social acceptance* – Any proposals for changes in the way of life and livelihoods of the residents and local peoples, including indigenous communities, need to be acceptable to them.
7. *Solid and comprehensive information* – All stakeholders have access to critical information prepared to facilitate planning and management. Geographic Information System technology is used to help stakeholders envision their region and its distinctive features clearly. GIS also helps them model options and scenarios for the future.
8. *Research and monitoring* – Research and inquiries focus on people/environment interactions, the development of innovative methods for the managing of natural resources, and the long-term monitoring of environmental factors and the impact of management practices.
9. *Use of knowledge* – Scientific, local, and traditional knowledge are employed in planning and management activities. Biology, anthropology, economics, engineering and other related fields are all tapped.

Table 3.1 *(cont'd)*

---

10. *Adaptive management* – Bioregional programmes operate on an experimental basis, drawing lessons from real-world experience, and responding appropriately.
11. *Restoration* – Restoration is pursued where the viability of some habitats or ecological functions have been impaired through excessive or inappropriate use.
12. *Co-operative skills development* – Communities and public and private organizations together locate and mobilize the skills, knowledge, and information needed to manage the area.
13. *Institutional integration* – Alliances with other institutions and with local organizations are forged to close gaps, minimize overlap, and make management and investment in the region more efficient.
14. *International co-operation* – Because some ecosystems cross international boundaries and, in some cases, extend globally along animal migration routes, international co-operation may be required.

Source: Adapted from Miller (1996)

of social and economic information in planning, and an emphasis on providing appropriate economic and social incentives to support both livelihoods and conservation goals. Finally, the importance of co-operative leadership and management skills at all levels, sharing of knowledge, technology, and information, and the use of formal and informal agreements and alliances can help to create a flexible institutional policy environment that translates national or state/provincial goals into local actions.

Obstacles to effective local participation in bioregional management

Conservation experiences worldwide show that bioregional management approaches can encourage local communities and institutions to improve the sustainability of natural resource management practices. However, numerous obstacles frequently limit the effective participation of local interests in bioregional management. Most of these obstacles fall into one of three categories:

- a lack of institutional co-operation and political will from centralized institutions;
- a lack of authority, tenure, or incentives for local communities and institutions;

■ a lack of financial, human, and institutional capacity, especially at a local level.

In addition, limited time on the part of resource managers and users combined with incomplete access to available information can hinder the engagement of local interests. If these obstacles are not addressed, local interests may limit their contribution toward bioregional conservation goals. They may even actively resist bioregional programmes whose goals and methods are uncertain and possibly counter to their own priorities.

Lack of institutional co-operation and political will

In any bioregion, the landscape is already filled with an array of public and private organizations and institutions that control, influence, or affect the management of natural resources. The institutional arrangements already in place are often organized around traditional sectoral concerns including forestry, agriculture, fisheries, water, transportation, wildlife, and tourism. These sectoral agencies often operate separately, except where differing goals create conflict. In addition to a lack of co-operation amongst sectoral interests, conflict between local, state/provincial, and national authorities can complicate efforts to catalyse co-operative efforts toward bioregional conservation goals. Sudden reversals in national policy can destroy the co-operation and trust built to reach bioregional goals defined in partnership with local interests.

The inability of governments to plan and carry out national policy that supports conservation action at the local level is a widely cited obstacle to co-operative resource management and conservation efforts in both developing and developed countries (e.g. Western et al. 1994; Keystone Center 1996). This lack of political will often manifests itself in a failure to revise policies, provide resources, or to ensure co-operation between sectoral institutions, and between local, state/provincial and national organizations. When this happens, promising initial efforts to develop bioregional management projects become compromised, delayed, or totally derailed.

Australia's first co-management experience in Kakadu National Park illustrates how conflicting interests delayed the establishment of an effective management regime for the park (Hill and Press 1994). First proposed in the mid-1960s, it took over two decades before conflicts between Aboriginal, conservation and mining interests in Kakadu National Park were settled and permanent management arrangements put in place. From the beginning, the fate of the national park has been complicated by competing rather than co-ordinated government activities. In the late 1960s, as the Australia National Parks and Wildlife Service began planning the park, the Commonwealth government issued leases for exploratory uranium mining for most of the proposed park area. After the Commonwealth government granted self-government to the Northern Territory in 1978, the territorial government sided with mining interests and has consistently opposed the park. As various proposals for defining park boundaries and management plans were developed by the Commonwealth government, Aboriginal interests were largely ignored despite the fact that Aboriginal peoples had lived in the region for as long as 50 000 years (Hill and Press 1994). Meanwhile, conservationists who opposed the mining interests were also sceptical of Aboriginal proposals for greater involvement in park management.

It was only after years of slowly integrating Aboriginal views into the development of policy and management arrangements that some level of certainty in the management Kakadu National Park was established. By the late 1980s, mining in the park was banned by the Commonwealth government, conservation activities were being widely implemented, and Aboriginal participation had become integral to park management activities. Although the various conflicts nearly killed the park on several occasions, Kakadu National Park is now viewed as a successful model for co-management between Aboriginal people and the national government in Australia (Hill and Press 1994).

Another common obstacle to local engagement is the lack of commitment by national institutions to follow through on specific policy reforms promised in connection with a bioregional management effort. Sharing or decentralizing authority and responsibility is often fundamental to such reforms and provides a key incentive for local engagement, but as Feldmann (1994) noted, the failure to implement reforms generates dangerous gaps between the promises of central authorities and the expectations of local people.

For example, to compensate Masai pastoralists for the establishment of Amboseli National Park, the Kenyan government promised a set of benefits and incentives, including a pipeline to provide water for livestock outside the park, direct financial compensation for loss of game previously hunted within park boundaries, organization of hunting concessions outside the park, and assistance in developing wildlife tourism (Wells and Brandon 1992). However, many promises made by the government were not kept, with disruptive effects for both people and wildlife. A promised water pipeline, for example, did not provide adequate water supplies for livestock, forcing the pastoralists to enter the park illegally to meet the needs of their livestock, and crowding wildlife out of critical food and water sources during the dry season. As a result, festering conflicts over cattle grazing in protected areas, wildlife damage to crops and livestock outside protected areas, and the question of who benefits from the tourism revenue generated by wildlife conservation have lengthened the odds for biodiversity conservation success in the long term (Miller 1996).

In some cases, national agencies perceive local players as possessing little information of value for conservation objectives. This 'we know best' attitude can undermine efforts to stimulate community involvement. For example, the Coalition for Unified Recreation in the Eastern Sierra (CURES) attempted to address recreation planning and management in the California Sierras. However, the project encountered barriers when local players were drawn into the debate, mostly because some Forest Service personnel found it difficult to open the management process to outsiders. As one agency representative stated, 'We may feel that we are giving away power or authority' (Wondolleck and Yaffee 1994). Indeed, many national agencies are not accustomed to allowing others into the debate on natural resources management. This may be partially a result of legal and policy barriers that limit the participation of federal agency members in non-governmental or even inter-agency initiatives (Keystone Center 1996).

While national agencies must be prepared to involve other players, bioregional approaches also require co-ordination between and among a vast array of players at the local level. Not only do governments need to foster co-operation among agencies and institutions, but agreements and negotiations will need to be forged among a variety of local, regional, and national interests. Such efforts require considerable trust, effort and time, all of which may be lost should any members of the original group need to be replaced with newcomers. For example, the Deerlodge Forest Plan in Montana required considerable negotiation among a variety of players before all were willing to sign the plan. This process was seriously hampered by a relatively high turnover of forest managers and personnel who had participated in the original negotiations (Wondolleck and Yaffee 1994). Because effective co-operation requires establishing relationships of trust among individuals and institutions, a turnover of personnel in an agency or a simple change in landownership within the bioregion can significantly alter the will to implement previous agreements.

Bioregional efforts may never have a chance to get off the ground if central authorities make no provision for voluntary local participation.

For example, in Niger, widespread deforestation, erosion, and desertification undermine the country's ability to feed itself and maintain already meagre standards of living. Political conditions in Niger restrict the organization of inhabitants in rural communities to undertake amongst other goals the revitalization of the country's depleted forests. Moreover, the central government controls most forms of communication, monitoring, natural resource management activity, and regulation (Otto and Elbow 1994). As a result, successful natural forest management projects in Niger are difficult to nurture.

On the other hand, decentralizing authority too far can have detrimental consequences for national conservation and resource management objectives. In Korea's Mt Sorak National Park, for example, local village leaders planned to develop infrastructure for tourism that would have essentially destroyed the basis for creating the park in the first place (Ishwaran 1995). In the United States, the creation of most national parks and wildlife refuges has encountered stiff local opposition by residents more interested in the certain benefits that mining, logging, or grazing would bring than in the uncertain benefits from conservation and tourism. Striking a balance in the amount of authority awarded to communities and redefining the role of central authority provide perplexing challenges to decision makers charged with managing resources on a bioregional scale.

Lack of tenure, authority and incentives

Implementing more sustainable natural resource management practices frequently means local people must incur short-term costs in the hope of generating longer-term benefits. In many rural areas, however, this trade-off is difficult or impossible to make because abilities to capture or control future benefits are highly uncertain. For example, a lack of legal tenure to forest or fisheries resources means that an investment in sustainable practices made today may yield benefits to someone else tomorrow (e.g. Lynch and Talbott 1996; Zerner 1994). Or, a local agency may be handed additional responsibilities for managing natural resources without being given authority to collect revenues that could be used to reinvest in resource management. Finally, even if appropriate legal and tenurial authorities are in place, local people and institutions may have few incentives to undertake conservation actions that provide uncertain benefits in comparison to maintaining status quo management practices.

In many developing countries, a lack of secure land or resource tenure by local communities exacerbates and accelerates the destruction of forest resources (Sharma 1992). In Indonesia, traditional *adat* or community tenure rights to local forest resources have been widely displaced by state assertions of national interests in the control and disposition of forest lands (Barber *et al.* 1994). The central government has thus granted logging concessions on millions of hectares of land that already had well-established *adat* rights claimed by local communities. The resulting uncertainty about the fate of their forests in Sumatra, for example, has led some communities to join the destructive rush to exploit the increasingly rare lowland forests:

> . . . for the large majority of local people, securing their rights through obtaining [government land] certificates is not a realistic option. They have but one possibility left: to force the traditional land tenure system to its bitter end, hoping that at least some kind of recognition will be given to them when the land is expropriated. Thus, their strategy is to clear as much land as possible within previously uncleared forest before somebody else does so. 'We know we are destroying our forests, but it is a race and whoever does not join it will lose' is the fear expressed by villagers as they move to new forest areas. (Ostergaard 1993)

Where conservation is yielding government revenues and economic benefits, agencies may lack the authority to capture those benefits for local use. This is a common constraint in the management of protected areas. Wells and Brandon (1992) found that in 23 protected areas, the vast majority of visitor fees and revenue from lodging, meals, and other services went to central treasuries and concessionaires, not to overworked and underfunded parks departments. For example, in Ecuador, the Galapagos National Park attracts more visitors and generates more revenue than the rest of the country's national parks combined (Lindberg 1991). Park managers have no authority to limit visitation or to use revenues from visitor fees to manage the islands' globally unique ecosystems and wildlife populations, despite threats posed by rapidly growing numbers of tourists. On the small Caribbean island of Saba, half of the island's income – over $23 million – is generated by visits to Bonaire Marine Park, while the park receives only $150 000 for management (Brandon 1996). In the United States, visitors to world-class national parks such as Yosemite or Yellowstone pay just $20 for one car to have unlimited access to the park for one week. At the same time, research in the national park system has been slashed, the system has a $2 billion maintenance backlog, and fewer rangers have to manage crowds that swell by 5–10 per cent annually. In other words, managers are not equipped with sufficient authority to enable them to capture even the available income to their management units, and thereby seek to manage their resources sustainably.

A lack of incentives for local people and institutions to undertake conservation activities or to engage in a co-operative bioregional management project is a common obstacle to local involvement (UNEP 1995). This is especially problematic since there are usually numerous and more valuable incentives – or perverse incentives – that encourage local people to pursue activities that conflict with conservation goals. For example, in the areas surrounding Thailand's Khao Yai National Park, farmers have low levels of education, relatively little farm income, relatively high debt burdens from the purchase of seeds and fertilizer on credit, and poor health and social services (McNeely 1988). Incentives for poaching wildlife in the park, on the other hand, were substantial and encroachment in the park was widespread. In several Latin American countries, governments have invested hundreds of millions of dollars in subsidies for converting forests to pastures while spending virtually nothing on incentives that would encourage landowners to retain at least some forest cover (UNEP 1995). In the Galapagos Islands of Ecuador, local residents have relatively few opportunities to be involved in the highly organized and controlled tourism industry. Since they derive few benefits from tourism and the natural features and wildlife that attract it, residents have few incentives to co-perate with conservation efforts and indeed conflicts between park managers and local residents are common (MacFarland and Cifuentes 1996).

Lack of financial, human, and institutional capacity

Even when bioregional management projects successfully address the above challenges, they may also face additional obstacles, such as a lack of capacity, experience, adequate information, or time commitments from community residents. If central authorities devolve power and authority to the community level, residents and decision makers must be prepared to manage natural resources, in many cases without adequate financial, human or institutional capacity. Acquiring finances and trained personnel which previously may have been provided by central authorities has proved a serious obstacle for many bioregional management projects.

Due to the uncertainties of the global economy, as well as national governments' commitments to bioregional management efforts, funding can be elusive. In the case of La Amistad Biosphere Reserve in Costa Rica, the project suffered as a result of changing administrations and a resulting reduced financial and human capacity. NGOs and international donors came to the rescue, providing $12 million in project funding. However, as Miller (1996) notes, reliance on outside funding is risky and, ultimately, project managers will need to find sustainable ways of generating funds.

In some cases, funding for bioregional projects depends on year-to-year budget planning. For example, in the USA, funding for federal agencies is allocated by Congress on a yearly basis, making the survival of long-term bioregional projects hang precariously on the generosity of national legislators.

Because central authorities have usurped authority that previously existed at the local or regional level, communities have become reliant on top-down management regimes for direction on natural resource stewardship. Thus, providing adequate information and experience for local or regional managers is a significant challenge to stimulating appropriate community involvement. This problem is particularly exacerbated in areas with indigenous communities, especially when governments perceive indigenous stakeholders to be ignorant or poor candidates for training. In La Amistad, indigenous communities were at a disadvantage in negotiations due to a lack of appropriate and intelligible information.

Proponents of bioregional approaches may also need to convince community members that conservation efforts will not exclude local players from resource use. In the ACE basin project in South Carolina, community members needed to be convinced that despite the project's conservation objectives, private and public land would not be 'locked up' from traditional uses, such as hunting and fishing (Frentz et al. 1995). Similarly, local residents in the Annapurna region of Nepal feared that designation of their land as a 'conservation area' would exclude them from practising traditional resource-extractive activities. Given Nepal's previous record of excluding local residents from its national parks, supporters of the Annapurna Conservation Area faced considerable obstacles in assuring residents that the project would be different from previously established national parks (Wells 1994). A lack of awareness and inadequate information are obstacles that leaders of bioregional projects must overcome if communities are to be involved in the effort.

From a practical standpoint, increased community participation in natural resources management also requires substantial time commitments from community members, above and beyond their habitual activities. This can sometimes present difficulties, especially if negotiations and meetings are scheduled at times that conflict with stakeholders' day-to-day activities. Aborigines in Kakadu National Park complained of the constraints they faced when they were required to attend daily meetings (Hill and Press 1994). To the degree that community participation is sought, bioregional management approaches will necessarily require a greater time commitment from all stakeholders. Indeed, the Kakadu management plan stipulated that park personnel seek Aboriginal input on all park management decisions. However, at times excessive bureaucratic requirements can create confusion and hinder the management process, as well as alienating certain players for whom committing enormous amounts of time is impractical.

Strategies to engage local communities in bioregional management

From Kakadu National Park in Australia to La Amistad Biosphere Reserve in Costa Rica, bioregional efforts around the world are responding to barriers that frequently prevent or discourage local engagement. The strategies and tools being used are diverse – a recognition that there are no universal recipes for promoting local participation in conservation efforts. Still, in most situations, proponents face similar challenges in devising strategies that effectively engage local institutions and communities in the planning, implementation, and governance of bioregional conservation efforts. In particular, strategies to encourage and enable local participation need to address four sets of issues. First, the role of local communities in management arrangements must be defined with their participation. Second, incentives need to be created that motivate individuals and organizations to be involved and engaged in the bioregional effort. Third, local institutions need to have tools and information to evaluate available options. Fourth, mechanisms must be established that allow for changes in management plans to reflect changing realities. From a local perspective, one of the benefits of addressing these issues is a reduction in actual or perceived uncertainty.

Define local roles in management arrangements

Local participation in an effective bioregional approach to conservation cannot be forced or bought. Rather, as experiences from around the world increasingly show (see Miller 1996; Keystone Center 1996; Wells 1995), it must be invited, negotiated, and planned. Sorting out management arrangements typically involves three steps. The first step is inviting all stakeholders to be at the table as goals are defined, management strategies are planned, and implementation responsibilities are considered. Second, since stakeholders will come with a range of values and perspectives, mechanisms to negotiate differences will be needed. Third, creating open and equal access to information is vital to building trust amongst stakeholders and between local institutions and outside agencies and researchers. Together these steps reduce local uncertainties about what goals are being considered, who is responsible and accountable, and what information is available and how it is to be used.

As a first step in establishing management responsibilities, leaders, planners, and policy makers should get to know the stakeholders, their concerns, interests, and perspectives (Miller 1996). This may seem trite, but often it does not happen early enough to avert major conflicts and controversies over planning and implementing bioregional management programmes. For example, when the US National Park Service and the US Forest Service released a report on their vision for co-operative management in the Greater Yellowstone Ecosystem, it was simultaneously hailed as a model for integrating scientific information and ecological concepts into joint agency management planning and assailed for being largely ignorant of local knowledge, interests, and aspirations. The report (GYCC 1987) created a storm of local controversy that effectively ended federal leadership in co-ordinating a regional approach to conservation and natural resource management in the 7.3 million hectare bioregion surrounding Yellowstone National Park. According to Clark and Lichtman (1994), one of the key lessons from this experience is that 'It is in a policy or plan's infancy that it is possible to mitigate or eliminate obstacles most successfully' and that once a new policy initiative begins, 'unpredicted responses can be much more potent, undermining the entire policy process'.

By contrast, when the Great Barrier Reef Marine Park was established along a 2000 kilometre stretch of the Queensland coast in 1975, a permanent consultative committee representing a wide cross-section of public and private interests including tourism, fishing, science, and conservation was put in place by the same Act that created the park (Great Barrier Reef Marine Park Authority 1993).

Unless stakeholders have the opportunity to become full partners in planning and implementing bioregional conservation programmes, one group or another is likely to find or perceive threats to its self-interest. This can quickly lead to non-co-operation or active resistance to the effort, which could deprive the larger effort of important knowledge that group might have had and its ability to seek co-operation essential to programme goals. In some countries, such as Niger, soliciting participation from all stakeholders may involve dealing with ethnic conflicts or traditional customs. Islamic customs in Niger prevent women from organizing or participating in co-operatives, and grazing activities are not fully incorporated in forest management plans, although both women and herders are among the primary forest resource users (Otto and Elbow 1994). Similarly, in eastern India, women do not participate in forest community groups, although they are among the primary fuel wood harvesters (Poffenberger 1994). At this level, including stakeholders who have been marginalized due to cultural norms or ethnic conflicts means addressing fundamental cultural and social traditions, some of which have long-standing histories. However, a lack of consideration of certain stakeholders can prove devastating for long-term conservation.

At the same time, governments must establish their role in local bioregional efforts. While local players may have more influence in bioregional projects than in traditional protected areas approaches, central authorities will still need to play a role in the management process. In many cases, the role of central governments will be to define the goals and values of all of society and to ensure that local bioregional efforts are consistent with national objectives. Thus, governments will need to establish mechanisms to safeguard national interests within local bioregional efforts. In some cases, this will mean establishing national environmental protection policies, such as the *Endangered Species Act* in the United States, which guarantees protection for a nationally determined list of endangered species. Occasionally, policies forged at the national level can catalyse communities into action. For example, the Deer Creek community in northern California moved to restore their watershed in response to the threat of an endangered species listing for Chinook salmon as well as the possible designation of the watershed to the protected status of 'wild and scenic' (Bingham 1996).

Use incentives to motivate local engagement

Even where interest in conservation goals is widespread, few stakeholders can afford to do more than attend a few public meetings or respond to documents and questionnaires circulated for public comments. Bringing about changes in farming, fishing, logging, or tourism practices to reach bioregional management goals and objectives will require incentives.

These incentives can take different forms and will need to be tailored to the group from which changes are being sought. While incentives are most often thought of in economic and financial terms, they can also take shape in the form of policy or administrative changes. In Australia's Kakadu National Park, the Commonwealth government provided incentives to local Aborigines by legalizing their land claims, enabling and encouraging participation, and devolving power to the community level. Administrative incentives were offered to tour operators in the Great Barrier Reef for voluntary codes of conduct (Miller 1996). In the North York Moors National Park in the United Kingdom, local farmers were given direct economic incentives for managing their land in a way that restores the natural landscape (Stratham 1994). In some cases, economic incentives can turn out to be even more profitable than previous activities. For example, in Malawi's Kasungu National Park, local people were allowed to harvest caterpillars and establish beehives within park boundaries in exchange for the cessation of other ecologically damaging activities. This incentive proved to be more profitable than traditional agricultural activities (UNEP 1995).

Incentives can also take the form of compensating local people for the loss of resource use, or for the encroachment of wildlife on their property. In Zimbabwe's Matobo National Park, local residents were allowed to harvest thatch for construction of their houses. This symbiotic exchange between park rangers and local residents allowed villagers to collect thatching materials that were scarce outside the park, while also reducing work for park managers, who would have burned the thatch anyway (McNeely 1988). In Nigeria, compensation was provided to local villagers whose crops were damaged by elephants roaming outside the boundaries of the Cross River National Park (UNEP 1995).

Ensure access to information for all stakeholder groups

All players at the negotiating table must be provided with the same information, in a form that can be easily understood by all stakeholders. Those involved in building local capacity must provide appropriate information in a way that is meaningful to local communities. While this seems intuitively obvious, experience from La Amistad Biosphere Reserve suggests that local participation was hindered by a lack of access to appropriate information. As a result, indigenous groups at La Amistad formed their own NGO to help the community develop negotiation skills and to facilitate access to information (Talamanca 1992). Where information is available only through sophisticated technical mechanisms, such as Geographic Information Systems (GIS), bioregional managers must be prepared either to train locals in the use of sophisticated information systems, or to disseminate information in a way that is readily accessible to all stakeholders.

Local communities can gain access to information by requiring that researchers provide them with the findings of any work conducted in their area. For example, the Kuna Indians in Panama have attempted to regulate the activity of researchers in their region by requiring copies of reports in Spanish, as well as copies of photographic materials and samples of plant and animal species, and by training and hiring local individuals as guides (Laird 1993). Likewise, in Kakadu National Park, park managers are required to train and hire a percentage of the Aboriginal community to work for Kakadu's park management (Hill and Press 1994). At times, managers can help local groups systematize local knowledge so that it can be made available for a larger community. For example, in the Sierra Nevada de Santa Marta in Colombia indigenous groups used GIS to map out physical features in the landscape surrounding their homes. Indigenous peoples then modelled cause and effect relationships to determine the ramifications of different land use options and scenarios (Mayr, personal communication). These measures simultaneously promote the participation of local people, foster the development of new skills and information, and maximize existing local knowledge.

Support the engagement of local groups in adaptive management

Environmental change, evolving knowledge of biophysical systems, shifts in economic conditions, and changing attitudes amongst constituents are part of the context of any bioregional management programme. The capacity to anticipate such changes and to respond appropriately is therefore a vital asset to any bioregional management effort. Adaptive management not only requires appropriate monitoring systems, but planning and decision-making processes that are capable of analysing the implications of new information and flexible enough to revise plans and management practices when needed.

Co-operative information exchanges contribute to adaptive management. Recognizing that uncertainties exist, adaptive management efforts promote flexibility in establishing and implementing plans. In this way, new information can be incorporated more effectively into the management plan. In 1991, an earthquake devastated the Talamanca Region in La Amistad Biosphere Reserve, damaging forests and the nearby Cahuita reefs. Information obtained from this experience could help prepare communities and natural resource managers for future natural occurrences, enabling them to manage their bioregion better.

Monitoring the impact of changes on the bioregion is an important aspect of adaptive management. Without the ability to monitor impact, bioregional managers are unable to ascertain the results of resource decisions made by community groups. In many cases, central government agencies are not in a position to provide funds for research and monitoring activities, or too little importance is ascribed to these activities. Budgets for bioregional projects must include funds for monitoring and research. For example, in Louisiana the Coastal Wetlands Planning, Protection, and Restoration Act requires that funds for monitoring be included in the overall programme budget (IEMTF 1995).

Managing ecosystems through adaptive processes will also require that mechanisms be established for dealing with shifting players. Ideally, those stakeholders taking part in initial project planning stages should follow the project through its implementation and monitoring stages. In reality, however, residents move out of the community and agency employees can be transferred to other regions. Agencies should try to ensure project continuity by appointing their employees to serve for the duration of the project. At the very least, communities and agencies should agree on a key 'point person' who can commit to the entire process, training newcomers as they enter the process. Such a tactic was successfully employed in Oregon's Silverspot Butterfly Recovery Project, in which a research scientist from a local university accompanied the project from planning to implementation stages, training new staff as they replaced those who left the project (Wondolleck and Yaffee 1994).

Strengthen the role of the research community in bioregional management

As bioregional management projects are forged, local communities will need support and information from the research community, especially because risk and uncertainty are inherent in management at the local level. The earthquake that devastated the La Amistad Biosphere Reserve was certainly unexpected, and while natural disasters are only one of many types of uncertainty, scientists must be prepared to help communities anticipate, plan and prepare for any kind of disaster, be it natural, financial, ecological or economic. In the case of natural disasters, this means supplying scientific information to help predict future occurrences when possible, providing recommendations for policy responses, and helping develop community emergency preparedness plans.

Researchers are particularly well positioned to contribute to bioregional management efforts. Through their highly specialized training, researchers are capable of articulating how ecosystems function, as well as the risks involved in undertaking specific activities. Thus, researchers can act as guides, advisors, and players in the debate over resources management, closing information gaps that might otherwise remain unresolved. Therefore, researchers must be prepared to provide technical support and build informational capacity to enable communities to make the most informed decisions possible. In this sense, researchers interested in contributing to bioregional management efforts should view communities and local organizations as an active audience, keeping in mind the potential impact of scientific studies at the local level.

One way to reach this strategy is through the establishment of resource centres that are readily accessible to the local community. For example, in 1990 the California Academy for Sciences, a private natural history museum, began a preliminary biodiversity resource centre to provide information on biodiversity to researchers and to the public. In its experimental phase, the centre consisted of conventional reading materials, journal and newspaper clippings, CD-ROMs, multi-media materials, Internet access, and video capabilities. Among other successes, the resource centre provided information on the environmental implications of the nearby Sacramento River Toxic Spill (Moritz 1996).

In addition, information necessary for innovative research is often most readily available at the community level. Researchers investigating biotechnology options in Panama certainly benefited from the Kuna's indigenous knowledge of flora and fauna. Because traditional investigations have been conducted at the local level, researchers are particularly suited to monitoring trends and establishing indicators of change that affect the ecosystem, such as soil quality, distance to water sources, population growth, or abundance of fish. Local people are crucial players in establishing indicators because in many cases they have direct access to natural resources and are best equipped to describe changes in ecological, economic, and social processes. For example, researchers from the University of Zimbabwe helped the CAMPFIRE project establish indicators by enlisting the community in surveys measuring sociological and ecological trends (Metcalfe 1994).

For most countries, bioregional approaches to conservation are new and untested. These co-operative approaches require collaboration between a variety of interests, but many governments still approach natural resources management in a sectoral manner (i.e. separate departments of agriculture, forestry, wildlife, and housing). Researchers can help bridge these sectors by approaching their work in an interdisciplinary manner. For instance, research on the degradation of a watershed may lead to identifying perverse agricultural policies that encourage activities destructive to the river's health.

Finally, researchers would do well to consider their own stake in programmes that focus upon entire bioregions. By joining the process, researchers can gain a seat at the table, access information, knowledge and wisdom, and contribute their own insights that will guide conservation and resource use strategies. The fact that bioregional programmes feature adaptive management procedures provides a comfortable home for scientists and researchers accustomed to experimental approaches and procedures.

This same environment is also a fertile field for science and information, and carries the particularly powerful opportunity for both managers and scientists to learn from experience. Indeed, the experimental nature of bioregional management may be the key to its success.

References

Barber, C.V., Johnson, N.C. and Hafild, E. (1994) *Breaking the Logjam: Obstacles to Forest Policy Reform in Indonesia and the United States*. Washington, DC: World Resources Institute.

Bingham, N. (1996) A voice from Deer Creek. *Cultural Survival Quarterly*. Spring 1996.

Brandon, K. (1996) *Ecotourism and Conservation. A Review of the Key Issues*. Washington, DC: The World Bank.

Ceballos, G., Rodriguez, P. and Medellin, R.A. (1998) Assessing conservation priorities in megadiverse Mexico: mammalian diversity, endemicity, and endangerment. *Ecological Applications* 8: 8–17.

Clark, T. and Lichtman, P. (1994) Rethinking the 'Vision' exercise in the Greater Yellowstone Ecosystem. *Society and Natural Resources* 1(4): 459–78.

Dovers, S.R. (1997) Sustainability: demands on policy. *Journal of Public Policy* 16: 303–18.

Feldmann, F. (1994) Community environmental action: the national policy context. In: Western, D., Wright, R.M. and Strum, S.C. (eds) *Natural Connections: Perspectives in Community-based Conservation*. Washington, DC: Island Press.

Frentz, I., Hardy, P., Maleki, S., Phillips, A. and Thorpe, B. (1995) *Ecosystem Management in the U.S.: An Inventory and Assessment of Current Experience*. Master's project, Natural Resources and Environment Program, University of Michigan.

Great Barrier Reef Marine Park Authority (1993) *1993–1998 Corporate Plan.* Townsville, Australia: Great Barrier Reef Marine Park Authority.

GYCC (1987) *The Greater Yellowstone Area: An Aggregation of National Park and National Forest Management Plans.* Report of the Greater Yellowstone Co-ordinating Committee. USDA Forest Service, Intermountain Region, Ogden, Ut.

Hill, M.A. and Press, A.J. (1994) Kakadu National Park: an Australian experience in co-management. In: Western, D., Wright, R.M. and Strum, S.C. (eds) *Natural Connections: Perspectives in Community-based Conservation.* Washington, DC: Island Press.

Interagency Ecosystem Management Task Force (IEMTF) (1995) *The Ecosystem Approach: Healthy Ecosystems and Sustainable Economies.* Volume II: Implementation Issues. Springfield, Va: US Department of Commerce.

Ishwaran, J. (1995) Biodiversity Regimes for Protected Area Management. Paper presented at the *Global Biodiversity Forum*, Convention on Biological Diversity, Conference of the Parties, Jakarta, Indonesia, November 1995.

Keystone Center (1996) *Final Report of the Keystone Dialogue on Ecosystem Management.* Keystone, Co: The Keystone Center.

Laird, S. (1993) Contracts for biodiversity prospecting. In: Reid, W.V. *et al.* (eds) *Biodiversity Prospecting: Using Genetic Resources for Sustainable Development.* Washington, DC: World Resources Institute.

Lindberg, K. (1991) *Policies for Maximizing Nature Tourism's Ecological and Economic Benefits.* Washington, DC: World Resources Institute.

Lynch, O.J. and Talbott, K. (1996) *Balancing Acts: Community-Based Forest Management and National Law in Asia and the Pacific.* Washington, DC: World Resources Institute.

MacFarland, C. and Cifuentes, M. (1996) Ecuador case study. In: Dompka, V. (ed.) *Human Populations, Biodiversity and Protected Areas: Science and Policy Issues,* Washington, DC: AAAS Press.

Mayr, J. (1996) Personal communication. July 11.

McNeely, J.A. (1988) *Economics and Biological Diversity: Developing and Using Economic Incentives to Conserve Biological Resources.* Gland, Switzerland: IUCN.

Metcalfe, S. (1994) The Zimbabwe Communal Areas Management Programme for Indigenous Resources (CAMPFIRE). In: Western, D., Wright, R.M. and Strum, S.C. (eds) *Natural Connections: Perspectives in Community-based Conservation.* Washington, DC: Island Press.

Miller, K.R. (1996) *Balancing the Scales: Managing Biodiversity at the Bioregional Level.* Washington, DC: World Resources Institute.

Moritz, C. (1996) Biodiversity Resource Centers: Appropriate Information Technology to Empower Local Communities. Paper presented at the *Global Biodiversity Forum*, Convention on Biological Diversity, Conference of the Parties, Jakarta, Indonesia, November 1995.

Ostergaard, L. (1993) Traditional Swidden Cultivators and Forces of Deforestation in Sumatra: the Significance of Local Land Tenure Systems. Paper presented at the Second Asia-Pacific Consultative Meeting on Biodiversity Conservation, 2–6 February, 1993, Bangkok, Thailand.

Otto, A. and Elbow, D.I. (1994) Profile of national policy: natural forest management in Niger. In: Western, D., Wright, R.M. and Strum, S.C. (eds) *Natural Connections: Perspectives in Community-based Conservation.* Washington, DC: Island Press.

Poffenberger, M. (1994) The resurgence of community forest management in eastern India. In: Western, D., Wright, R.M. and Strum, S.C. (eds) *Natural Connections: Perspectives in Community-based Conservation.* Washington, DC: Island Press.

Repetto, R. and Gillis, M. (1988) *Public Policies and the Misuse of Forest Resources.* Cambridge: Cambridge University Press.

Riebsame, W.E. (1996) Ending the range wars? *Environment* 38(4): 4–9, 27–29.

Sharma, N.P. (1992) (ed.) *Managing the World's Forests: Looking for Balance between Conservation and Development.* Dubuque, Ia: Kendall/Hunt Publishing Co.

Stratham, D.C. (1994) The farm scheme of North York Moors National Park, United Kingdom. In: Western, D., Wright, R.M. and Strum, S.C. (eds) *Natural Connections: Perspectives in Community-based Conservation.* Washington, DC: Island Press.

Talamanca, T. (1992) Reserva de la Biosfera La Amistad. Boletin Informative 5. October, San José, Costa Rica.

Tilman, D., Wedin, D. and Knops, J. (1996) Productivity and sustainability influenced by biodiversity in grassland ecosystems. *Nature* 379 (6567): 718.

UNEP (1995) Global Biodiversity Assessment. United Nations Environment Programme, Nairobi. Cambridge: Cambridge University Press.

Wells, M. (1995) A profile and interim assessment of the Annapurna Conservation Area Project, Nepal. In: Western, D., Wright, R.M. and Strum, S.C. (eds) *Natural Connections: Perspectives in Community-based Conservation.* Washington, DC: Island Press.

Wells, M. and Brandon, K. (1992) *People and Parks: Linking Protected Area Management with Local Communities.* Washington, DC: The World Bank.

Western, D., Wright, R.M. and Strum, S.C. (1994) (eds) *Natural Connections: Perspectives in Community-based Conservation.* Washington, DC: Island Press.

Wondolleck, J.M. and Yaffee, S.L. (1994) *Building Bridges Across Agency Boundaries: In Search of Excellence in the United States Forest Service.* Research report submitted to the USDA-Forest Service, Pacific Northwest Research Station.

Wright, R.G. (1999) Wildlife management in the national parks: questions in search of answers. *Ecological Applications* 9: 30–36.

Zerner, C. (1994) Transforming customary law and coastal management practices in the Maluku Islands, Indonesia, 1870–1992. In: Western, D., Wright, R.M. and Strum, S.C. (eds) *Natural Connections: Perspectives in Community-based Conservation.* Washington, DC: Island Press.

4 | Sentimental ecology, science and sustainable ecosystem management

Robert J. Mason and Sarah Michaels

The Adirondack Park

Sustainability in principle and sustainability in practice can differ dramatically. Indeed, even the definition of sustainability is highly contested, with the term being appropriated by divergent interests to advance and serve their needs. Similarly, 'ecosystem management' raises a host of questions about if, how, and when human intervention should occur; the state to which an ecosystem should be restored; and how to balance political considerations with ecosystem integrity. New York State's Adirondack Park has been struggling with these issues – if not always framing them in the current parlance – for more than a century (Erickson 1998). The Adirondack experience provides a compelling illustration of the complications that can arise in attempting to define and implement sustainable ecosystem management in an economically marginal, but highly valued, region situated in close proximity to one of the wealthiest, most mobile populations on the planet.

The largest park in the United States outside Alaska, the Adirondack Park is an important, accessible recreation resource that accounts for almost one-fifth of New York State's land mass (see Figure 4.1). Located in northern New York and managed by New York State as a park and forest reserve, the Adirondacks consist of a 2.4-million-hectare patchwork of privately and publicly owned lands. Much of the land cover consists of second growth forest dating back to the late nineteenth century.

Bill McKibben (1995: 30), a best-selling author and resident champion of the Park, considers the Adirondacks 'the world's first experiment in restoring an entire ecosystem'. He considers it 'just about the most

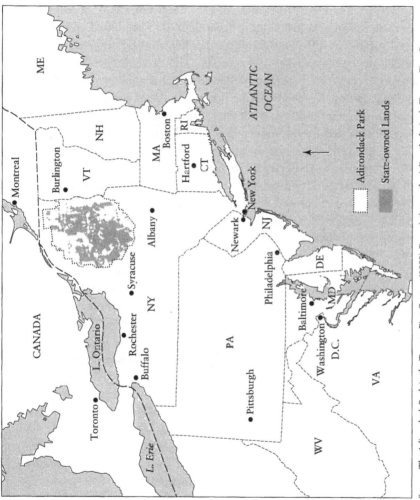

Source: The Adirondack Park in the Twenty-First Century (1990) Commission on the Adirondacks in the Twenty-First Century, Albany, State of New York.

Figure 4.1 The Adirondack Park

heartening spot on earth' (McKibben 1995: 31) because it is possible to imagine conserving what is still pristine and restoring the viability of habitats and ecological functions impaired through excessive or inappropriate use (Miller 1996).

In the Adirondacks, as in many other places, 'sustainability' is a concept that has been adopted by stakeholders with widely divergent interests and conceptions of just what is meant by this term. Sustainability, it seems, is relative rather than absolute, highly dependent on place-based considerations and subject to different meanings ascribed to it by different groups with their own perspectives and agendas. The now classic Brundtland Commission definition of sustainable development is often invoked: 'development that meets the needs of the present without compromising the ability of future generations to meet their own needs' (World Commission on Environment and Development 1987: ix). Even this definition of sustainability, inclusive as it is, can be used to cover a wide variety of environmental management perspectives. So, too, can some of the prescriptions, offered by advocates of comprehensive protected-area management, that speak to both ecological and economic sustainability.

Biosphere reserve planning is one such perspective. The Adirondack Park comprises a distinct subunit of about 2.4 million hectares within the 4-million-hectare Champlain-Adirondack Biosphere Reserve. These reserves are designated by the United Nations Man and the Biosphere Program, but are managed by national (or subnational) governments. In ideal form, the biosphere reserve consists of a core area, where ecosystem protection is paramount; a buffer zone, where a wide range of carefully regulated land uses is allowed, so that the core is protected; and a transition zone, where human use (especially 'traditional use') is meant to peacefully coexist with natural systems (Batisse 1982, 1985; Lucas 1992; Wells and Brandon 1992; Solecki 1994). While the Adirondacks may not be a clear fit for the core–buffer–transition model, these concepts are invoked by at least some Adirondacks regional planning advocates. Indeed, the contentious report of the Commission on the Adirondacks in the Twenty-First Century (1990), discussed below, called for the creation of a narrow transition zone around the park's perimeter (Mason 1995).

The Adirondack Park also is viewed as a model 'greenline park' (Corbett 1983; Hirner and Mertes 1986; Belcher and Wellman 1991; Little 1992; Mason 1994). Greenline parks, which consist of a combination of public and privately owned lands, are meant to sustain a mix of land uses such as forestry, agriculture, tourism, and wilderness protection (Mason 1994). Partnerships involving various levels of government, as well as individual citizens and citizens' organizations, are the ideal models for greenline management. In the Adirondacks, as elsewhere, the objectives are multiple and often conflicting (West and Brechin 1991). They include, according to greenline advocates, promotion of viable local economies, protection of local cultures, ecological preservation, and provision of recreational opportunities for urban populations outside the parks. Greenline planning and biosphere reserve planning are conceptually compatible with broad definitions – such as the Brundtland definition – of sustainability. The devil is in the details of implementing these concepts.

Kemmis (1990) describes a neighbourly citizenship, one which stems from people who find themselves connected to the same or nearly adjoining places. Yet, typically, conflicts over Adirondacks management have polarized park residents and non-residents, both of whom proclaim a strong attachment to the Adirondack region. These differences were brought into extreme relief with the 1990 release of the report of the Commission on the Adirondacks in the Twenty-First Century (1990). Among other things, the report called for more stringent regulation of regional land use and more state land acquisition. Interest groups from all quarters responded to the report's recommendations, with strong opposition coming mostly from locally based citizens and citizens' groups (Mason 1995). While most Adirondacks interest groups – supporters as well as detractors of regional planning as it is currently practised – would likely embrace the concept of sustainability at some level, their conceptions of a sustainable regional environment and economy differ rather markedly.

These varied groups have different concepts of 'nature', with some favouring 'wise use' of resources (often a guise for supremacy of private property rights, see Brick and Cawley 1996), human benefits, and economic sustainability, while others give precedence to protection of biodiversity and ecological sustainability. Different stakeholders have different conceptions of crisis: environmentalists argued that the Adirondacks of the early 1990s (as in the early 1970s) were experiencing a development crisis; many local residents were much more concerned about what they viewed as an economic crisis. Different conceptions of appropriate economic development also enter into the debate: how much emphasis should be placed on resource-based activities, such as farming, forestry, and fishing, and what role should tourism and recreation play? What constitutes an acceptable level of local participation in regional decision making and how important is the voice of the much larger statewide public in Adirondacks management? Much of the debate comes back to a rather appropriate, if oversimplified, conception of privileged outsiders versus economically disadvantaged residents who are forced to accommodate the needs and interests that are articulated by influential outside groups (Terrie 1997; Knott 1998).

Adirondacks management, officially at least, focuses on open space preservation, water quality protection, local planning, and maintenance of 'working landscapes'. Regional management is not necessarily to be equated with ecosystem management. While some influential interest groups do promote ecosystem management (Mason 1995), a holistic conception of what this entails does not seem to be in place. Elements of ecosystem management include habitat protection, wilderness preservation, and promotion of biodiversity. But these broad conceptual guideposts are not easily applied in specific management situations, as will be illustrated later in this chapter with the discussion of two recent issues. In part, this situation is the outcome of a complex history of regional management efforts that began more than a century ago. This history, sketched below, is critical to appreciating the challenges that arise when ecosystem management is juxtaposed with long-standing place-based political tensions.

Park history

The Adirondack Park experience is representative of a number of American and worldwide trends that are relevant to ecosystem management. At the same time, it is distinctive because of its particular context and place-based considerations.

As with most places on earth, the fate of the Adirondacks is intimately tied to the direct and indirect revenue-generating potential of the land and land cover. Within five years of New York State's 1779 appropriation of what had been British Crown lands prior to the War of Independence in 1776, the state legislature – desperate for funds – began selling off the land (Graham 1978; McMartin 1994).

In response to ravenous exploitation of Adirondack forests in the nineteenth century, utilitarian resource conservation became the rationale for first restricting further transfer of public property to private ownership, and then introducing wilderness preservation (Heiman 1988). This concept was championed by an effective outside-of-the-park constituency, a constituency which has been highly influential in managing the Adirondacks over the past century. Tension between state and local-level interests takes the form of competing efforts by park residents and non-residents to shape the state-level political debate about managing the park.

In 1883, the legislature halted further sales of lands in the six Adirondack counties (Donaldson 1921). A coalition of business interests based outside the park had lobbied for the cessation of sales because of concern that rainfall and stream flow would be adversely affected by destroying the Adirondack forest cover. It was thought that water transportation on the Hudson River and the Erie Canal, essential arteries of state commerce, would be harmed by the reduced flow (Heiman 1988; Halper 1992).

The state legislature designated the state-owned forest land in the Adirondacks as a Forest Preserve in 1885. Its creation was made possible by a single unidistrict referendum, which gave equal weight to urban downstate voters and rural voters. The unidistrict referendum is noteworthy because it was uncharacteristic of the electoral procedures of the times, which tended to favor rural voters through malapportionment (Halper 1992).

In 1892, the state created the Adirondack Park, including in it public and private lands. The limited protection afforded first by the preserve and then by the park designation was then reduced by legislative amendment which permitted the ongoing sale of state timber and also through the State Forest Commission's inefficient management, which led to speculators purchasing state lands (Heiman 1988). Public indignation over state management practices led to a public vote in 1894 favouring a constitutional amendment to prevent the sale, removal, or destruction of timber on state lands and prohibit public land sales or swaps without additional constitutional amendments. New York became the first state in the country to provide constitutional protection to a natural resources area (Heiman 1988).

Over the past century, the Adirondacks, in common with other parts of the United States and the world, experienced explosive growth in development. An increasingly affluent and mobile post-World War II middle-class population sought out the accessible recreational amenities of the Adirondack Park (Heiman 1988). Much of the pressure has come from improved access, such as the development in the late 1950s and 1960s of New York State's limited-access interstate highway, the Northway, that links Montreal with Albany and New York City. In response to these pressures, Laurence Rockefeller – noted conservationist, brother of New York State governor Nelson Rockefeller, and chair of the State Council of Parks – released a report recommending that a new national park be created in the core of the Adirondacks (Booth 1987; Halper 1995).

State, local and private interests were so opposed to the idea of a national park that a Temporary Study Commission on the Future of the Adirondacks was appointed by the governor in 1968 to propose a more politically palatable option (Graham 1978; Liroff and Davis 1981). The final report, produced two years later, weighted local economic considerations less than preservation, aesthetics, and ecological integrity (Heiman 1988).

Second home subdivisions, the Temporary Study Commission (1970) concluded, threatened the ecological and social integrity of the park. At the time, New York State led the nation in the number of households and in the percentage (10.1) of households owning second homes. Land speculators were attracted to the Adirondacks because only 10 per cent of private lands were subject to local land use regulation (Heiman 1988). Between 1967 and 1987, 21 000 single family dwellings and 65 000 vacant lots were created. The sale of subdivided property increased 300 per cent between 1982 and 1985 and doubled again by 1988 (Diffenderfer and Birch 1997). Much of the land converted into recreational estates is working forestland (Klyza and Trombulak 1994). This transformation of landscapes of production into landscapes of consumption – with the latter usually imposing greater ecological impacts – is a challenge to sustainable ecosystem management in many parts of the world.

In 1971, the state Adirondack Park Agency (APA) was created as an outcome of the Temporary Study Commission. The APA was given broad authority to control private land development in the park. The 1972 draft land use plan was strongly supported at public hearings outside the park and vocally opposed at in-park hearings (Liroff and Davis 1981).

The final version of the Adirondack Park Land Use and Development Plan was approved in 1973, extending state oversight to the entire park. A zoning approach is used as a means to reconcile development demands with the open space character of the park.

Historically, land ownership in the Adirondacks has been quite concentrated. In 1970, 32 individuals or corporations owned parcels over 4000 ha in size, which made up 30 per cent of all Adirondack Park land (Halper 1992). Wood products companies, which have been major landholders in the park, were able to generate revenue from standing timberland because of the Adirondack tradition of private recreational use. Leasing out hunting and fishing rights to private clubs and organizations helped ensure that uncut forestlands did not drain profits (Halper 1995). Restructuring of the forest industry in the last part of the twentieth century caused wood products companies to rethink practices of retaining properties not currently used for timber extraction. Property sale for residential development continues to be an economically attractive option.

Given that acquisition of significant natural areas through public purchase is becoming increasingly costly and difficult (Liroff and Davis 1981), and in many cases less desirable than it may have been in the past, the

Adirondack Park provides a longstanding alternative model for public/ private resource management. The Adirondacks is a peopled wilderness where multiple use has been an integral feature of park management from the outset. The park is approximately 40 per cent publicly owned and 60 per cent privately owned (Commission on the Adirondacks in the Twenty-First Century 1990). Consequently, conservation on private lands and preservation on state lands could not have been achieved separately (Halper 1995). Regional management of privately and publicly held lands is essential if the park is to retain its distinctive ecological character and continue to satisfy place-based economic, political, social, and cultural demands. 'Ecostewardship partnerships', involving government agencies, non-profit organizations and private enterprises are a means to address the imperfect match between dynamic ecological processes and highly defended private property rights (Michaels, Mason and Solecki 1999).

Two recent issues illustrate the sorts of dilemmas that arise when ecosystem management, economic interests, conservation concerns, private property interests, and scientific questions all come to bear on Adirondacks management. The concept of sustainable ecological management in these instances becomes very difficult to put into practice.

The blowdown

On July 15 1995 a huge windstorm swept across the Adirondack region, leaving in its wake many fallen trees (Folwell 1995; Meade 2000). As much as 25 per cent of the total land area of the park – approximately 600 000 ha in total – was affected. Fifty million trees were downed, in a patchy configuration. In the wake of the storm, debate ensued about whether or not to allow harvest of this timber. Although salvage would have created edge habitat suitable for some species, it would have reduced habitat for others.

Under the 'forever wild' amendment to the state constitution, timber is not to be destroyed, sold, or removed from state lands (Terrie 1985). Yet following the 1950 hurricane, which took place during a rather different era in terms of environmental awareness and influence of environmental organizations, salvage did take place. The salvage operation and subsequent questions about further timber operations on state land prompted a concerted response from the conservation community, known as the 'big blowup' (Terrie 1997). In the wake of the 1995 event, timber salvage would not even be discussed without instant challenge. Some Adirondack residents, in concert with timber companies, argued that salvage was needed to reduce the potential for dangerous forest fires. Environmentalists held firm against salvage and they were especially adamant because some of the worst damage occurred in wilderness and near-wilderness areas in the northwest portion of the park. The 'preservationist' position was aided by wet conditions that reduced the fire potential in the summer and autumn of 1995, and also by prevailing low timber prices (Meade 2000).

When viewed in the longer term (longer than the past few years), the Adirondacks region is a highly disturbed one, shaped as much by human modification of the landscape as by natural events. Thus, the question of how to respond to a natural disaster like the massive storm of 1995 eludes simple answers. Yet preservation is the only option viewed as viable by most environmental groups. From their perspective, they must hold firm on one of the constants in Adirondacks management: constitutionally enshrined protection of state owned lands. The debate that followed the 1995 blowdown was highly charged politically, and the only strategy that most groups were willing to adopt was one of adhering to their traditional positions. Local politicians – along with industry, agriculture, and landowner groups – tended to favour conservation/wise use of resources, while environmentalists advocated preservation (see Halper 1995). There was to be no negotiated middle ground on the timber salvage question. But as we shall see with the case of species reintroduction, such hard and fast positions are not the most strategic responses to all controversies.

Species reintroduction

Comparatively passive interventions, such as wilderness designation, tend to be highly contentious in the Adirondacks and elsewhere in the United States. But proposals for more active and intrusive measures – species reintroduction key among them – can bring about much more heated debate (Fascione and Kendrot 1998; Hutchens 1998).

Any attempt at ecosystem restoration raises questions about baseline ecological knowledge, possible unforeseen and undesirable ecological consequences, and questions of political acceptability. Advocates of restoration often must balance questions of scientific uncertainty with their desire to take action when and where political conditions seem most favourable. Proposals for reintroducing specific species may be as much prompted by timeliness and righteous convictions about what is ecologically correct as they are by scientific detachment and considered assessments of short-term and long-term ecological consequences of such interventions. Indeed, widespread public support – even in the face of significant local opposition – is generally present for reintroduction of 'poster species'. In the Adirondacks, such species as wolf, moose, elk, and lynx are on this list; historically, even such non-indigenous species as caribou, mountain goats, bighorn sheep, and wild boar have been considered for reintroduction (Terrie 1993). Much of the support for reintroduction has been rooted in 'anthropocentric pride' (e.g. moose as 'king'), the belief that reintroduction is akin to stocking a stream with fish, and the prospect of economic benefits (Terrie 1993). Economic arguments have included everything from increased revenue from sportsmen to visitors who would come to join in 'wolf howls'. Although the more recent reintroduction campaigns are far less anthropocentric in perspective than earlier efforts, proponents of reintroduction still find it in their strategic interest to capitalize on public sentiment that supports the 'poster species' of the day.

An unsuccessful moose reintroduction took place in the late 1800s, while beaver were successfully restocked in the early 1900s (Terrie 1993). Elk reintroduction was attempted, unsuccessfully, in the early 1900s and the 1930s, and the possibility has been raised again in recent years by the Rocky Mountain Elk Foundation (Harris 1997). Even though it is not clear that elk have ever inhabited the Adirondack region (Terrie 1993), the early – and limited – efforts were supported by wealthy private individuals and sportsmen's organizations. They had virtually no knowledge of and gave no attention to the ecological ramifications – at least as we understand them today – of these actions.

More recently, efforts to reintroduce moose foundered in the face of local concerns in the early 1990s. These concerns principally centred around risks of collisions with vehicles, the belief that local residents should have a greater say in reintroduction plans, and the belief that moose would slowly reintroduce themselves (Hicks 1995, Lauber and Knuth 1998). And indeed, moose did find their own way back in the region. Canada lynx, introduced in 1989, have suffered heavy casualties from motor vehicles. The lynx, transplanted from the Yukon, seemed to experience difficulty adapting to the much more densely populated Adirondacks. Though their status remains uncertain, apparently there still is an Adirondacks lynx population.

Although wolf reintroduction was proposed in 1970 by a consultant to the Temporary Study Commission on the Future of the Adirondacks (1970), serious debate would not get underway until the mid-1990s. Once demonized, wolves are now romanticized, with popular literature, film, and a host of web sites giving testimony to the noble character and proud disposition of these magnificent creatures. While such anthropomorphizing may have no place in a purely ecological assessment, it clearly does play a role in the management of a peopled landscape located in a populous state where wolves have not been present since the late 1800s. At that time, a bounty was offered for wolves. A century later, by contrast, Defenders of Wildlife described 1996 as the 'year of the wolf'. (Defenders of Wildlife 1996). Reintroductions had taken place in Yellowstone National Park and central Idaho, and Alaska's voters passed a ballot initiative restricting aerial hunting of wolves. It was an opportune time to launch an Adirondacks campaign.

Wolf reintroduction is a very current, very divisive issue in the Adirondacks, adjacent northern New England, the Yellowstone region, and elsewhere in North America (Wilson 1997; Fascione and Kendrot 1998). Wolves still are being demonized, mainly by local officials, as well as deified, principally by environmental organizations. The ongoing debate, with both sides marshalling evidence from a variety of popular and scientific sources as they seek to rally various constituencies, raises important questions about scientific evidence and uncertainty (Wilson 1997).

Defenders of Wildlife, a non-governmental organization, has been actively pressing for restoration of wolf populations across their North American ranges. While moose have reintroduced themselves to the Adirondacks, it appears that wolves are more readily deterred by the highways, waterways, and extensive agricultural lands that surround the region. If a wolf population is to be re-established, active intervention is almost certainly required.

Defenders of Wildlife has worked methodically and inclusively in building its case for wolf reintroduction and was behind the creation of the 'Adirondack Citizens' Advisory Committee on the Feasibility of Wolf Reintroduction'. Formed in 1997, the advisory committee has voting members representing various landowner groups, resource-based industries, environmental groups, and other statewide interest groups, along with observers and ex-officio members from the US Fish and Wildlife Service, Defenders of Wildlife, and New York State Department of Environmental Conservation.

Despite their limited co-operation as members of the advisory committee, advocacy groups tend to line up predictably on the wolf question. Most Adirondack local governments and property rights groups, including the New York State Farm Bureau, are hostile to reintroduction. Environmental NGOs, such as the Adirondack Council and the Sierra Club, are at least cautiously supportive. Limited data on public sentiment indicate

Table 4.1 1996 attitudes of Adirondack residents and New York State residents to wolf reintroduction.

Attitude	Adirondack residents (% of respondents)	New York State residents (% of respondents)
Strong support	34	38
Moderate support	42	42
Neither support nor opposition or did not know	5	10
Moderate opposition	8	6
Strong opposition	11	4

Source: Responsive Management, Inc. (2000)

that there is considerable public support for wolf reintroduction, even within the Adirondacks.

Table 4.1, based on a 1996 survey funded by Defenders of Wildlife, summarizes attitudes of Adirondack residents and residents of New York State as a whole toward wolf reintroduction.

A follow-up 1997 survey of Adirondacks residents by Responsive Management, Inc. (2000) indicated that overall support for reintroduction had decreased from 76 per cent to 46 per cent while opposition increased from 19 per cent to 42 per cent. Strong opposition had risen from 11 per cent to 26 per cent. Preliminary results from a 1999 Cornell University survey reveal that a majority of New York State residents favour reintroduction (though about one-third are neither in favour nor against), while Adirondackers are about evenly split on the question (Human Dimensions Research Unit 2000).

The differences in results over a period of two or three years seem to suggest that support for reintroduction is broad, but not necessarily very deep. The 1999 survey was conducted before the release of the biological study, noted below, that concludes that there should be no reintroduction. Within the Adirondacks, at least, it seems that negative publicity about wolf reintroduction may tilt large numbers of citizens toward a negative view. Similarly, perhaps, new information about reintroduction and new public campaigns may bring large numbers toward the positive view. In the end, what may matter most in the debate are the voices of local residents for whom wolf reintroduction is a very salient concern. Their vocal opposition stems from concerns about dangers that wolves may pose to crops, livestock, and children; concerns that reductions in deer populations negatively affect hunting in the region; and the continuing local aversion to having decisions about environmental management made by outsiders (in this case, environmental groups based outside the region).

The Citizens' Advisory Committee agreed, in March 1998, to the commissioning of a biological study by the Conservation Biology Institute – a respected organization whose philosophical underpinnings clearly favour ecological preservation. Funded by Defenders of Wildlife, the study was released in December 1998 (Paquet *et al.* 1999). What is striking about the study, and what perhaps no one anticipated, are its very cautious findings. Immediate action is not recommended. Indeed, the study raises questions about the adequacy of undisturbed habitat in the Adirondacks and threats from disease, and views as a key issue human attitudes toward reintroduction. The habitat question was also raised in an earlier study, by the US Fish and Wildlife Service, that looked at recovery sites in northern New England and New York (US Fish and Wildlife Service 1998).

Most interesting, though, is the report's recommendation that the grey wolf should not be introduced in the Adirondacks. The eastern timber wolf, which Defenders had proposed introducing, is a subspecies of grey wolf. Because of habitat unsuitability and lack of linkages to other areas that the grey wolf inhabits, it is concluded that introduction of the grey wolf could not be sustained. This is especially likely to be the case in the absence of an unambiguous government commitment to protecting the wolves both inside and outside the park.

The study also questions whether grey wolves were ever present in the Adirondacks. Much of the evidence indicates that eastern Canadian wolf, a red wolf, may have once been endemic. And the coyotes currently in the park may actually be a Canadian wolf–coyote hybrid. If eastern Canadian wolves were reintroduced to the Adirondacks, there would be potential for more interbreeding and hybridization with the existing coyote population. Grey wolves, unlike red wolves, prey on coyotes. But introduction of grey wolves may well be ecologically inappropriate, if, as it now appears, they were not the dominant species. All this uncertainty seems to have taken some of the wind out of the Defenders of Wildlife's sails and, curiously enough, may allow the opponents of reintroduction to incorporate ecological arguments into their case. Clearly, proponents of wolf reintroduction are frustrated, given that there is no guarantee that the momentum that has built up in favour of wolves will last over the span of years that may have to be devoted to further research. Ecological uncertainty and political uncertainty synergistically create strategic despair!

The precedent for reintroducing predators into an ecosystem came from the much-publicized wolf reintroduction scheme in the Greater

Yellowstone region (Wilson 1997). Indeed, the approach taken by Defenders of Wildlife in the northeastern United States reflects the lessons learned by western advocates of species reintroduction. Defenders of Wildlife worked hard to build the kind of change in public attitude that can directly open a policy window. They worked very methodically towards leveraging a growing openness to ecosystem management and its many ramifications. In their campaign, Defenders of Wildlife and their allies assumed that the limiting factor for reintroduction would be attitude. The ecological rationalization was taken as a given at the outset. When this assumption proved false, Defenders of Wildlife found themselves in the uncomfortable and embarrassing position of not having the science to underpin their sentimental ecology. Science, not politics or public perception, was the leading factor in closing the policy window. The very basic lesson here is that, without convincing and compelling science, there is little basis for creating public support to undertake ecological intervention. Although sentiment remains important – perhaps even critical – it cannot be the sole basis for ecological management, as was the case with the attempts at elk and moose reintroduction that took place so many decades ago.

Conclusion

Experience in the Adirondacks illustrates what can happen when ecological planning and advocacy confront scientific and political reality. Recent controversies over the 1995 blowdown and reintroduction of wolves (or, more accurately, 'introduction', if the eastern timber wolves under consideration were never there) have not been cases of all-out war among interest groups; indeed, the wolf reintroduction effort is a model for introducing contentious issues in a non-confrontational, consensus-building way. The aftermath of the blowdown highlights how constitutional law delineated the parameters of debate. The reintroduction of the wolves was halted by the weight of scientific evidence. These two cases illustrate how sustainable ecosystem management requires a robust legal framework and unfettered scientific inquiry.

Recent events in the Adirondacks point to the complexity of putting the concept of sustainable ecosystem management into practice. Within the local context, at least, these events may have had something of a humbling effect on some of the more narrowly focused ecocentrists and defenders of property rights. More generally, they highlight the utility

– for advocates and resource managers alike – of incorporating an adaptive management approach (see Holling 1978; Gunderson *et al.* 1995) into proposals that involve ecosystem management. When what is 'best' is not known, ecosystem management requires both a willingness to respect what is known and an experimental approach. These principles necessitate greater co-operation among divergent interests, as well as the abandonment of righteous convictions about how best to manage regional resources.

References

Batisse, M. (1982) The biosphere reserve: a tool for environmental conservation and management. *Environmental Conservation* 9: 101–11.

Batisse, M. (1985) Action plan for biosphere reserves. *Environmental Conservation* 12: 17–27.

Belcher, E.H. and Wellman, J.D. (1991) Confronting the challenge of greenline parks: limits of the traditional administrative approach. *Environmental Management* 15: 321–8.

Booth, R. (1987) New York's Adirondack Park Agency. In Brower, D.J. and Carol, D.S. (eds) *Managing Land-Use Conflicts: Case Studies in Special Area Management.* Durham, NC: Duke University Press, 140–84.

Brick, P.D. and Cawley, R.M. (1996) (eds) *A Wolf in the Garden: the Land Rights Movement and the New Environmental Debate.* Lanham, MD: Rowman & Littlefield.

Commission on the Adirondacks in the Twenty-First Century (1990) *The Adirondack Park in the Twenty-First Century.* Albany: State of New York.

Corbett, M.R. (1983) (ed.) *Greenline Parks: Land Conservation Trends for the 1980s and Beyond.* Washington, DC: National Parks and Conservation Association.

Defenders of Wildlife (1996) Animal of the Year: New Poll Shows Widespread Support for Wolves. News release, December 31. www.defenders.org/releases/pr1996/pr123096.html

Diffenderfer, M. and Birch, D. (1997) Bioregionalism: a comparative study of the Adirondacks and the Sierra Nevada. *Society and Natural Resources* 10(1): 3–16.

Donaldson, A.L. (1921) *A History of the Adirondacks.* New York: The Century Co.

Erickson, J.D. (1998) Sustainable development and the Adirondack Experience. *Adirondack Journal of Environmental Studies* 5(2): 24–32.

Fascione, N. and Kendrot, S. (1998) Wolves for the Adirondacks? *Adirondack Journal of Environmental Studies* 5(1): 7–10.

Folwell, E. (1995) Lowdown on the blowdown. *Adirondack Life* 26(4): 48–53.

Graham, F.J., Jr. (1978) *The Adirondack Park: A Political History.* New York: Knopf.

Gunderson, L.H., Holling, C.S. and Light, S.S. (1995) (eds) *Bridges and Barriers to the Renewal of Ecosystems and Institutions.* New York: Columbia University Press.

Halper, L.A. (1992) A rich man's paradise: Constitutional preservation of New York State's Adirondack Forest, a centenary consideration. *Ecology Law Quarterly* 19(2): 193–267.

Halper, L.A. (1995) The Adirondack Park and the Northern Forest: an essay on conservation and preservation. *Vermont Law Review* 19: 335–62.

Harris, G. (1997) A brief history of elk introduction in the Adirondacks. *Adirondack Journal of Environmental Studies* 4(1): 13–18.

Heiman, M.K. (1988) *The Quiet Evolution: Power, Planning, and Profits in New York State.* New York: Praeger.

Hicks, A. (1995) Moose in New York – Past, present, and future. *Adirondack Journal of Environmental Studies* 2(1): 26–31.

Hirner, D.K. and Mertes, J.D. (1986) Greenlining for landscape preservation. *Parks and Recreation* 21(11): 30–4, 59.

Holling, C.S. (1978) *Adaptive Environmental Assessment and Management.* London: John Wiley.

Human Dimensions Research Unit, Department of Natural Resources, College of Agriculture and Life Sciences, Cornell University (2000) *Annual Report: Preliminary Assessment of Social Feasibility for Reintroducing Gray Wolves to the Adirondack Park in Northern New York.* www.dnr.cornell.edu/hdru

Hutchens, W.D. (1998) It doesn't make sense: reintroduction of wolves is not supported by Adirondack communities. *Adirondack Journal of Environmental Studies* 5(1): 11–13.

Kemmis, D. (1990) *Community and the Politics of Place.* Norman, Okla: University of Oklahoma Press.

Klyza, C.M. and Trombulak, S.C. (1994) (eds) *The Future of the Northern Forest.* Hanover, NH: University Press of New England.

Knott, C.H. (1998) *Living with the Adirondack Forest: Local Perspectives on Land Use Conflicts.* Ithaca and London: Cornell University Press.

Lauber, T.B. and Knuth, B.A. (1998) Refining our vision of citizen participation: lessons from a moose reintroduction proposal. *Society and Natural Resources* 11: 411–24.

Liroff, R.A. and Davis, G.G. (1981) *Protecting Open Space: Land Use Controls in the Adirondack Park.* Cambridge, Mass: Ballinger.

Little, C.E. (1992) *Hope for the Land.* New Brunswick, NJ: Rutgers University Press.

Lucas, P.H.C. (1992) *Protected Landscapes: A Guide for Policy-makers and Planners.* London: Chapman & Hall.

McKibben, B. (1995) *Hope, Human and Wild.* Boston: Little, Brown & Company.

McMartin, B.A. (1994) *The Great Forest of the Adirondacks*. Utica, NY: North Country Books.

Mason, R.J. (1994) The greenlining of America: managing private lands for public purposes. *Land Use Policy* 11: 208–21.

Mason, R.J. (1995) Sustainability, regional planning and the future of New York's Adirondack Park. *Progress in Rural Policy and Planning* 5: 15–28.

Meade, J. (2000) The 1995 Adirondack blowdown: An analysis of the ecological and sociological phenomena. *Adirondack Journal of Environmental Studies* 7(1): 13–18.

Michaels, S., Mason, R.J. and Solecki, W.D. (1999) Motivations for eco-stewardship partnerships: Examples from the Adirondack Park. *Land Use Policy* 16(1): 1–9.

Miller, K.R. (1996) *Balancing the Scales*. Washington, DC: World Resources Institute.

Paquet, P.C., Strittholt, J.R. and Staus, N.L. (1999) *Wolf Reintroduction Feasibility in the Adirondack Park*. Corvallis, Oreg: Conservation Biology Institute.

Responsive Management, Inc. (2000) *Public Attitudes Toward the Reintroduction of Large Predators*. www.responsivemanagement.com/wolves_bears.html

Solecki, W.D. (1994) Putting the biosphere reserve concept into practice: Some evidence of impacts in rural communities in the United States. *Environmental Conservation* 21(3): 242–7.

Temporary Study Commission on the Future of the Adirondacks (1970) *The Future of the Adirondack Park*. Albany, NY: Temporary Study Commission on the Future of the Adirondacks.

Terrie, P.G. (1985) *Forever Wild: A Cultural History of Wilderness in the Adirondacks*. Philadelphia: Temple University Press.

Terrie, P.G. (1993) *Wildlife and Wilderness: A History of Adirondack Mammals*. Fleischmanns, NY: Purple Mountain Press.

Terrie, P.G. (1997) *Contested Terrain: A New History of Nature and People in the Adirondacks*. Syracuse: Syracuse University Press.

US Fish and Wildlife Service (1998) *Recovery Plan for the Eastern Timber Wolf*. Washington, DC: US Fish and Wildlife Service.

Wells, M. and Brandon, K. (1992) *People and Parks: Linking Protected Area Management with Local Communities*. Washington, DC: The World Bank, The World Wildlife Fund, and US Agency for International Development.

West, P.C. and Brechin, S.R. (eds) (1991) *Resident Peoples and National Parks: Social Dilemmas and Strategies in International Conservation*. Tucson, Ariz: University of Arizona Press.

Wilson, M.A. (1997) The wolf in Yellowstone: Science, symbol, or politics? Deconstructing the conflict between environmentalism and wise use. *Society & Natural Resources* 10: 453–68.

World Commission on Environment and Development (1987) *Our Common Future*. Oxford: Oxford University Press.

5 | Limitless lands and limited knowledge: coping with uncertainty and ignorance in northern Australia

John Woinarski and Freya Dawson

Introduction

After more than a century of European settlement, northern Australia remains a frontier land. Its strange landscapes and vastness have repeatedly attracted developers with grand vision. Typically these developers have ignored the constraints imposed by the environment, the visions have failed, and the remnants of forsaken development have been left to be reabsorbed by a resilient land. This chapter examines some of the reasons for this pattern, and notes characteristics which may be general to other frontier regions (e.g. Holmes 1992). This contribution builds on some major, but more sectoral, analyses of development attempts in northern Australia (most notably Davidson 1965; Lacey 1979; Bauer 1964, 1977; Mollah 1980).

Essentially, the argument developed links several features:

- The environment is poorly known and dissimilar to that from which its developers originated. This strangeness has contributed negatively to the attribution of value to that environment and its component parts.
- The environment is extensive and this scale engenders the perception that successful use of the lands can be achieved only by large-scale development.
- The environment is perceived to be so extensive and of so little value that little safeguard needs to be built into development proposals.
- Repeated development failure reinforces the perception that the land is of limited value, and hence few resources should be directed towards understanding it (or rehabilitating it after development failure).

- The environment is perceived to be so marginally productive that the only route to substantial profit is through intensive modification of the environment and large-scale development.
- The economic framework supporting the settlers is so tenuous that developers and their supporters (government) consider that development shouldn't be burdened by substantial imposed conservation regulations.

Despite (or perhaps ironically because of) the developers' approach, northern Australia continues to have vast areas of relatively unmodified landscapes. Gradually this is being recognized as an asset rather than an affront. This change in attitude stems from a range of factors including: increased power of Aboriginal landowners; ongoing uncertainty about land tenure; the rise of the tourist industry; the inclusion of the frontier lands within national environmental strategies; a larger and more stable settler population; and, belatedly, some learning from the mistakes of the recent past.

Geographic and environmental setting

Northern Australia as defined here (Figure 5.1) occupies about 1.5 million square kilometres of mainly open forest and savanna lands. In this chapter, we concentrate on one political component of this region, the 'Top End' of the Northern Territory (NT), as this is the area with which we are most familiar. The patterns apparent in the Northern Territory have been general across northern Australia, with the exception that environments of tropical Queensland have been subjected to more intensive modification.

Across this region, soils are almost universally low in nutrients (Davidson 1965; Christian 1977). The climate is characterized by a short wet season (November to March) alternating with a long period of little or no rain. This general pattern varies considerably from one year to another (Christian 1977; Taylor and Tulloch 1985). Frequent cyclones add a further dimension to the climatic capriciousness (Lourensz 1981).

Many of the environmental values of northern Australia are obvious and substantial, if largely taken for granted. Its vast landscapes retain the most extensive eucalypt forests in the world (Woinarski et al. 2000). Its savannas are less intensively modified and support a vastly smaller human population than all other tropical savannas. Its ecological processes remain relatively undisturbed and its wildlife remains relatively undiminished: for example, in contrast to the catastrophic fate of native

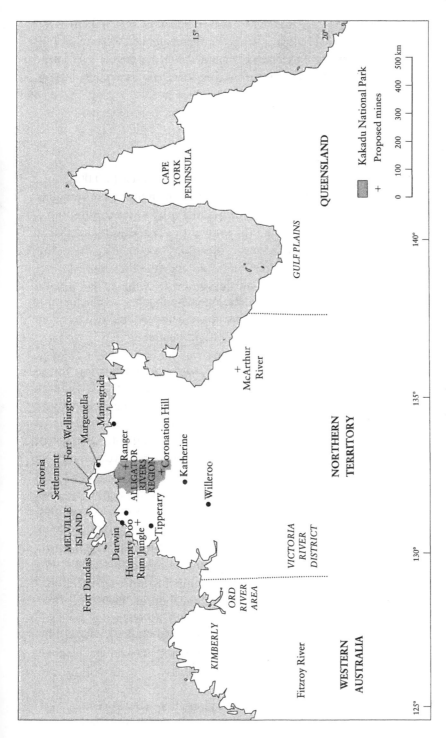

Figure 5.1 Northern Australia, showing place names referred to in text

mammals elsewhere in Australia, the mammal fauna of northern Australia has lost no species since European management (Woinarski and Braithwaite 1990). The extent and perceived 'naturalness' of its landscapes provide the main drawing card for the region's second largest industry, tourism.

Development philosophy

Aboriginal people entered Australia from the north at least 60 000 years ago. The extent of their modification of Australian environments remains hotly disputed (e.g. Flannery 1994; Bowman 1998). However, substantial environmental change following the removal of their land management in parts of northern Australia over the last 50 years (e.g. Price and Bowman 1994) suggests that they imposed considerable control on vegetation dynamics. Aboriginal people developed a complex and intimate relationship with the north Australian environment (e.g. Russell-Smith et al. 1997), with survival depending upon detailed knowledge of its components, and appreciation of their requirements.

European settlement of the area began in the 1820s, and consisted initially of small military outposts on the northern coastal rim, whose purpose was to protect the English colonial settlements of southern Australia from invasions from the north. Setting a recurring pattern, these 'forsaken settlements' (Spillett 1972) were typically underresourced and defeated by failures to appreciate their environment. Of the first three attempts at coastal settlement, Fort Dundas lasted five years before being abandoned due to hostility from local Aboriginal people; Fort Wellington lasted two years before being abandoned for logistic reasons, and Victoria Settlement was abandoned after 11 years due to high mortality, 'despondency', and general failure to find exploitable resources. The Aboriginal peoples' bountiful home became 'the White Man's frontier of adventure, isolation, difficulty and forbidding nature' (Jull 1991).

Penetration of the interior by pastoralists quickly followed accounts of explorers such as Leichhardt, Gregory and Stuart during the period 1844–1862. Many of these explorers, beguiled by the apparent luxuriance of grass at the end of the wet season and/or the demands to reward their financial backers with significant discoveries, falsely interpreted the potential of the northern Australian landscape. Thus,

> I have no hesitation in saying that the country I have discovered on and around the banks of the Adelaide River (near present-day Darwin) is

more favourable than any other part of the continent . . . I feel confident that, in a few years, it will become one of the brightest gems in the British crown. (Stuart 1865)

Stuart (1863) also repeatedly noted that the landscape was 'beautifully grassed' with 'the country of excellent quality and great extent'. Even when their eyes told them otherwise, imagination got the better of them: for example, Stokes in 1846 called the Gulf Plains of Queensland 'the Plains of Promise', as

. . . even in these deserted plains, equally wanting in the redundance of animal, as in the luxuriance of vegetable life, I could discover the rudiments of future prosperity, and ample justification of the name which I bestowed upon them . . . I could not refrain from breathing a prayer that ere long the now level horizon would be broken by a succession of tapering spires rising from the many Christian hamlets that must ultimately stud this country. (quoted in Bauer 1959)

Such overblown praise was generally greeted more warmly by sponsors than the assessments of the more perceptive Cassandras, such as Cadell, who described the previously recommended Victoria River District as 'a most wretched, rocky, barren, and waterless country . . . if the Elysian fields had been beyond it I should have felt it to have been a duty to report against its selection' (cited in Bauer 1964).

In the century since the hopeful vision of most of these early explorers, northern Australia has continued to be viewed as a land of promise, waiting for the right key to release its riches. This desire to use the land is linked to a broader national context. The development philosophy of northern Australia is still very much tied to the perceived need to protect the nation from invasion from the north, explicitly with the military as a major user of the northern environments (Barton and McDonald 1996) or implicitly with federal support of projects in northern Australia which can help populate the lands to the deterrence of invaders.

The more enthusiastic developers rail against constraints imposed by land tenure, concerned that lands retained as Aboriginal living space or as conservation areas ('locked up forever if possible, in Aboriginal lands with unique occupancy rights or in wilderness areas protected by United Nations conventions': Blainey 1992) thwart their promise of riches (Ewing 1996). The Northern Territory is often singled out as the place where most land has been 'quarantined' from major development because of the greater influence of the national government.

Who owns the land?

Land tenure is a significant factor for development and conservation. Its significance lies in the connection between property rights in land and the ownership and control over resources which are connected to or dependent on that land. Land tenure in the Northern Territory clearly reflects the history of European settlement. The tropical savannas of the north were one of the last frontiers of settlement for European pastoralists in Australia aided by the doctrine of *terra nullius* as a basis for the dispossession of Aboriginal people. Vast tracts of land were occupied under leasehold title, many of which remain under pastoral lease even though they are barely able to support one family (Holmes and Mott 1993). Despite the low productivity of these pastoral properties, at the time they were established the land they covered represented the resources of greatest value in the region. Pastoral leases still cover almost half of the Northern Territory.

Almost all of the Northern Territory not under pastoral lease is owned by Aboriginal people under the *Aboriginal Land Rights (Northern Territory) Act 1976* (Cth) (henceforth, the Land Rights Act). The tenure of Aboriginal land differs radically from the transferable title that is granted to non-Aboriginal pastoralists and holders of a 'normal' freehold. The main difference is that the tenure created by the Land Rights Act incorporates a partial recognition of Aboriginal custom and law (Neate 1989). The Aboriginal traditional owners of this land are entitled to occupy the land and use the biological resources on that land in accordance with Aboriginal tradition governing the particular group of people. The extent to which the general laws of the Northern Territory governing such things as wildlife conservation, fire management, planning and water resources can apply to Aboriginal land without interfering with these rights is a matter of considerable legal uncertainty (Dawson 1996). This creates two very different regimes of property rights in biological resources in the Northern Territory.

The land tenure regime in the Northern Territory is currently under pressure stemming from the recognition of native title under common law, by the High Court in *Mabo v. Queensland* (No. 2), and subsequently in legislation by the *Native Title Act 1993* (Cth). 'Native title' means the communal, group or individual rights and interests of Aboriginal people or Torres Strait Islanders to land or waters under their own laws and custom. It is not yet clear how much land or sea in the Northern Territory will be able to be claimed under the *Native Title Act*, or what the exact nature of this interest will be. The full impact of the recognition of native title on the land tenure system in the Northern Territory is unlikely to be seen in the near future. The recognition of native title has been vigorously opposed by many sections of the pastoral and mining industries because they feel it creates uncertainty over land tenure and the access they might have to the land to pursue commercial development (e.g. Ewing 1992).

Land tenure in the Northern Territory creates a challenge for conservation measures which seek to deal with ecological processes across the whole landscape (Hughes 1995; Dawson 1996). Native title simply adds a further layer to an already complex situation. The mix of jurisdictions across northern Australia further complicates conservation planning. The relatively continuous and homogeneous environments of northern Australia are partitioned between two states and the Northern Territory, with different laws and strategies for land use and conservation planning. Within the Northern Territory, some conservation reserves are managed by a federal government agency and others by the Territory authority, with inconsistent and, at times, conflicting practice and objectives. As an example of the distortion of information and planning due to the political dissection of northern Australia, each of the states and the Northern Territory has their own vegetation mapping scheme, and these are mutually incompatible and stop at the state/territory borders. Such parochialism hampers conservation planning across the broad environments of northern Australia.

Case studies

We introduce some of the main development and conservation players and issues of northern Australia by way of a series of examples which provide the flavour of, and indicate the flaws in, the pervasive development philosophy and its attempts to deal with (or paper over) the environmental ignorance and uncertainty within settler society.

Case study 1: Forestry

The history of forestry in the Northern Territory provides an expensive but telling lesson in the costs of environmental ignorance (Lacey 1979) . The relatively limited supply of prime durable timbers, principally ironwood (*Erythrophleum chlorostachys*) and northern cypress-pine (*Callitris intratropica*), was rapidly depleted for railway sleepers and construction around Darwin within the first 40 years of settlement, and well before any substantial resource inventory or investigation of recruitment and growth rates (Hanssen and Wigston 1989). Subsequent early investment in plantations of exotic timber species generally failed as most species proved susceptible to the termite *Mastotermes darwiniensis*.

The next major phase of forestry activity in the NT occurred from the late 1950s to the late 1970s, and involved a more concerted campaign of timber plantations (principally on Melville Island) and prospecting for major areas of native forest production. The philosophical change underlying this development was described by Hanssen and Wigston (1989):

> The preceding century of exploitative disregard for NT forest resources was displaced by a welcome enthusiasm but a disastrous optimism.

While there were sober assessments of the limited availability of timber (Bateman 1955), these were largely ignored by misplaced confidence in forestry technology. Lack of information on forest characteristics was belatedly recognized to be a problem, which was addressed in 1958 by the establishment of a forest research station, charged initially with establishment of plantations and investigation of tree growth. The research station grew into a government department with a clear vested interest in the maintenance and expansion of forestry. Their 'assessments of forestry resources and their potential to supply timber became progressively more hopeful as the demand for timber rose and the actual production of sawn timber progressively fell' (Lacey 1979), and 'any research findings that were counter to the spirit of the plantation program were suppressed or discouraged' (Cameron 1985). For both plantations and native forests, the projected growth rates, timber yields, projected economic returns and potential for pulpwood turned out to be wildly exaggerated, if not duplicitous (e.g. Higgins and Phillips 1973), at least partly due to facile extrapolation of data from temperate Australia. For example, estimates given to the Forwood Conference (1974) suggested an increase in future availability of forest product from an actual yield in 1971 of 3300 cubic metres to 58 000 cubic metres by 1980, a hopeful estimate raised in 1977 to 82 000 cubic metres. The eventual tally for 1980 was zero. Estimates given for yield in the Maningrida area were 13 cubic metres per hectare, whereas these proved to be only 0.6 cubic metres per hectare. Estimates of the surveyed area of high-quality forests on Melville Island were 20 per cent greater than the total land area of that island (Anon. 1978).

Contemporary but less ambitious timber harvesting projects, such as the establishment of sawmills at Murganella and Maningrida, also quickly foundered, due to inaccurate assessments of the resource and/or conflict over ownership of the land and its products. The more than AUD30 million outlaid (Lacey 1979) for negligible returns led to a federal government inquiry in 1978, which effectively closed down the NT Forestry Program.

This inquiry concluded that the plantation and native forests programmes had failed largely because of inadequate research: 'fundamental data on the type, volumes and accessibility of native timber stands which should have been the starting point for an exploitation program was not available in any meaningful form' (House of Representatives Committee on Expenditure 1978), and that at least ten years of detailed research on such topics was required before addressing the feasibility of any future timber industry in northern Australia.

Just over ten years after this inquiry, and flying in the face of its main recommendation, a major new export-oriented forestry operation was announced for the NT, based on lancewood (*Acacia shirleyi*) and, less importantly, gutta-percha (*Excoecaria parvifolia*). That this major development came out of left field is indicated by a complete lack of mention of it in a major outlook paper for forestry and forest research in the NT published four years before its commencement (Cameron 1985). The sorry history of this project is described in the Forest and Timber Inquiry of the Australian Resource Assessment Commission (RAC 1992) and submissions to that inquiry. An initial licence was granted in 1989 by the NT government to harvest 360 000 tonnes of these timbers over a five-year period, mostly from pastoral leases. At the time of licence issue there were no estimates of timber volume nor of growth rates for these species, nor was there any information on the wildlife values of lancewood-dominated vegetation (Bowman 1991). Following the issue of licences to cut, the federal government granted export licences on condition that:

- a report should be prepared on the conservation status of lancewood and gutta-percha;
- an agreement should be reached with the company to ensure the sustainable utilization of the two species;
- programmes should be introduced to monitor and report on the operations; and
- pre-logging surveys should identify significant cultural, biological and geological data.

Research into the conservation values of this environment and sustainability of the industry were to be supported by service fees and royalties paid by the company to the NT government.

Within two years of commencing logging, the company reduced their projected cut from 72 000 tonnes per year to 3000 tonnes per year. No pre-logging biological surveys were conducted and there was no reporting on the conservation status of lancewood or the associated bullwaddy (*Macropteranthes kekwickii*) communities. By the following year, the operation had been abandoned, with licences rescinded by the NT government, a total of only 200 tonnes cut over the life of the industry, complaints of large areas of cut and left logs, and failure of the industry to pay any service charges.

Failure of the industry was due partly to unrealized market projections, but, more importantly, to completely unrealistic estimates of timber availability and quality. The first detailed estimates of wood volume were produced only after the folding of the industry, and indicated that the total amount of lancewood (pole and sawlog equivalent) across the entire Northern Territory was only 5.2

million tonnes, of which most was unsuitable for logging due to inappropriate growth form or defects (NFI 1993). The government's role in this doomed industry was not atypical – enterprises were to be facilitated, rather than be hampered by any need to ground assessment of their viability, sustainability or impact in relevant information. The first examination of the wildlife of lancewood communities (Woinarski and Fisher 1995a,b) did not appear until well after the closure of the industry. There is still no lancewood within conservation reserves in the Northern Territory.

Case study 2: Agriculture

It is in agricultural ventures that northern development has had its most spectacular failures. The causes and economic costs of these debacles have been widely reviewed (notably by Davidson 1965; Bauer 1977; Mollah 1980; MacKenzie 1980). Davidson (1965) argued that large-scale agricultural developments in northern Australia were almost inevitably doomed by the small population base and (hence) distance from markets and large transportation costs, compounded by poor soils and erratic climate. Ignorance about the environment and the absence of risk assessment were also major contributors to failure. While the environments of a relatively benign temperate Australia were strange and initially an impediment to the transfer of the agricultural customs of European settlers, those of tropical Australia were doubly so, and proved completely unsuitable for most of their traditional stock and produce.

Agricultural failures became a regular feature soon after the first settlement of Darwin in 1869. Large-scale projects began and failed between 1875 and 1884 (sugar cane around Darwin), and continued intermittently in northern Australia (e.g. cotton plantations at Derby in the 1920s). However, such ventures grew grander in scale after the Second World War, when the threat of invasion from the north encouraged a strong development push (with enthusiastic support or compliance from the federal government). Beginning in 1955, massive American capital was injected into rice production at Humpty Doo, a site considered by the local government agricultural agency 'a more suitable track (sic) of land (than) anywhere in the world for mechanised rice growing' (Curteis 1961). This scheme aimed to produce 400 000 tonnes of rice annually from 200 000 hectares of intensive cultivation. In the eight years before it was abandoned, it produced a total of only 3000 tonnes. Although consumption and trampling of crops by magpie geese (*Anseranas semipalmata*) was blamed for the failure, the real cause was inadequate knowledge of environmental constraints, most notably the availability of water (Fisher *et al.* 1977). One benefit of the scheme was the establishment of a government-sponsored research station focusing mainly on agronomy, but also examining the ecology of floodplain environments. The latter included detailed studies of the native potential 'pest' species, magpie goose and dusky rat (*Rattus colletti*) (e.g. Frith and Davies 1961; Redhead 1979). These remain some of the few detailed animal autecological studies in northern Australia.

Detailed knowledge about the environment was belatedly recognized to be a prerequisite for the successful establishment of major agricultural developments (e.g. Forster *et al.* 1960). To this end, a government agency, the North Australian Development Committee, directed a major series of land capability studies describing soils, vegetation and agricultural potential (CSIRO 1952, 1965, 1969, 1970, 1976; Christian and Stewart 1953; Perry 1960). Notwithstanding such information about land capability (CSIRO 1965), and substantial relevant agronomic research (Basinski *et al.* 1985), the next major development scheme, grain sorghum production at Tipperary, was another disaster. With an investment of more than AUD20 million (much of it American capital), this scheme was established in 1967 and predicted annual production of 300 000 tonnes of grain sorghum from almost 10 000 square kilometres. To facilitate the scheme and its projected township of 15 000 people, the NT Legislative Council made substantial changes in leasehold conditions to allow broad-scale land clearance (Mollah 1980). By the abandonment of the scheme in 1973, only 16 000 tonnes of sorghum had been harvested (and almost all of this was below export standard): but at least 10 000 hectares of forests had been destroyed. Tipperary failed partly because it was directed from so far away, partly because its planners had neglected to consider the unpredictability of climate, partly because too few of the techniques and crop strains had been adequately trialled on a small scale, but largely because of a misplaced confidence in technology: 'There is no comparable region in Australia where the white man has so consistently overestimated the power of his technology in the field of primary industry to draw forth bounty from the land' (Lacey 1979).

Despite this obvious failure, but spurred on by the land clearance incentives offered in the Tipperary legislative amendments, a similar scheme commenced at the 6 000 square kilometres of Willeroo in 1971. This venture proposed to clear 120 000 hectares within five years for crop sorghum and high-quality beef (Fisher *et al.* 1977). Unlike the Tipperary scheme, no detailed land capability surveys preceded decisions about clearing and cropping, although land resource survey scientists apparently accompanied or followed the bulldozers (Bauer 1977). The venture was abandoned within four years. Of about 50 000 hectares cleared, only 16 000 hectares was used for cultivation. Failure was due to 'a fundamental lack of understanding of the limitations of soils and climate' and 'the Company grossly overestimated the savings of large-scale operations' (Fisher *et al.* 1977).

At least recognizing the limitations of soil nutrients across much of the north Australian landscape and the problem of inadequate knowledge, the next major venture picked an area with more fertile (though still nutrient-deficient) soils and preceded large development with research and pilot farms. However, this next development was linked to the fancy that irrigation was the key to unlocking the resources of northern Australia. Haigh (1963), for example, championed this vision, claiming that northern Australia had 3.2 million hectares of potentially irrigable lands (of which only 5 per cent was developed), compared to half that amount in southern Australia (of which 90 per cent was developed), and the north had 27 billion cubic metres of 'uncommitted water potentially available for irrigation'. Such beguiling figures readily convinced politicians keen to establish monuments. The Ord River Irrigation Area Scheme was established in 1963, primarily to produce cotton on irrigated blacksoil plains in the north of Western Australia. Within ten years the cotton venture collapsed, despite public expenditure of almost AUD100 million (1978 value: Anon 1978). It failed partly due to higher transport and infrastructure costs relative to other cotton-producing areas, inappropriate cultivars, and inability to control insect pests (Fisher *et al.* 1977; Anon 1978; Graham-Taylor 1978). The detailed agronomic research had helped indicate what crops had highest potential and suggested that insect pests would be a problem in extensive crops (CSIRO 1978), but its caveats were ignored and it had not provided solutions which were economical for the scale of development which followed. As with research associated with the Humpty Doo rice venture, some of the studies associated with the Ord provided major contributions to our understanding of the ecology of northern Australia, including some detailed autecological studies of vertebrate species (e.g. Beeton 1977); although these studies were always coloured by the framework of 'pest' designation.

The development of these agricultural ventures has been characterized by disregard for environmental consequences, with the occasional exception of some concern for land erosion. There were no assessments of the conservation values of the areas developed. With minor exceptions (e.g. the Ord scheme left five chains from each side of the river: Graham-Taylor 1978), no natural areas were reserved from development to maintain representative undisturbed habitat. Typically, there was no monitoring of the ongoing impacts of the agricultural activities. For example, after 15 years of the Ord scheme, the Government Review Committee rather lamely noted that

> Generally speaking, the Review Committee concludes that known environmental effects of the development of the ORIA to date do not appear to be very significant. However, in the absence of adequate data few detailed conclusions can be drawn. (Anon. 1978)

Twenty years further on, there are still no adequate data to assess the environmental impacts of this scheme.

The agricultural developments have typically lacked risk assessment, despite the now well-recognized climatic vagaries and the repeatedly catastrophic impacts of lack of rain or too much rain during critical phases of crop production. Regard to resource conflicts and environmental sustainability has generally not been part of the development planning either. Whitehead (1991) noted that the development of extensive tropical fruit orchards near the Top End floodplains would inevitably lead to crop damage by magpie geese and other vertebrates (reviewed in Lim *et al.* 1993), and that such likely lost production (or the expenses associated with preventative measures) should be included at initial planning and project justification. Belatedly, planning for agriculture within such an environmental framework is now being undertaken, at least in some cases (e.g. Whitehead *et al.* 1990; Anon. 1995).

Subsequent to the failure of cotton, the Ord scheme has persisted on a generally less ambitious scale and with more diversified cropping. However, over the last five years, grand plans have resurfaced to develop the lower Ord for intensive sugar plantations (Sinclair Knight Merz 1996). Tipperary development returned in the late 1980s, repeating the grand vision and accompanied by even more extensive forest destruction. Horticultural production has increased in the Darwin-Katherine area, and a major intensive farming scheme for the Douglas-Daly catchment is again proposed.

Case study 3: Pastoralism

Pastoralism is the dominant industry across northern Australia, monopolizing land use in almost all environments. It has succeeded, or persisted, where agriculture has failed, largely because its establishment and operational costs are generally low.

Pastoral intrusion into northern Australia quickly followed the generally enthusiastic, if less than perceptive, reports of the first European explorers (Bauer 1959, 1964). Favourable reports led, in 1858, to a land grab of what is now the Northern Territory by the colony of South Australia, which then promptly sold over 200 000 hectares to pastoral investors sight unseen; in many cases, the actual location of the lots was unknown (MacKenzie 1980). This action set the tone of distant management which still characterizes much of the pastoral industry. Fiascos followed this first land trade, with problems in survey and unsuitability of lands leading to the first years of South Australian governance being described as 'a decade of ineptitude' by MacKenzie (1980). Despite administrative bungling, north and west Queensland were 'opened up' in the 1860s; and by the 1880s, cattle and sheep had reached the Victoria River District and the Kimberley (Bauer 1984). In many areas it was clear that the promises painted by the explorers would not be readily realized. Bauer (1984) reported the surveyor McKinlay's description of his livestock's first experience of the wet season: '. . . stock did not fatten, even when standing in grass higher than they; several horses died of plant poisoning and most of the others lost their hair from the continual drenching. The sheep did even more poorly . . .'

Notwithstanding such reports (and the very limited experience of pastoralism in the region), by 1881 the 'fantastic rush for pastoral land' (Bauer 1984) resulted in pastoral leases granted for over 95 per cent of the Northern Territory (despite the fact that most of the area had never experienced cattle nor Europeans). Speaking of the Gulf Plains area, but apposite generally, Bauer (1959) noted that 'there seems to have been what amounts to a refusal on the part of many settlers to recognise that (the lands) had definite limitations'.

In some areas, the livestock and pastoralists prospered. However, in many parts, the more attractive and less resilient plants declined and unsustainable stocking pressure quickly produced soil erosion, degradation of water bodies, and vegetation change. Even proponents of the pastoral industry began talking about 'abuse of our rich pastoral areas' (Wise (1929), cited in Riddett (1990)). After little more than 50 years of pastoral use, Medcalf (1944) reported that 10 per cent of the area he surveyed in the Ord valley was affected by erosion to varying degrees, with the formation of deeply eroded gullies, and progressive deterioration proceeding rapidly in many other areas. By 1976, over 30 per cent of the Fitzroy River catchment was degraded to poor or very poor condition (Payne et al. 1978). Such degradation was not seriously addressed until its impact was felt on other land uses. When overgrazing led to an annual deposition of 24 million tonnes of sediment into the Ord River, the well-founded concerns about effects on the irrigation scheme on the lower Ord, and the massive capital investment spent on dams (Winter 1990) forced reduction in stocking rates and stock removal. However, much of the damage has persisted, and there is little knowledge about recovery from degradation (Foran et al. 1985). The impacts of pastoralism upon biodiversity have not been examined systematically or in any detail. For many environments, there has been substantial, and possibly irreversible, change in vegetation composition due to grazing or trampling by livestock and/or feral stock (Stocker 1970; Winter 1990). The impact upon wildlife of pastoral hegemony in northern Australia is almost unknown, but it has clearly contributed to the decline and local extinction of riparian birds in East Kimberley (Smith and Johnstone 1977; Woinarski 1993) and in Queensland (Barnard 1925). Increased grazing pressure has been shown to be associated with decreased abundance of granivorous birds in the Victoria River District (Tidemann 1990). Medium-sized mammals have also declined precipitously in most areas subjected to extensive grazing (Kitchener 1978; McKenzie 1981).

The environmental impacts of pastoralism are magnified when pastoral managers transform the land to increase its suitability (at least over the short term) for stock. In northern Australia, such 'improvements' have been eagerly sought, and have included removal of trees, modification of the fire regime, alteration of water supply and, particularly, introduction of exotic plants. While there has been a very large investment in research aimed at such pastoral 'improvements' (e.g. Eyles et al. 1985), little of this research has considered the environmental costs of such transformation (though there are a few notable recent exceptions: Gillard et al. 1989; Glanznig 1995; Scanlan and Turner 1995). Lonsdale (1994) illustrated the problem with such an unbalanced research agenda. He found that, of nearly 500 plant species introduced for pasture improvement in northern Australia, only 21 were eventually listed as useful, whereas 60 became listed as weeds (incidentally including all but four of the 'useful' species). Many of the plants introduced to improve pasture are now major threats to conservation (and other) values and will cost millions of dollars to control (Whitehead and Dawson 2000).

While some environments in northern Australia clearly support economically sustainable pastoralism (and the economic viability of many holdings has been substantially improved by the recent burgeoning live cattle export trade: Stewart 1996), much of the land remains marginal or submarginal (e.g. Holmes 1990). Governments have shown a marked reluctance to remove pastoralism from such areas (Anon. 1991), presumably because the retention of nonviable pastoralism is sufficient to stake a claim to the area rather than to leave a land use vacuum, which may provide an opportunity for Aboriginal people to reclaim part of their lost estate.

Case study 4: Mining

The mining industry in northern Australia provides clear examples of the difficulties associated with decision making in the context of environmental uncertainty and ignorance and the added challenge of strongly conflicting cultural values.

The impact of the Rum Jungle uranium mine should provide a clear lesson of the dangers of ignoring environmental considerations. The Rum Jungle mine, 80 km south of Darwin, was Australia's first major uranium mine. Development and operation of the project was carried out by a subsidiary of Consolidated Zinc Pty Ltd (now known as C.R.A. Pty Ltd.) as an agent for the Commonwealth government. Mining operations began in 1953 and finished in 1971.

Towards the end of the life of the mine the Australian Atomic Energy Commission undertook a series of studies aimed at identifying the extent and degree of environmental damage (Davy 1975). This review found the overburden heaps at the mine were oxidizing and producing acid mine drainage containing heavy metals, sulphates and acid. Pollution was entering the East Finniss River, ground water and surrounding ground. The open cut pits had filled with water, which had become polluted with heavy metals and acid to the extent that one of the pits had a pH of approximately 2.4. The tailings dam was acidic, contaminated with heavy metals and was also a low-level source of radiation. By 1974, 150 000 tonnes of these tailings had been eroded from the tailings dam and had entered the river system. Estimates of the time for the environment to recover naturally from the effects ranged from 100 to 1000 years (NT Department of Mines and Energy 1986).

The Commonwealth government eventually agreed to fund a rehabilitation project to reduce the environmental effects of the mine. The project cost AUD16.2 million in 1982 values. The long-term effects on the flora and fauna of the Finniss River and its radiological safety are not fully known (NT Department of Mines and Energy 1986).

Since the operation of the Rum Jungle mine, much has changed in terms of the attitudes, expectations and requirements of society with respect to environmental issues. By 1974, much had also changed in terms of the development of environmental law. Arguably, the most significant of these changes in relation to the mining industry was the enactment of the *Environment Protection (Impact of Proposals) Act 1974* (Cth). The Administrative Procedures under this Act deal in detail with the requirements for the environmental assessment of projects likely to have a significant effect on the environment.

In July 1975 the Commonwealth directed that an inquiry should be conducted pursuant to the *Environment Protection (Impact of Proposals) Act 1974* (Cth) in relation to a proposal for the development of uranium deposits in the Alligator Rivers Region of the Northern Territory by the Australian Atomic Energy Commission in association with Ranger Uranium Mines Pty Ltd. Many submissions were received, most of which opposed the project proposal. The Commissioners made an extensive inquiry into the natural features of the Alligator Rivers Region and the potential environmental impact of the Ranger proposal. Their findings are set out in the *Ranger Uranium Environmental Inquiry Second Report* (known as the Second Fox Report) (Fox *et al.* 1977). No reference is made in the Fox Report to the Rum Jungle mine and the associated environmental disaster.

What emerges from the Second Fox Report is an approach to dealing with uncertainty and ignorance that has become typical for mining projects in the region in the era of environmental assessment. The report accepts the contention of a number of biologists that the then existing information was not sufficient to enable the ecological effects of mining to be predicted, especially the long-term effects on aquatic ecosystems. This was not viewed as a sufficient reason to recommend that the project should not proceed, or that it should be delayed until more information is collected. Instead, the report recommends that the best practicable technology should be used and that standards for contaminant releases should be strictly defined (Fox *et al.* 1977). Since these standards are being defined against unknown baseline data they must, in fact, be determined arbitrarily.

The federal government decided to allow the development of the Ranger Uranium mine. To ensure that something was being done to reduce the ignorance and uncertainty surrounding the environmental impact of the mine they established the Office of the Supervising Scientist (now known as the Environmental Research Institute of the Supervising Scientist) for the Alligator Rivers Region to coordinate research and monitoring operations and generally to supervise the performance of the mining company in the environmental field. The Supervising Scientist does not have any powers of enforcement if the environmental requirements established for the project are not complied with (Fry 1980). The impact of the Ranger mine on the wetlands downstream of the mine has been the subject of fierce debate. Releases of contaminated water from the mine have caused considerable distress to Aboriginal traditional owners, uncertain of the effects these releases may have on the aquatic life which forms a large component of their traditional food supply. The traditional owners were unsuccessful in their attempts to challenge these releases in court in the case *Northern Land Council, Big Bill Neidjie and Ors v. Energy Resources of Australia Ltd and the Minister for Mines and Energy* (Supreme Court of the Northern Territory of Australia, Martin CJ, 24 March 1995).

The timing of the Fox Report was significant in that it coincided with the passage of the *Aboriginal Land Rights (Northern Territory) Act 1976* (Cth) (the Land Rights Act). The Fox Report also made recommendations in relation to a land claim made by the traditional owners of the Alligator Rivers Region (Fox *et al.* 1977). Ultimately, the recognition of traditional ownership of most of the

Alligator Rivers Region has had a significant impact on its development. This is particularly so in relation to Kakadu National Park which now surrounds the Ranger mine, and which is jointly managed by the traditional owners and the Australian Nature Conservation Agency (Press *et al.* 1995).

One of the effects of the legislative recognition of the traditional Aboriginal relationship with the land has been to highlight the great differences between the cultural values of Aboriginal and non-Aboriginal people in Australia. White Australians generally have a poor understanding of Aboriginal culture, and yet the impact that mining may have on the exercise of Aboriginal tradition, including spiritual matters, must now be considered in relation to many development proposals in Northern Australia. This change has been perceived by the mining industry to add a further element of uncertainty to that already created by environmental impact assessment.

The combined effect of concern over environmental impact and opposition from Aboriginal people to disturbance of their land by mining was seen most acutely in relation to the Coronation Hill Joint Venture project. This project proposed to mine gold and palladium in a window within Kakadu National Park known as the Conservation Zone. The normal process of environmental assessment under the *Environment Protection (Impact of Proposals) Act 1974* (Cth) was completed and yet the information so obtained was seen as inadequate to provide a basis for decision making. In particular, scientific research conducted by CSIRO was almost completely at odds with the draft environmental impact statement produced by consultants Dames and Moore for the joint venture partners (Toyne 1994). A land claim had also been made under the Land Rights Act and the area registered as a sacred site under the *Aboriginal Sacred Sites Act 1978* (NT). The elders of the Jawoyn people (the traditional owners) opposed the Coronation Hill project.

The response of the Commonwealth government to this difficult situation was to order an inquiry under the *Resource Assessment Commission Act 1989* (Cth). The Resource Assessment Commission was a primarily research-oriented body, collecting and analysing information in order to make recommendations about resource development issues. The inquiry ran for just over one year, received 199 written submissions and held numerous public hearings. Although the inquiry acknowledged the outstanding conservation value of the region and the fact that environmental risk could not be eliminated, it concluded that a single mine, properly managed and monitored, would have a small and geographically limited impact on the known biological resources of the Conservation Zone (RAC 1991). The finding of the inquiry which most influenced the ultimate decision makers was the conclusion that mining would adversely affect the ability of the Jawoyn people to sustain cultural and religious values, beliefs and practices that are important to them. In June 1991 Federal Cabinet decided that the mine should not proceed. The Conservation Zone was then included in Kakadu National Park.

The Coronation Hill case shows an attempt by the federal government to come to terms with the full range of issues relevant to sustainable development in a systematic way in relation to a particular mining project. However, there has been a considerable backlash against the decision. The range of matters dealt with in the inquiry served to highlight the deficiencies of the regular environmental assessment process, and yet this process was subsequently weakened by the announcement in 1992 by the Prime Minister that development projects of AUD50 million or more would be 'fast-tracked' through the approvals process (Toyne 1994).

The impact of this policy decision fell first on the Northern Territory where it was applied to the McArthur River mine on the Gulf of Carpentaria and the Mt Todd mine near Katherine. The result of 'fast-tracking' was that the environmental assessment process was carried out to a timetable designed to meet the convenience of the mining company, and only the absolute minimum requirements of the Commonwealth and Northern Territory environmental assessment legislation (Dawson 1993). In the case of McArthur River, Aboriginal traditional owners in the area objected to the fast-tracking process and the negotiating techniques of the mining company (Land Rights News 1993). The environmental impact statement frankly admitted that the baseline environmental data were inadequate and that further studies were necessary to properly assess the impact of the project and yet this had no bearing on the hasty decision to allow the project to proceed. The weight of the environmental controls applied to the McArthur River mine are encompassed in an environment management plan (EMP), thereby shifting the responsibility of dealing with the uncertainty and ignorance associated with any environmental impact to the Northern Territory government as part of the overall regulation of the mine.

The Mt Todd mine was another notable case where assessment of the impacts of the mine was substantially clouded by uncertainty over the conservation values of the proposed mine site. The mineral deposit was located within or adjacent to a major breeding colony of the Northern Territory's only endangered bird species, the Gouldian finch (*Erythrura gouldiae*). The information on which the assessment of the mine's impact was based was insufficient to clarify the bounds of the breeding colony relative to the mine site, the number of breeding birds likely to be affected, the impacts that could be expected on the breeding colony, or the significance of the breeding colony relative to the total population of the species (Buckley 1993). In the absence of such data, the mine proposal was approved in 1992; but any impacts of the mine upon the Gouldian finch were to be ameliorated by a substantial research contribution from the mine proponents and the establishment of a monitoring programme to assess the mine's impacts. Given the risks associated with the location of a very large mine in or adjacent to a significant breeding colony of an endangered species, much of the protection of that colony was dependent upon the efficacy of the monitoring programme, and the ability to ensure that remedial action was taken if serious impacts were detected (Clancy 1995). However, the monitoring programme adopted (Zapopan 1993) had substantial limitations, and has now been abandoned. No relevant pre-impact baseline data were collected; the power of the programme to detect any change was very low (and never quantified); just one part of the life history and resource requirements of the species was considered; the programme measured impact on an annual basis such that rapid remedial response may have been foreclosed, the threshold changes required to trigger remedial action were not specified, and remedial actions were not defined. This tolerance of uncertainty in impact assessment works heavily in the developer's favour (Green 1989) as 'low power research has a low probability of rejecting the idea of no effect, so the use of the environment is more likely to be approved without expensive environmental safeguards' (Leis 1992). In situations such as this where there is limited information but a risk of substantial conservation costs, it has been argued that the possibility of failing to detect a decline when one in fact occurs should be explicitly stated and minimized (Shrader-Frechette and McCoy 1992).

Case study 5: Sustainable harvesting

For at least 60 000 years, Aboriginal people of northern Australia have relied upon harvesting wildlife for sustenance. A few wildlife species (notably the two crocodile species *Crocodylus porosus* and *Crocodylus johnstoni*) are now being harvested commercially, and the economic value of this trade is being used as an argument to conserve the habitat of these species (Webb and Manolis 1993). Recently, the NT government has proposed a generic strategy for the sustainable use of wildlife (PWCNT 1995a), and specific trial sustainable use management programmes, for the red-tailed black cockatoo (*Calyptorhynchus banksii*) (PWCNT 1995b) and cycads (PWCNT 1995c). The establishment of sustainability in these industries is contingent upon adequate information about population size, reproductive rates, harvest rates, costs of surveillance and market demands. For red-tailed black cockatoos in the Northern Territory, no such information is currently available. Lacking such information, the proponents propose a monitoring programme linked to adaptive management (Walters 1986). The required monitoring and research into the ecology of the species would be funded by service charges from the commercial harvesters, in a manner similar to those proposed for lancewood harvesting. For monitoring to effectively assess the impact of harvesting, it must be capable of detecting changes in population size and demographic structure.

Surveying red-tailed black cockatoos has a number of inherent problems which reduce the statistical power needed to meet this objective (Pollock 1995; Marsh 1995). The species is highly mobile and may occur in large flocks, hence error bars are likely to be large relative to population estimates, and distributional responses to spatial variation in rainfall or food availability may mask impacts of harvesting (Ludwig *et al.* 1993). The species may be inconspicuous, particularly in its favoured recently burnt areas, hence population estimates may be inaccurate. The species has a very low reproductive rate, is long-lived and has a long period to maturity (Forshaw 1981), factors which would render it especially sensitive to over-harvesting (Beissinger and Bucher 1992) as substantial changes in demographic structure resulting from harvesting may not be apparent until well after such changes occur. The rapid decline of the red-tailed black cockatoo to the point of endangerment elsewhere in Australia (e.g. Joseph *et al.* 1991) should mandate that if commercial harvesting of this species is to be permitted in the Northern Territory, then the supervisory authorities must adopt a particularly cautious and well-informed approach, and clearly state and address the uncertainties involved in all aspects of the industry (Clark 1996).

Lack of information does not necessarily preclude the establishment of a sustainable harvesting strategy, although it does magnify the risks. Beissinger and Bucher (1992) provided an approach to establishing sustainable harvesting from an initially inadequate information base: however, this strategy was dependent upon landowners taking only the additional population resulting from environmental modification which had improved habitat suitability. Such is not a component of the proposed NT scheme.

The effectiveness of these programmes in delivering their conservation goals (Anon. 1994) will be dependent upon, and measurable by, the extent to which the industry leads to a reduction in the rate of land clearing. In the case of the red-tailed black cockatoo, the financial inducements to landowners offered by the development of a sustainable harvesting regime may be difficult to translate into land protection and species conservation. This bird undertakes extensive landscape-scale movements in response to spatial variability in resources, rendering protection of only part of its used area an insufficient guarantee of persistence.

Case study 6: Land management and conservation planning

Uncertainty bedevils conservation planning in northern Australia. For most of this century, environmental concerns were not taken into account in development proposals (Frith 1961). Now, where there is some consideration of environmental impacts, the information base for assessing conservation values is generally very limited: for example, despite a concerted effort to aggregate all available distributional data for vertebrates in the Northern Territory, the average density of records is 0.6 records per square kilometre (for all species combined), decreasing to < 0.05 records per square kilometre in the more remote bioregions (Connors *et al.* 1996). This is not an adequate information base to assess likely impacts of development, nor to undertake regional land use planning.

Further, any conservation assessment is difficult to set against the more clinical (if demonstrably often unrealistic) economic values ascribed to developments. In contrast to most Australian states, comprehensive regional land use planning has been undertaken on only one occasion in the Northern Territory (Anon. 1991), and in that case conservation values were generally subsumed to development goals. In many more cases, land use planning in northern Australia has been sectoral, seeking only to identify development opportunities and constraints (e.g. Forster *et al.* 1960). (Though in the Cape York Peninsula of far north Queensland, very detailed and relatively comprehensive land use planning has been attempted (Holmes 1992).) Without strategic and regional planning, the assessment of environmental impacts of development proposals has been piece-meal and has evaded the issue of cumulative effects.

The conservation reserve system in northern Australia is unrepresentative of many environments (Woinarski 1992), especially those with high pastoral values. For example, pastoral leases cover more than 99 per cent of the NT's extensive Mitchell grasslands (Woinarski et al. 1996). While there has been a great deal of research aimed at increasing the productivity of these environments (e.g. Orr and Holmes 1984), almost nothing is known of their native fauna and its responses to grazing or other management practices, an imbalance explicitly justified by the pastoral scientists Orr and Holmes (1984):

> No large area of these grasslands is currently reserved in any form and at present there are no firm plans for any such reserve. The situation arises because the need and pressure for reservation is low. No information on fauna native to these grasslands has been collected.

While little is known of the effects of land use upon those wildlife species occurring mainly in poorly reserved environments, there is also negligible information on the management requirements of species more fortunate to live in conservation reserves. For many reserves, there has been no wildlife inventory and only superficial consideration of how the wildlife present can be maintained and prosper. This uncertainty about the identification and management of conservation values within northern Australia's national park system is accentuated by the fledgling state of joint management of many of these reserves by Aboriginal landowners and government conservation agencies. While such co-operation may bring many benefits, there is also much scope for discord in the establishment of management goals. This discord may produce a lack of conservation security for biota within national parks, a result which is foreign to long-held beliefs (at least within the settler society) about the purpose of such reserves. For example, Aboriginal traditional owners of Kakadu National Park, who form the majority of the management board of that park, and are now advocating extensive areas of buffalo farms within the park and the return of relatively high densities of feral animals, despite clear evidence of the conservation costs (Kakadu Board of Management and Australian Nature Conservation Agency 1996). The paramouncy of Aboriginal interests is explicit: 'Any conservation partnership must be based upon the premise that indigenous cultural objectives of a conservation program have priority over environmental issues', and 'Indigenous cultural practices, including the use of land and natural resources, should not be limited by formal conservation requirements set by governments' (Smyth 1995). The implications of this prioritization for the conservation reserve system in northern Australia are yet to be determined.

However, even sympathetic and informed management of reserves will not by itself ensure the persistence of 'reserved' species within northern Australia. While the vast open forests and savanna woodlands of northern Australia appear superficially homogeneous, there are many patches of distinct habitat embedded within them, and substantial spatial variation in rainfall patterning leads to marked heterogeneity in resource availability, even within extensive environments. In response to this patchiness, many of northern Australia's animal species undertake extensive geographic dispersal and habitat shifts (Woinarski *et al.* 1992). Examples include: intercontinental migrants, notably shore birds (for which coasts and wetlands of northern Australia are of major significance: Lane 1987); continental migrants, including many bird species which breed in temperate Australia; long-distance nomads, notably including waterfowl (the northern Australian wetlands may occasionally hold virtually all of Australia's waterfowl population: Frith and Davies 1961); and species whose movements are largely restricted to the arena of north Australia. The magpie goose is perhaps the classic example of the latter, with the location of breeding colonies varying between years in response to a range of local rainfall factors, with substantial shifts in preferred feeding habitat depending upon gosling age, and with strongly differential use of habitat (and hence location) between breeding and non-breeding periods (including the occasional concentration of the bulk of the regional population within very limited areas) (Whitehead *et al.* 1992). Despite almost four decades of research directed at this species, it is still possible to make only loose predictions about the impacts upon it of alienation or modification of particular areas within its broad environment. Even where knowledge of the requirements of this species can be translated to explicit identification of threatening processes, it has proved difficult to have these considerations accepted by developers or government regulatory agencies (Whitehead 1991). It is far harder to guess the set of required areas (or resources) for the vast majority of unstudied vertebrates, let alone to ensure that these can be protected within the prevailing culture of facilitating development.

Conclusions

Our examples of attempts by European settlers to use the land of northern Australia suggest a pattern of general disregard for information and scant concern for environmental consequences of success (or failure). Although these developments have inevitably led to personal and environmental casualties, such losses have been deemed bearable in the context of a government drive to dominate or stake a claim on these lands, and the pervasive perception that environmental costs weigh little against the land's limited value and its excessive extent.

We have generally selected cases of failure, at least partly because these may better illuminate the importance of uncertainty, and perhaps because they outnumber the cases of clear success. However, not all development attempts in northern Australia have failed. Northern Australia has supported a moderately high density of Aboriginal people for at least

60 000 years, and for much of that period these people have relied upon careful management of resources, rooted in an intimate environmental knowledge (Hiscock and Kershaw 1992). The settler population in northern Australia has increased and become less transient, allowing greater knowledge of the environment by locally based decision makers. Darwin now houses 80 000 people and is a prosperous city: however, it is far from being self-supporting, with much of its income and population base derived from distant funding (Committee on Darwin 1995). Northern Australia has some large and magnificent national parks (most notably Kakadu: Press *et al.* 1995), which protect much of the region's biota. The significant rise of the tourist industry in northern Australia has forced a greater appreciation of the value of natural areas, wildness and national parks, leading to a gradually increasing recognition that extensive relatively unmodified environments are an asset and not an affront. The Northern Territory (and northern Australia generally) has also caught up with at least some environmental policy operating elsewhere in Australia, through its endorsement of a wide range of national strategies and agreements, such as the National Forest Policy Statement, National Strategy for the Conservation of Australia's Biodiversity, and the National Strategy for the Conservation of Australian Species and Communities Threatened with Extinction.

Many of the examples we have presented are now historic, and the evidence suggests that some of the lessons have now been learned, that the frontier mentality is gradually fading. Conservation issues are now generally included in reviews of development prospects (e.g. Moffatt and Webb 1992; Anon. 1995), and environmental research (including conservation issues) is being seen as an essential precursor of strategic development (ASTEC 1993). Although the number of researchers in northern Australia remains low relative to the rest of the nation (ASTEC 1993), there is now a substantial body of western scientific knowledge about the ecology of northern Australia (e.g. Ridpath and Corbett 1985; Haynes *et al.* 1991). Some of this research has now been of sufficient length to assess responses to climatic variability, and of sufficient relevance to provide clear warning of the environmental constraints to development. The limits to which lands can be pushed have been exceeded so frequently now that these limits are recognizable, and some land users have begun to operate sustainably, and search for indicators of sustainability (Winter 1990; Stewart 1996). In some cases (e.g. the use of exotic pasture grasses, and tree clearing) the conservation impacts of land use practices are now known, at least dimly, and, at times, factored into management advice (e.g. Scanlan and Turner 1995; Anon. 1995). There is also a growing awareness of the importance of understanding the environmental consequences of pervasive land management practices, such as fire.

There is a growing diversity of interest groups in northern Australia, such that the land management agenda is now less monopolized by the previously dominant sectoral interest, pastoralism. There is also substantially increased communication between a range of researchers, landowners and other stakeholders, not least through the recently established Cooperative Research Centre for Sustainable Development of Tropical Savannas, which spans all jurisdictions across northern Australia and all main land user groups. Aboriginal people have far greater power than they did a generation ago, especially where land ownership is concerned. In some cases, their detailed environmental knowledge is now more widely respected and sought.

However, some of the ignorance, limited vision and/or hubris remains, at least partly because the history of European entry to this region has been so brief. Calls for large developments continue, with often little prospect of (or concern for) sustainability and little regard for other values. In some areas (notably in the Darwin rural area), the large developments have been replaced by a multiplicity of more modest enterprises, each of more limited impact, but together providing possibly greater cumulative change. In most of northern Australia, land use planning, if it occurs at all, is neither integrated nor strategic, so much decision making is piecemeal and impacts are incremental. The vastness of the region continues to daunt research: it is far more tractable to figure out the workings of a remnant woodland patch than of an environment in which there are no obvious limits to ecological processes. While that vastness remains a comfort (we can get things wrong here and there without fatal environmental consequences), it is very much an attribute which must be carefully protected.

There is still a need in settler society for far greater knowledge of the north Australian environments, and the impacts of development upon them. But development will continue to be pursued regardless of the adequacy of the information base. How then can such development be constrained within prudent environmental limits? Comprehensive regional land use planning, linked to environmental planning principles, may provide the most effective mechanism. A framework and principles should include: regionalization based on social and environmental criteria; collection or collation of sufficient environmental information to identify within the region sites of greatest conservation significance and to predict the environmental costs of possible development; an inclusive process to identify regional land use objectives and to identify sites of conservation significance within the region; the incorporation of these land use object-ives into legislation and/or planning instruments; the establishment and wise management of a comprehensive and adequate reserve network and the incorporation of ecologically sustainable management in unreserved lands; the participation of all stakeholders in land use decisions; and the staged expansion and monitoring of development such that management may adapt to unforeseen impacts before these become excessive and/or irreversible.

There is also a broader context from which environmental research requirements should be considered. Much of the Australian continent, and the world in general, has been substantially modified over the last 200 years, with fragmentation and alteration of previously extensive landscapes in most temperate areas. In much of these altered environments the workings of ecological processes have been distorted or lost. It may be that patching up such broken environments will require the knowledge of how intact landscapes operate. The relatively unmodified northern Australian environment can provide such an example. It is perhaps ironic that this opportunity might well have been foreclosed if many of the development schemes initiated in the north had achieved their ambitions; for ultimately, the failures have probably had fewer environmental impacts than the successes.

Acknowledgments

We thank Andy Chapman, David Bowman, Penny Wurm, John Childs and Peter Whitehead for comments on drafts of this paper, and Sam Lake for some unpublished information. Some of the land use planning principles noted in this paper were distilled from a fuller listing compiled by Peter Whitehead. We thank the Tropical Savannas Cooperative Research Centre for support.

References

Anon. (1978) Editorial: professional and organisational loyalty, *Australian Forestry* 41: 136–7.

Anon. (1979) *Ord River Irrigation Area. Review: 1978*. Canberra: Australian Government Publishing Service.

Anon. (1991) *Gulf Region Land Use and Development Study*, Darwin: Northern Territory Department of Lands and Housing.

Anon. (1994) *A Conservation Strategy for the Northern Territory*. Darwin: Government Printer.

Anon. (1995) *Report into Matters Relating to Environmental Protection and Multiple Use of Wetlands Associated with the Mary River System*. Sessional Committee on the Environment, Legislative Assembly of the Northern Territory. Darwin: NT Government Printer.

ASTEC (Australian Science and Technology Council) (1993) *Research and Technology in Tropical Australia and their Application to the Development of the Region. Final report*. Canberra: Australian Government Publishing Service.

Barnard, C.A. (1925) A review of the bird life on Coomooboolaroo Station, Duaringa district, Queensland, during the past fifty years. *Emu* 24: 252–65.

Barton, A. and McDonald, J. (1996) The Australian Defence Force and the future of tropical savannas. In: Ash, A. (ed.) *The Future of Tropical Savannas: an Australian Perspective*. Melbourne: CSIRO.

Basinski, J.J., Wood, I.M. and Hacker, J.B. (1985) *The Northern Challenge: a History of CSIRO Crop Research in Northern Australia*. Research report no. 3, CSIRO Division of Tropical Crops and Pastures. St. Lucia: CSIRO.

Bateman, W. (1955) Forestry in the Northern Territory. *Forestry and Timber Bureau Leaflet no. 72*.

Bauer, F.H. (1959) *Historical Geographic Survey of part of Northern Australia. Part 1: Introduction and the Eastern Gulf Region*. CSIRO Division of Land Research and Regional Survey report no. 59/2. Canberra: CSIRO.

Bauer, F.H. (1964) *Historical Geography of White Settlement in Northern Australia. Part 2: The Katherine-Darwin Region*. CSIRO Division of Land Research and Regional Survey report no. 64/1. Canberra: CSIRO.

Bauer, F.H. (1977) (ed.) *Cropping in North Australia: Anatomy of Success and Failure*. Canberra: Australian National University.

Bauer, F.H. (1984) What man hath wrought: geography and change in northern Australia. In: Parkes, D. (ed.) *Northern Australia: the Arenas of Life and Ecosystems on Half a Continent*. Sydney: Academic Press.

Beeton, R.J.S. (1977) The impact and management of birds on the Ord River development in Western Australia. M.Nat.Res. thesis, University of New England, Armidale.

Beissinger, S.R. and Bucher, E.H. (1992) Can parrots be conserved through sustainable harvesting? A new model for sustainable harvesting regimes when biological data are incomplete. *BioScience* 42: 164–73.

Blainey, G. (1992) Overcoming southern apathy towards Australia's northern regions. *The Mining Review* 16: 22–5.

Bowman, D.M.J.S. (1991) Aims and achievements in Northern Territory forest wildlife biology. In: Lunney, D. (ed.) *Conservation of Australia's Forest Fauna*. Mosman: Royal Zoological Society of NSW.

Bowman, D.M.J.S. (1998) The impact of Aboriginal landscape burning on the Australian biota. *New Phytologist* 140: 385–410.

Buckley, R. (1993) How well does the EIA process protect biodiversity? *Australian Environmental Law News* 2: 42–52.

Cameron, D.M. (1985) Forest crops. In: Muchow, R.C. (ed.) *Agro-research for the Semi-arid Tropics: North-West Australia*. St. Lucia: University of Queensland Press.

Christian, C.S. (1977) Agricultural cropping in northern Australia: a general review. In: Bauer, F.H. (ed.) *Cropping in North Australia: Anatomy of Success and Failure*. Canberra: Australian National University.

Christian, C.S. and Stewart, G.A. (1953) General report on survey of Katherine-Darwin region, 1946. *CSIRO Land Research Series no. 1*. Canberra: CSIRO.

Clancy, T. (1995) Workshop report. Evidence of change – when and how should the manager respond? In: Grigg, G.C., Hale, P.T. and Lunney, D. (eds) *Conservation through Sustainable Use of Wildlife*. Brisbane: University of Queensland.

Clark, C.W. (1996) Marine reserves and the precautionary management of fisheries. *Ecological Applications* 6: 369–70.

Committee on Darwin (1995) *Report of the Committee on Darwin*. Canberra: Australian Government Publishing Service.

Connors, G., Oliver, B. and Woinarski, J. (1996) *Bioregions in the Northern Territory: Conservation Values, Reservation Status and Information Gaps*. Report to ANCA National Reserves System Cooperative Program. Darwin: Parks and Wildlife Commission of the Northern Territory.

CSIRO (1952) Survey of Barkly region, 1947–48. *Land Research Series no. 3*. Canberra: CSIRO.

CSIRO (1965) General report on lands of the Tipperary area, Northern Territory. 1961. *Land Research Series no. 13*. Canberra: CSIRO.

CSIRO (1969) Lands of the Adelaide-Alligator area, Northern Territory. *Land Research Series no. 25*. Canberra: CSIRO.

CSIRO (1970) Lands of the Ord-Victoria area, W.A. and N.T. *Land Research Series no. 28*. Canberra: CSIRO.

CSIRO (1976) Lands of the Alligator Rivers area, Northern Territory. *Land Research Series no. 38*. Canberra: CSIRO.

CSIRO (1978) CSIRO's research in and for the Ord River Irrigation Area. Submission to the Ord River Irrigation Area Review Committee.

Curteis, W.M. (1961) Prospects for Rice in the Northern Territory. In: Papers of the Northern Territory Scientific Liaison Conference. Melbourne: CSIRO.

Davidson, B.R. (1965) *The Northern Myth: a Study of the Physical and Economic Limits to Agricultural and Pastoral Development in Tropical Australia*. Melbourne: Melbourne University Press.

Davy, D.R. (1975) *Rum Jungle Environmental Studies*. Sydney: Australian Atomic Energy Commission.

Dawson, F. (1993) A major mining project is fast tracked in the Northern Territory, but at what cost? *Impact* 30: 6.

Dawson, F. (1996) *The Significance of Property Rights for Biodiversity Conservation in the Northern Territory*. Unpublished report to the Cooperative Research Centre for the Sustainable Development of Tropical Savannas, Northern Territory University, Darwin.

Ewing, G. (1992) The likely impact of the Mabo case on Aboriginal land rights claims. *The Mining Review* 16: 8–12.

Ewing, G. (1996) Sustainable mining in Australia's tropical savannas. In: Ash, A. (ed.) *The Future of Tropical Savannas: an Australian Perspective*. Melbourne: CSIRO.

Eyles, A.G., Cameron, D.G. and Hacker, J.B. (1985) *Pasture Research in Northern Australia – its History, Achievements and Future Emphasis*. Research report no. 3, CSIRO Division of Tropical Crops and Pastures. St Lucia: CSIRO.

Fisher, M.J., Garside, A.L., Skerman, P.J., Chapman, A.L., Strickland, R.W., Myers, R.J.K., Wood, I.M.W., Beech, D.F. and Henzell, E.F. (1977) The role of technical and related problems in the failure of some agricultural development schemes in northern Australia. In: Bauer, F.H. (ed.) *Cropping in North Australia: Anatomy of Success and Failure*. Canberra: Australian National University.

Flannery, T.F. (1994) *The Future-eaters: an Ecological History of the Australasian Lands and People*. Sydney: Reed.

Foran, B.D., Bastin, G. and Hill, B. (1985) The pasture dynamics and management of two rangeland communities in the Victoria River District of Northern Territory. *Australian Rangeland Journal* 7: 107–13.

Forshaw, J.M. (1981) *Australian Parrots*. 2nd (revised) edn. Melbourne: Lansdowne.

Forster, H.C., Kelly, C.R. and Williams, D.B. (1960) *Prospects of Agriculture in the Northern Territory*. Report of the Forster Committee. Canberra: Department of Territories.

Fox, R.W., Kelleher, G.G., Kerr, C.B. (1977) *Ranger Uranium Environmental Inquiry Second Report*, Canberra: Australian Government Publishing Service.

Frith, H.J. and Davies, S.J.J.F. (1961) Ecology of the magpie goose, *Anseranas semipalmata* Latham (*Anatidae*). *CSIRO Wildlife Research*, 6: 91–141.

Fry, R.M. (1980) Environmental protection and uranium mining in the Alligator Rivers Region. In: Harris, S. (ed.) *Social and Environmental Choice, the Impact of Uranium Mining in the Northern Territory*. CRES Monograph 3. Canberra: Australian National University.

Gillard, P., Williams, J. and Monypenny, R. (1989) Clearing trees from Australia's semi-arid tropics, *Agricultural Science* 2: 34–9.

Glanznig, A. (1995) *Native Vegetation Clearance, Habitat Loss and Biodiversity Decline: an Overview of Recent Native Vegetation Clearance in Australia and its Implication for Biodiversity*. Biodiversity series, paper no. 6. Canberra: Department of Environment, Sport, and Territories.

Graham-Taylor, S. (1978) A History of the Ord River Scheme: a Study in Incrementalism. PhD thesis. Murdoch University.

Green, R.H. (1989) Power analysis and practical strategies for environmental monitoring. *Environmental Research* 50: 195–205.

Haigh, F.B. (1963) Irrigation potential and problems. In: Tomlinson, J. and Walker, P. *Water Resources: Use and Management*. Melbourne: Melbourne University Press.

Hanssen, N.L. and Wigston, D.L. (1989) Approaches to a forest history of the Northern Territory. In: Frawley, K.J. and Semple, N. (eds) *Australia's Ever-changing Forests*. Campbell, ACT: Department of Geography and Oceanography, University College, Australian Defence Force Academy.

Haynes, C.D., Ridpath, M.G. and Williams, M.A.J. (1991) *Monsoonal Australia: Landscape, Ecology and Man in the Northern Lowlands*. Rotterdam: Balkema.

Higgins, H.G. and Phillips, F.H. (1973) Technical and economic factors in the export of wood chips from Australia and Papua New Guinea. *Australian Forest Industry Journal* 39: 47–53.

Hiscock, P. and Kershaw, A.P. (1992) Palaeoenvironments and prehistory of Australia's tropical Top End. In: Dodson, J. (ed.) *The Naive Lands: Prehistory and Environmental Change in Australia and the South-west Pacific*. Melbourne: Longman Cheshire.

Holmes, J.H. (1990) Ricardo revisited: submarginal land and non-viable cattle enterprises in the Northern Territory Gulf District. *Journal of Rural Studies* 6, 45–65.

Holmes, J. (1992) Strategic regional planning on the northern frontiers. Discussion paper no. 4. Darwin: North Australian Research Unit.

Holmes, J.H. and Mott, J.J. (1993) Towards a diversified use of Australia's savannas. In: Young, M.D. and Solbrig, O.T. (eds) *The World's Savannas, Economic Driving Forces, Ecological Constraints and Policy Options for Sustainable Land Use*. UNESCO Man and the Biosphere Series Vol 12. Paris: UNESCO.

House of Representatives Committee on Expenditure (1978) *Report on Forestry Program*. Canberra: Government Printer.

Hughes, C.J. (1995) One land: two laws – Aboriginal fire management. *Environment and Planning Law Journal* 12, 37–49.

Joseph, L., Emison, W.B. and Bren, W.M. (1991) Critical assessment of the conservation status of red-tailed black-cockatoos in south-eastern Australia with special reference to nesting requirements. *Emu* 91: 46–50.

Jull, P. (1991) *The Politics of Northern Frontiers*. Darwin: Australian National University, North Australia Research Unit.

Jull, P. (1996) *Constitution-Making in Northern Territories: Legitimacy and Governance in Australia*. Alice Springs: Central Land Council.

Kakadu Board of Management and Australian Nature Conservation Agency (1996) *Kakadu National Park Draft Plan of Management 1996*. Jabiru: Australian Nature Conservation Agency.

Kitchener, D.J. (1978) Mammals of the Ord River area, Kimberley, Western Australia. *Records of the Western Australian Museum* 6: 182–219.

Lacey, C.J. (1979) Forestry in the Top End of the Northern Territory. Part of the Northern Myth. *Search* 10: 174–80.

Land Rights News (1993) Fast tracking ignores Aboriginal interests. Vol. 2, No. 11, p. 4.

Land Rights News (1993) Borroloola clans want mining agreement. Vol. 2, No. 27, p. 3.

Lane, B. (1987) *Shorebirds in Australia*. Melbourne: Nelson.

Leis, J.M. (1992) The dilemma of power. *Australian Natural History* 24: 72.

Lim, T.K., Bowman, L. and Tidemann, S. (1993) Winged vertebrate pest damage in the Northern Territory. Department of Primary Industries and Fisheries Technical Bulletin no. 206. Darwin: DPIF.

Lonsdale, W.M. (1994) Inviting trouble: introduced pasture species in northern Australia. *Australian Journal of Ecology* 19: 345–54.

Lourensz, R.S. (1981) Tropical cyclones in the Australian region July 1909 to June 1980. Canberra: Bureau of Meteorology, Australian Department of Science and Technology.

Ludwig, D., Hilborn, R. and Walters, C. (1993) Uncertainty, resource exploitation, and conservation: lessons from history. *Science* 260: 17, 36.

MacKenzie, I. (1980) European incursions and failures in northern Australia. In: Jones, R. (ed.) *Northern Australia: Options and Implications*. Canberra: Australian National University.

Marsh, H. (1995) The limits of detectable change. In: Grigg, G.C., Hale, P.T. and Lunney, D. (eds) *Conservation through Sustainable Use of Wildlife*. Brisbane: University of Queensland.

McKenzie, N.L. (1981) Mammals of the phanerozoic south-west Kimberley, Western Australia: biogeography and recent changes. *Journal of Biogeography* 8: 263–280.

Medcalf, F.G. (1944) *Soil Erosion Reconnaissance of the Ord River Valley and Watershed*. Perth: Department of Lands and Surveys.

Moffatt, I. and Webb, A. (1992) (eds) *Conservation and Development Issues in North Australia*. Darwin: North Australia Research Unit.

Mollah, W.S. (1980) The Tipperary story: an attempt at large-scale grain sorghum development in the Northern Territory. *North Australian Research Bulletin* 7: 59–183.

Neate, G. (1989) *Aboriginal Land Rights in the Northern Territory*, Vol. 1. Sydney: Alternative Publishing Cooperative.

NFI (National Forest Inventory) (1993) Lancewood communities. *Australian Forest Profiles* 2: 1–8.

Northern Territory Department of Mines and Energy (1986) *The Rum Jungle Rehabilitation Project Final Project Report*, Darwin: Northern Territory Department of Mines and Energy.

Orr, D.M. and Holmes, W.E. (1984) Mitchell grasslands. In: Harrington, G.N., Wilson, A.D. and Young, M.D. (eds) *Management of Australia's Rangelands*. Melbourne: CSIRO.

Payne, A.L., Kubicki, D.G., Wilcox, D.G. and Short, L.C. (1978) A report on erosion and range condition in the West Kimberley area of West Australia. *Western Australia Department of Agriculture Technical Bulletin* no. 42.

Perry, R.A. (1960) Pasture lands of the Northern Territory. *Land Research Series* no. 5. Canberra: CSIRO.

Pollock, K.H. (1995) The challenges of measuring change in wildlife populations: a biometrician's perspective. In: Grigg, G.C., Hale, P.T. and Lunney, D. (eds)

Conservation through Sustainable Use of Wildlife. Brisbane: University of Queensland.

Press, T., Lea, D., Webb, A., Graham, A. (1995) *Kakadu Natural and Cultural Heritage Management*. Darwin: ANCA and NARU.

Price, O. and Bowman, D.M.J.S. (1994) Fire-stick forestry: a matrix model in support of skilful fire management of *Callitris intratropica* R.T. Baker by north Australian Aborigines. *Journal of Biogeography* 21: 573–80.

PWCNT (1995a) *A Strategy for Conservation through the Sustainable Use of Wildlife in the Northern Territory of Australia*. Darwin: Parks and Wildlife Commission of the Northern Territory.

PWCNT (1995b) *A Trial Management Program for the Red-tailed Black Cockatoo* Calyptorhynchus banksii *in the Northern Territory of Australia*. Darwin: Parks and Wildlife Commission of the Northern Territory.

PWCNT (1995c) *A Trial Management Program for Cycads in the Northern Territory of Australia*. Darwin: Parks and Wildlife Commission of the Northern Territory.

Redhead, T.D. (1979) On the demography of *Rattus sordidus colletti* in monsoonal Australia. *Australian Journal of Ecology* 4: 115–36.

Resource Assessment Commission (RAC) (1991) *Kakadu Conservation Zone Inquiry. Final report*, Vol. 1. Canberra: Australian Government Publishing Service.

Resource Assessment Commission (RAC) (1992) *Forest and Timber Inquiry. Final report*. Canberra: Australian Government Publishing Service.

Riddett, I.A. (1990) *Kine, Kin and Country: the Victoria River District of the Northern Territory 1911–1966*. Darwin: North Australia Research Unit.

Ridpath, M.G. and Corbett, L.K. (1985) *Ecology of the Wet-Dry Tropics. Proceedings of the Ecological Society of Australia* 13.

Russell-Smith, J., Lucas, D., Gapindi, M., Gunbunuka, B., Kapirigi, N., Namingam, G., Lucas, K. and Chaloupka, G. (1997) Aboriginal resource utilisation and fire management practice in western Arnhem Land, monsoonal northern Australia: notes for prehistory, lessons for the future. *Human Ecology* 25: 159–95.

Scanlan, J.C. and Turner, E.J. (1995) *The Production, Economic and Environmental Impacts of Tree Clearing in Queensland*. A report to the working group of the Ministerial Consultative Committee on tree clearing. Brisbane: Department of Lands.

Schrader-Frechette, K.S. and McCoy, E.D. (1992) Statistics, costs and rationality in ecological inference. *Trends in Research on Ecology and Evolution* 7: 96–9.

Sinclair Knight Merz (1996) *Public Environment Review. Ord River Irrigation Area: Stage 2*. Perth: Sinclair Knight Merz.

Smith, L.A. and Johnstone, R.E. (1977) Status of the purple-crowned fairy-wren (*Malurus coronatus*) and buff-sided robin (*Poecilodryas superciliosus*) in Western Australia. *Western Australian Naturalist* 13: 185–8.

Smyth, D. (1995) *Protecting Country: Indigenous Protected Areas*. Report to Australian Nature Conservation Agency. Atherton: D. Smyth and C. Bahrdt, Consultants in Cultural Ecology.

Spillett, P.G. (1972) *Forsaken Settlement: an Illustrated History of the Settlement of Victoria, Port Essington North Australia 1838–1849.* Sydney: Landsdowne Press.

Stewart, J. (1996) Savanna users and their perspectives: grazing industry. In: Ash, A. (ed.) *The Future of Tropical Savannas: an Australian Perspective.* Melbourne: CSIRO.

Stocker, G.C. (1970) The effects of water buffaloes on paperbark forests in the Northern Territory. *Australian Forest Research* 4: 31–8.

Stuart, J. McDouall (1863) Diary of Mr John McDouall Stuart's explorations from Adelaide across the continent of Australia. *Journal of the Royal Geographical Society of London* 33: 276–321.

Stuart, J. McDouall (1865) *Explorations in Australia: the Journals of John McDouall Stuart during the Years 1858, 1859, 1860, 1861 and 1862.* W. Hardman (ed.). London: Saunders Otley & Co.

Taylor, J.A. and Tulloch, D. (1985) Rainfall in the wet-dry tropics: extreme events at Darwin and similarities between years during the period 1870–1983 inclusive. *Australian Journal of Ecology* 10: 281–96.

Tidemann, S.C. (1990) Relationships between finches and pastoral practices in northern Australia. In: Pinowski, J. and Summers-Smith, J.D. (eds) *Granivorous Birds in the Agricultural Landscape.* Warsaw: PWN – Polish Scientific Publishers.

Toyne, P. (1994) *The Reluctant Nation, Environment, Law and Politics in Australia.* Sydney: ABC Books.

Walters, C. (1986) *Adaptive Management of Renewable Resources.* New York: McGraw Hill.

Webb, G.J.W. and Manolis, S.C. (1993) Viewpoint: conserving Australia's crocodiles through commercial incentives. In: Lunney, D. and Ayers, D. (eds) *Herpetology in Australia: a Diverse Discipline.* Sydney: Royal Zoological Society of New South Wales.

Whitehead, P.J. (1991) Magpie geese, mangoes and sustainable development. *Australian Natural History* 23: 784–92.

Whitehead, P.J. and Dawson, T. (2000) Let them eat grass! *Nature Australia.* Autumn: 46–55.

Whitehead, P.J., Wilson, B.A. and Bowman, D.M.J.S. (1990) Conservation of coastal wetlands of the Northern Territory of Australia: the Mary River floodplain. *Biological Conservation* 52: 85–111.

Whitehead, P.J., Wilson, B.A. and Saalfeld, K. (1992) Managing the magpie goose in the Northern Territory: approaches to conservation of mobile fauna in a patchy environment. In: Moffatt, I. and Webb, A. (eds) *Conservation and Development Issues in North Australia.* Darwin: North Australia Research Unit.

Winter, W.H. (1990) Australia's northern savannas: a time for change in management philosophy. *Journal of Biogeography* 17: 525–9.

Woinarski, J.C.Z. (1992) Biogeography and conservation of reptiles, mammals, and birds across north-western Australia: an inventory and base for planning an ecological reserve system. *Wildlife Research* 19: 665–705.

Woinarski, J.C.Z. (1993) Australian tropical savannas, their avifauna, conservation status and threats. In: Catterall, C.P., Driscoll, P.V., Hulsman, K., Muir, D. and Taplin, A. (eds) *Birds and their Habitats: Status and Conservation in Queensland*. Brisbane: Queensland Ornithological Society.

Woinarski, J.C.Z. and Braithwaite, R.W. (1990) Conservation foci for Australian birds and mammals. *Search* 21: 65–8.

Woinarski, J.C.Z. and Fisher, A. (1995a) Wildlife of lancewood (*Acacia shirleyi*) thickets and woodlands in northern Australia: 1. Variation in vertebrate species composition across the environmental range occupied by lancewood vegetation in the Northern Territory. *Wildlife Research* 22: 379–411.

Woinarski, J.C.Z. and Fisher, A. (1995b) Wildlife of lancewood (*Acacia shirleyi*) thickets and woodlands in northern Australia: 2. Comparisons with other environments of the region (Acacia woodlands, Eucalyptus savanna woodlands and monsoon rainforests). *Wildlife Research* 22: 413–43.

Woinarski, J.C.Z., Whitehead, P.J., Bowman, D.M.J.S. and Russell-Smith, J. (1992) The conservation of mobile species in a variable environment: the problem of reserve design in the Northern Territory. *Global Ecology and Biogeography Letters* 2: 1–10.

Woinarski, J.C.Z., Connors, G. and Oliver, B. (1996) The reservation status of plant species and vegetation types in the Northern Territory. *Australian Journal of Botany* 44: 673–89.

Woinarski, J.C.Z., Connors, G. and Franklin, D.C. (2000) Thinking honeyeater: nectar maps for the Northern Territory showing seasonal variation in nectar supply from the monsoon tropics to arid central Australia. *Pacific Conservation Biology* 6: 61–80.

Zapopan N.L. (1993) *Mt Todd Gold Mine Environmental Management Plan*. Perth: Zapopan.

6 | Global warming:
science as a legitimator of politics and trade?[1]

Sonja Boehmer-Christiansen

Introduction

Scientific research bodies have long given unclear advice on climate change, though many believe there now is 'scientific consensus'. A German physicist working on climate models recently agreed that there has been

> a shift in the consensus view of the experts in this field from predominantly doubtfully negative to predominantly doubtfully positive (Hasselmann 1997),

but others remain doubtful, seeking answers for observed changes from outside the terrestrial system and certainly not in human activity. The policy implications of the controversy are staggering, as are the implications for research and research strategy, with new scientific findings said to completely undermine the

> sincerity and intellectual conviction of John Houghton (leader of IPCC Working Group I) and his group in Bracknell (Calder 1997: 199).

This scientific challenge and widespread unease is in stark contrast with statements made by politicians, who rarely exhibit such lack of caution. Tony Blair, (former) Chancellor Kohl, President Clinton and members of the Commission of the European Community have all been widely quoted as claiming that anthropogenic global warming caused by the burning of fossil fuels is now a fact, rather than one among several competing hypotheses. Yet scientific doubts about the alleged 'global' disaster only increase when one looks beyond the scientific discussion about the validity of mathematical models based on calculated average global temperature increases, to those concerned with local or regional impacts of assumed climatic warming. At this scale it is at least recognized that climatic changes amount to very much more than predicted changes in average global temperature, constituting complex ecological phenomena rather than a single statistical abstraction.

Political behaviour suggests that the doubters have had considerable influence, for few governments – irrespective of their rhetoric – have so far done much beyond what was politically advantageous or economically promising anyway. With some modification this applies also to the Protocol to the Framework Convention on Climate Change (FCCC) agreed in late 1997 in Kyoto. Indeed, the legally binding status of this document remains in doubt 'even amongst some OECD government delegates' (WEC 1998: 5) – a clear hint nevertheless about who expects to be the primary beneficiary of its proposed 'no regrets' policies.

This chapter attempts to explain both the achievements of global climate science and the reluctance of policy makers to go further. It is argued that objectively – but not politically – the outcome of climate change negotiations hinges upon the predictions of dangerous climate change at regional levels being both true and subject to a credible solution. A group of scientists, acting as experts serving political interests who fund these vast research programmes, has delivered such predictions. The group is the Intergovernmental Panel on Climate Change (IPCC). Should organized science subject to governmental selection and pressure be relied upon as a global policy guide?

The argument is made that three 'global' – that is, organized and resourced worldwide – interests have now sufficiently embedded themselves in intergovernmental and major national bureaucracies and political systems to implement measures against carbon-intensive technologies and activities to make the scientific basis for such action largely irrelevant. This has been achieved by making greenhouse gas emission mitigation and sequestration, especially of carbon dioxide, attractive to international finance and bureaucracies seeking to control 'market forces'. If the truth of anthropogenic global warming is becoming irrelevant, what does this tell us about the power of modern science, including ecology?

The developing law of climate change: Kyoto and the sustainability of bureaucracy

FCCC and Kyoto

Efforts to prevent predicted climate change are underpinned by an international treaty, the FCCC (1992, 1996) and its associated national and international institutions and networks. The FCCC itself is the outcome of a long process of global and national bargaining initiated in the mid-1980s by US-based research bodies. Ironically, these were initially motivated primarily by the desire to protect nuclear power from the onslaughts of 1970s environmentalism. The FCCC was drafted by several networks of international experts working for a small number of governments. A close look at the hard substance of the FCCC reveals that its regime effectively codifies the research and data collection needs of the 'international scientific community' under the guidance of certain intergovernmental bureaucracies, which are in turn responsible for raising research funds and planning the global future (Boehmer-Christiansen 1993).

The objective of the FCCC remains the stabilization of greenhouse gas concentrations

> at a level that would prevent dangerous anthropogenic interference with the climate system [combined with the rider that stabilization is to be achieved] within a time frame sufficient to allow ecosystems to adapt naturally to climate change, to ensure that food supply is not threatened and to enable economic development to proceed in a sustainable manner. (All treaty texts are cited from Churchill and Freestone 1992: 240–290.)

This remains, of course, inoperable until a stabilization level is agreed and baselines and reduction targets are allocated to countries (and are measurable and enforceable). A great deal more remains to be understood about the behaviour of ecosystems before any agreement on needed reductions can be reached on purely scientific grounds. Not even global warming potentials can be derived from purely scientific criteria, no matter how politically desirable it would be to be able to aggregate the climate effects of the various greenhouse gases (Shackley and Wynne 1997). Under the FCCC, reduction and mitigation plans are to be drawn up to:

> formulate, implement, publish and regularly update national and, where appropriate, regional programmes containing measures to mitigate climate change by addressing anthropogenic emissions by sources and removals by sinks of all greenhouse gases . . . and measures to facilitate adequate adaptation to climate change; and promote and co-operate in the development, application and diffusion, including transfer, of technologies, practices and processes that control, reduce or prevent anthropogenic emissions . . . (FCCC Treaty, Article 4 1b).

The second Conference of the Parties met in Kyoto, but represented primarily environmental as distinct from energy interests, atmospheric science as distinct from geology, ecology, hydrology and solar science.

Why has a single science been so successful in persuading so many? Strenuous efforts to generate real commitments have been under way since the mid-1980s, but have still not entirely succeeded.

The interests involved in the negotiations on the proactive side – the planet must be saved from the global hothouse – consist of environmentalists, climate modellers, environmental bureaucracies and commercial foes of coal and coal-based technologies. They have made themselves dependent on further progress being made in the definition of precise rules underlying the emission reduction targets (or credits) agreed at Kyoto. As amended there, all Annex 1 (OECD countries and the former eastern bloc countries considered to be in 'transition'), are to reduce their net emissions of six greenhouse gases converted to carbon dioxide equivalents between 2008 and 2012. (The six gases are carbon dioxide, methane, nitrous oxide, hydrofluorocarbons, perfluorocarbons and sulphur hexafluoride – their equivalence to CO_2 can be decided but not derived definitively.) These cuts are to amount to a global average of 5.2 per cent below 1990 levels, provided that at least 55 governments whose 1990 emissions amount to 55 per cent of the world total, ratify the Kyoto Protocol (FCCC 1998). Significant measuring problems remain, but answers could undoubtedly be 'adjusted'. Unless checked and double-checked by several bodies, governmental statistics in this area cannot be assumed to be correct; being economical with the truth is very easy in the environmental area, where measurement is difficult, place- and technology-dependent – and where there may be no single 'truth'.

The Protocol contained surprises: an insignificant nuisance like Australia was allowed an 8 per cent increase (Iceland even 10 per cent), while the European Union faces an average reduction of 8 per cent – which amounts to a reduction of 15–20 per cent compared to the projected growth – and the USA eventually agreed to cut by 7 per cent. The big 'outrage' to environmentalists was Russia, which need merely stabilize its 1990 emissions, something it has already overachieved thanks to its economic collapse, also called restructuring. Given that the USA increased its CO_2 emissions between 1990 and 1996 by 8.3 per cent, the US signature must at first surprise, despite the 'flexibility' of the Kyoto Protocol, unless three points are remembered:

- The US Congress may refuse ratification unless the 'details' are acceptable; these will undoubtedly take into account various trading and investment offers made by the World Bank (World Bank 1997).
- The US government has already won a major victory over the EU by reportedly forming a 'secret cartel' (UK environment minister reported in ENDS 278: 40) by capturing for itself the right to buy the emission

reductions already 'achieved' by Russia due to its industrial collapse. This would allow the USA a de facto real increase from 1990 values of 10 per cent (Pearce 1998), while Europe – now reportedly refusing to sign the protocol until loopholes are closed – would have significantly to cut emissions. The developing world is likely to be subjected to World Bank 'facilitation' in this matter.

- The complex and flexible emission trading arrangements of the Protocol have prepared the path for the World Bank's bid to become a major administrator of this trade with non-OECD countries (Jepma 1998: 1). This would substantially strengthen its role in 'development'.

The last point deserves elaboration in view of the fact that the Convention does not define technology transfer operationally or even theoretically. Article 2 of the Protocol states that each party included in Annex 1 should, apart from achieving quantified emission limits 'in order to promote sustainable development', do so by enhancing:

> energy efficiency, promoting forestry, innovating environmentally sound technologies and by reducing or phasing out market imperfections, fiscal incentives, tax and duty exemptions, and subsidies in all greenhouse emitting sectors that run counter to the objective of the Convention and apply market instruments. (UN 1997)

In other words, subsidies to nuclear power and solar power but not coal, would be allowed. Market instruments as generally advocated by the World Bank are supported, but energy subsidies to farmers in India, for example, should stop. Adverse effects on international trade are also to be minimized. The parties must be able by 2005 to demonstrate progress under the Protocol.

As the FCCC states, vast amounts of information will have to be processed and digested before administration can begin. For the purpose of calculating their assigned amounts, Annex 1 countries for whom land use changes and forestry constituted a net source of greenhouse gas emission in 1990, would have to include in their 1990 emission base year or period the aggregate anthropogenic carbon dioxide equivalent emissions minus removals in 1990 from land use change (Article 3, paragraph 7), and national systems for the estimation of anthropogenic emissions by sources and removal by sinks of all greenhouse gases will have to be in place no later than one year prior to the start of the first commitment period (Article 5, paragraph 1). Many experts and officials will be kept busy and probably away from what some might consider more urgent national and scientific tasks. As demanded by the USA and World

Bank, any emission reduction 'which a Party acquired from another Party ... shall be added to the assigned amount for that party', and transfers will be subtracted. Several types of emission trading will be permitted, though the details remain to be worked out, including funding, insurance and transfer of technology (UN 1997).

Indeed, the parties to the FCCC have agreed to general measures which go far beyond traditional international law to establish institutions and financial mechanisms for implementation in eligible countries. These cover 'enabling activities', 'capacity building' (including research, data collection, surveillance), as well as 'innovative' means of financing emission-reduction technologies and raising money for industrializing countries by emission trading by credits or allowances. Joint implementation is recognized as the main method. As the FCCC was prevented from having its own financing body it came to be served by the Global Environment Facility (GEF) or rather the World Bank as one of the GEF's Implementing Agencies. The Bank and its regional bodies are now busy pursuing climate change objectives through project developments using aid money (Young and Boehmer-Christiansen 1998). By applying its own brand of resource economics, subsidies are to be limited to agreed incremental costs calculated to correspond to global environmental benefits.

Yet, given the well-documented low priority of global warming actions in many poor or politically weak countries and the unknown human capacity for effective action, many governments have little incentive to comply with the Protocol. Carrots and sticks are clearly needed, with the Bank in possession of both. As most recipient countries are already subjected to macro-economic conditionalities, the FCCC adds to these and in the name of global environmental protection is likely to influence energy and land use policies in developing countries. The FCCC could therefore come to require major governmental interventions in the political ecology of the entire world, primarily through changes demanded in energy prices and markets, as well as forestry and land use.

What are the political forces that have driven climate change policy developments so far? It is argued that this was the achievement of a small range of the natural sciences promoting hypothetical global environmental threats that promised more immediate benefits to those offering solutions.

Comment on Hague 2000

The Hague conference of the FCCC Parties is widely seen as a failure. The seeds of failure had been sown at Kyoto in 1997. In future years,

Japan and the USA may be able to increase emissions significantly in return for aid to Russia, while the excluded EU will have to cut emissions by 5.2 per cent. China has stated that it will not reduce its emissions until it has become a moderately developed country. Rather than mitigating climate, emission reduction appears to serve aid and trade, and thus the globalizing political economy. This may be a more powerful cause for action than climate, and suggests that a deal will eventually be reached, driven by non-environmental motivations. Even the new Bush administration may yet be persuaded.

Selected sciences and politics work together

The science of climate change remains contested (Calder 1997; Mason 1996), which means that its science foundation remains inadequately understood. This applies to the physics of cloud formation, the effects of aerosols on the energy balance of the atmosphere, extraterrestrial influences and ecological feedbacks due to man-made and natural changes in the biosphere. Climate is an extremely complex set of phenomena that the physical and biological sciences are only beginning to understand. Scientific publications therefore place considerable emphasis on uncertainty. This protects legitimate research interests, but also has political implications.

Environmental bureaucracies and experts, as well as organized losers in the competitive world of fuel and energy technology supply, have a strong interest in the likelihood of anthropogenic warming and urge that science, through its organized bodies, comes to a consensus in their support. These two broad groups are, of course, supported by environmental non-governmental organizations (NGOs), which they have carefully cultivated since Rio and which have gained public confidence and media support.

Government decision-making bodies were therefore put in the unenviable situation of being called upon to act by environmentalists inside and outside government, as well as those commercial allies that stand to gain subsidies, regulations or market shares from the implementation of mitigation strategies. However, because of their broad range of commitments and responsibilities, governments in general, and treasuries in particular, are not readily persuaded, at least as long as they do not benefit from the proposed policy changes. These changes typically include increases in taxes on fossil fuels and subsidies to 'renewables', nuclear power, public transport and, above all, 'energy efficiency' (World Energy Council 1998).

Since the greatest resistance to such interventions comes from developing countries, the World Bank and aid ministries are called upon to deal with poor or weak governments who, if they want GEF money, are required to invest in 'cost-effective global environmental benefits'. This typically means the 'transfer' (i.e. subsidized provision) of cleaner technologies and fuels to industrializing countries such as India, China and Brazil, and solar power investment to Africa. The international debts of these countries are thereby increased, though steps are taken towards what is defined as 'sustainable growth'.

The climate protection regime has therefore already begun to develop as an aid and trade regime, with branches of physics and chemistry, as well as computers providing a credible threat. Environmentalism, emotion, rhetoric and the 'precautionary principle' required by politics, together with the energy technologies developed in response to high oil prices since the 1970s, as well as the recently discovered abundance of natural gas, promise to provide the needed greenhouse gas emission reductions. Bureaucracies and their experts will provide the labour and intelligence needed to integrate all this into policy by drafting rational plans and devising national strategies. Together they are to call upon politicians to provide the resources needed to translate the plan into action. We are about to reach this difficult stage; hence science deserves to be revisited.

Resistance to such plans was to be expected and is now clearly observable, for the administrative loads alone are enormous. As far as practical implementation is concerned, therefore, the treaty remains a package that is held together, in logic, only as long as a selected number of natural sciences lend credibility to the claim that human activities cause *controllable* net surface warming that is 'dangerous'. But who will define 'dangerous', what criteria will be used, and by whom will those criteria be selected? Before a definition can be agreed on the basis of sound science, vast quantities of data will be needed relating to the stabilization of greenhouse gas (GHG) concentrations, the stabilization of emissions, the definition of ecological limits or levels of tolerance to warming based on impact studies, and the methodologies for making national inventories of emissions and sinks.

So far, only the methodologies for greenhouse gas (GHG) inventories have been defined, with help from OECD and IPCC. Research bodies and administrations are collecting data sets and designing ambitious strategies that are of benefit to many institutions, especially those which tend to centralize control. Bureaucracies at all levels are being engaged in this enormous task. However, many countries are unable to comply without assistance. The GEF has therefore been called upon to fund such efforts under the label of 'capacity building'. The more pressing tasks of many poor countries have had to be postponed. It appears that there is now a battle under way between UN agencies and the World Bank about who will be the primary 'guardian' of the global environment.

An alliance against 'unsustainable' coal use

The outcomes of Kyoto might be compared with the 20 per cent reduction target proposed in Toronto in 1988 and later adopted in various places in Europe, and a 60 per cent cut that is often demanded by environmentalists with reference to IPCC scenarios. Many industries, treasuries and politicians have remained unpersuaded for about a decade that this was achievable, recognizing that the likely transition and compliance costs would fall largely on energy industries and energy users. However, given the fall in the price of fossil fuels in the mid-1980s, global warming came to serve both as a justification for closing relatively expensive coal mines, and for revenue-raising taxes. Both responses are clearly demonstrated in the UK (Boehmer-Christiansen 1996; ENDS 1998).

In general, most efforts made so far on paper and in real projects to limit GHG emissions involve restrictions on the use of coal, especially in electricity generation. From the late 1980s onward markets were sought for an often cheaper and abundant substitute – natural gas, and, though less wholeheartedly, efforts were made to increase 'sinks' of carbon by planting forests to fix carbon as wood. Policies opposed – for whatever reason – to coal mining, deforestation and in favour of tree planting could therefore benefit from appeals to 'global warming'. All economies, and especially growing ones, would have to become less carbon-intensive, 'cleaner' and 'sustainable'.

Table 6.1 hints at the reasons for the attack on coal and at the motives of various energy players.

Rising oil prices during the 1970s had created new winners in the world of energy supply and R&D: more efficient technologies, nuclear power and renewables. Gas was not yet available for electricity generation and energy demand was expected to rise steeply, a development that has become fact only in parts of the developing world. During this period, energy suppliers were not attracted to the idea of climate change, though the issue was already being debated in research circles (Kellogg and Schware 1981) and members of some national bureaucracies noted an opportunity for expansion of their role. The market did not yet need an ally in environmentalism – prices alone were moving energy options

Table 6.1 Carbon dioxide emission factor estimates for different fuels.

Natural gas	15
Oil	18
Wood	19
Coal	25.3 (very variable)
(Nuclear)	0 (excluding construction and transport)

Source: Organisation for Economic Co-operation and Development (OECD) (1991)
Note: Emission factor estimates ignore efficiency of conversion or release of gas during transmission. The thermal efficiency of coal-fired plants has increased from 25 per cent to 38 per cent since the 1950s; that of combined cycle gas plants is approaching 50 per cent, while gas burned directly has a much higher efficiency.

away from low-tech fossil fuels towards the options advocated in response to the 'energy crisis': nuclear power, 'renewables' and gas exploration. When fossil fuel prices fell again in the mid-1980s (and especially sharply in 1986), the situation reversed and the newcomers sought a green ally. Powerful commercial incentives operated to demand official measures to maintain the competitiveness of 'clean' energy in the name of 'sustainability'. The IPCC was conceptualized in 1985, planned in 1987 and began operating in 1988 in response to policy opportunities rather than science.

By the mid-1980s, non-fossil energy interests and several major governments committed to new technologies and fuels because of sunk investments felt sufficiently threatened to pay attention to environmentalists and some scientists disseminating worst-case scenarios: the global hothouse, floods and hurricanes. However, stagnant demand and falling prices are not an easy context for governments to subsidize higher-cost options. Further research to reduce uncertainties seemed more attractive and extant scientific knowledge appeared to justify this. Moreover, some emissions reductions were being achieved through the transition to natural gas (which also boosted the revenue of certain oil companies and countries). So the main economic losers (nuclear, 'renewables', 'energy efficiency' technologies) began to look for help to the well-prepared environmental lobby. Fossil fuel interests – the villains of the piece according to those who subscribe to the GCM models as explanatory devices – replied in kind, turning to the scientific uncertainties for support and pointing to population growth and China's rising energy demand, or consumption by the rich, as the real culprits.

Yet intervention in the 1990s was not as readily forthcoming as had once been hoped or feared. Fossil fuel prices have remained low and most countries failed to stabilize their carbon dioxide emissions by 2000. German compliance so far is the result of East Germany's industrial collapse. In the UK the 'dash to gas' and de-industrialization have ensured compliance, with economic and political benefits. While this gas can now be removed from plumes, and R&D concerned with carbon dioxide fixation and 'zero emission power generation' is booming in technologically advanced countries, subsequent containment is only economically feasible where CO_2 is not a pure waste product. Technology promises much but, as is so often the case, the constraints will not be technical but economic and political. Reducing GHG emissions per unit of energy generated is feasible over time but, because of energy demand growth, this is not likely to deliver overall reductions of the nature apparently needed to stabilize concentrations at predictable levels. This, according to the IPCC, would require a 60 per cent reduction of GHG emissions. The EU objective is now 8 per cent, and the UK is rapidly retreating from an earlier promise of 20 per cent, even though technical experts claim that this can be achieved (Green and Skea 1997). The date of this publication from a research council may be noted. The authors call upon government 'to co-ordinate and inspire action from every sector of society'. One might wonder what government could expect in return from such an effort.

Observable climate strategies are therefore better explained not as responses to environmental pressure, but as 'side effects' of energy policies or energy demand developments whose unpopularity may be 'greenwashed'. For example, Britain could not have come out so strongly in support of 'precaution' had not privatized industry been forced into price competition and therefore delivered emissions reductions by fuel switching and had not the Treasury agreed to subsidize nuclear power, and later impose a 'climate levy' on industrial energy use. Many UK coal-fired power stations were closed in the early 1990s and more are likely to follow for purely commercial reasons. It was shown that, for the UK, the 'cleanest' option for 2000 was a mix of only gas and uranium, which was also likely to be the cheapest. So governments may support GHG emission reduction for a variety of non-environmental reasons: to strengthen national nuclear industries (e.g. Germany), to enhance export potentials for gas or nuclear electricity (e.g. France and Norway), or to attract aid flows. The EC attempted to use the issue of 'carbon taxes' to strengthen its own competence.

Economies heavily dependent on coal for electricity generation would face the most serious difficulties if required to reduce emissions rapidly, and have therefore been noted to be among the most sceptical about global warming; they also happen to be the potentially most serious competitors of the OECD exporting countries, though technologically advanced importers of oil or coal may be sympathetic to a degree of emission control in order to encourage energy efficiency or to promote advanced energy technologies, e.g. Japan and Germany. The previous Bush administration's refusal to accept stronger commitments in the early 1990s is said to have stemmed from the wish to protect oil and coal interests. The change in attitude by the Clinton administration was probably related more to pro-gas changes in US energy legislation, an approaching election and declining reliance on coal than to new scientific understanding. As the USA became less willing to defend fossil fuel interests, Australia as a major coal exporter came to adopt this role. With no nuclear power and little gas, the Australian green vote was too small to keep the Labour government in power. To industrializing countries such as China the whole issue is primarily one of attracting cheaper money into their infrastructure developments. In 1992, the world's energy industries argued that world use and production of energy 'can only be changed marginally in the next thirty years because of weak administrative and institutional structures' (World Energy Council 1993: 20). This was clearly an invitation for 'capacity building' and joint implementation projects which are now being pursued.

A great deal of economic and political activity concerned with 'energy' is therefore justified with reference to 'science', relying largely on a limited understanding (because well funded, suddenly) of atmospheric physics, chemistry and oceanography. One may therefore wonder how the international institutions of science managed to raise and sustain the global warming threat in the first place. In the early 1970s the threat had been global cooling, a possibility discussed at the 1972 UN Stockholm Conference on Environment and Human Development with reference to aerosols and especially dust particles. This scenario was of very little interest to the energy world, least of all the nuclear lobby.

Earth systems research is Big Science and underpins IPCC science advice

The early efforts of the natural science research community, including the famous Villach meetings of the mid-1980s, are described elsewhere (Boehmer-Christiansen 1994). Only the attractions of climate change to scientific bureaucracies are summarized here.

In part because of the ease with which carbon dioxide is measured and modelled, climate change has come to constitute an umbrella label for large 'policy-relevant' or 'strategic' research packages, such as the International Geosphere Biosphere Programme (IGBP) which involves significant amounts of administrative supervision and assessment by bureaucracies: in the EU, most UN agencies and national research councils. The IPCC fits into this network as the body which best represents the interests of the 'hard' natural science in government or directly funded by it. Global modelling and earth observation, the tools of climate research, are 'big' science which takes place increasingly outside or separated from the communities of scholars and thinkers that constitute universities. Their financial foundations are precarious because they tend to be directly dependent on political goodwill. The danger is that science produced in this manner 'serves' rather than challenges what appeals most to prevailing political forces.

The World Meteorological Organisation's (WMO) interest in climate change research dates back to the mid-1970s, when its Climate Panel recommended a research programme on the subject, although even in 1979 delegates to the WMO, especially those from Britain and America, remained doubtful whether climate (as opposed to simply 'weather') should become a major focus for WMO-administered research. However, by 1981 a World Climate Research Programme (WCRP) was under way, albeit short of funding. It soon became a major part of the 'global change' research agenda which aims to model the physical Earth, including land use and emissions (WMO 1986). In this effort WCRP supplements the IGBP, as well as research on environmental monitoring and climate impacts undertaken under the auspices of the United Nations Environment Programme (UNEP). The IGBP is being implemented with the support of a number of national secretariats in the major research nations, UNEP, WCRP and UNESCO's oceanographic programmes. Further stages are planned. The IGBP builds on collaborations between atmospheric scientists which date back to the 1950s, e.g. the Global Atmospheric Research Programme managed by Bert Bolin, until recently chairman of the full IPCC, as well as UNESCO's Man and Biosphere Programme.

Behind these intergovernmental institutions is the International Council for Scientific Unions (ICSU), the scientific bureaucracy of academic natural science. The IGBP System for Analysis, Research and Training (START) programme is aimed at strengthening the scientific capacity of poor countries and is assisted by the GEF. Links between IPCC and IGBP, and national research bodies, can be demonstrated by cross-membership of senior research managers and scientists. A Human Dimension Program is now attached, which 'scientifically' studies institutional changes required to address 'global environmental change'. Economists modelling the global economy have been strong supporters of IPCC science and have become part of the IPCC process. They calculate the 'global' benefit of development projects.

All the above programmes were heavily influenced if not designed by US science administrators in the early 1980s. While the pure objective is the full understanding of the physical systems of the planet Earth, including the impacts of the human species as first envisaged by American scientists in the 1950s, the US agenda was written as 'a step in the evolving process of defining the scientific needs for understanding changes in the global environment that are of great concern' (US National Research Council 1990). This was disseminated globally for approval and implementation by ICSU in response to a concern that the climate threat had so miraculously created. The research is primarily defined by and for those working in the natural sciences and will take another decade to complete. Because carbon sinks are now being included, biologists have been drawn into the IPCC 'community' of government-funded strategic research. Achieving its objective will involve more attention to complex biotic feedbacks, often heavily influenced by human activity such as land use changes, but also to ocean circulation, the hydrological and the carbon cycles, as well as to 'socio-economic' assumptions of the emissions scenarios fed into the physical system. The commodification of nature and human nature is being promoted by this research agenda through the use of global models to forecast the impact of economic regulations, such as taxes, on world GNP and speculation about the costs and benefits of both global warming and its mitigation in dollar terms.

The functioning of the IPCC

The IPCC emerged from the WMO in 1988 with the help of UNEP and ICSU research networks and recently published its Third Assessment Report (TAR), which includes new scenarios and visions but little new science. It also periodically releases summaries specifically addressed to policy makers. Only the latter are read outside the scientific community and have attracted attention and resources to international research programmes. The former serve to focus the research carried out under these programmes and thereby help to ensure that subsequent funding follows the general pattern. The IPCC stresses available technical solutions and its famous 'scientific consensus' is designed by selected scientists and government officials united by a common interest to keep the IPCC ship afloat. The underlying IPCC promise to politicians is that climate change is sufficiently predictable to eventually allow rational decision making at the global level.

The individuals on the IPCC governing body primarily represent the interests of natural science research, although mainstream economic modellers and carbon accountants have now been added. IPCC working groups have paid little attention to why governments might find the implementation of technical options for reducing emissions difficult or even impossible. The need for understanding society, as distinct from collecting socio-economic data, was largely excluded from earlier deliberations. Research relevant to the IPCC concerns subjects that political leaders consider 'safe': technical matters, such as the diagnosis of future environmental problems, were to be explored with the aid of the latest developments in space and information technology. Most of the research upon which the IPCC draws is funded by government and prescribed by environment ministries. Some areas of science have been largely excluded, such as hydrology, solar forces, carbon dioxide chemistry and the aerosol impact though some of these have been allowed to join more recently.

The IPCC governing body consists of a small secretariat and bureau of about 50 people, with the former based inside the WMO in Geneva. The plenary body brings together leading government scientists and research managers with diplomats and government officials. The bureau is the facilitator of consensus and originator of policy makers' summaries which are released to the media and the world of politics. The scientific work of the panel takes place under the supervision of approved scientists whose main task is to co-ordinate the reports of working groups and attract funding for underpinning national research; the IPCC itself does not do research or fund it. The former NASA ozone scientist and chairman of one IPCC working group, Robert Watson, has also acted as advisor to the US president and is currently with the World Bank. He recently replaced Professor Bert Bolin as chair of the full IPCC, illustrating the fusion of science, governments and global governance in the IPCC.

The collection and writing up of information is in the hands of selected lead authors in charge of a large number of subgroups of three (original) working groups, WGI, WGII and WGIII, covering science, impacts and responses respectively. WGs I and II reflected the existing research interests of WMO and UNEP respectively. WGIII served the needs of governments more directly, allowing officials to meet policy advocates from NGOs and industry. WGs II and III were merged in 1994 and a third group on 'cross-cutting', i.e. socio-economic, issues was set up. In 1995, this group became enmeshed in controversy over the value of statistical life as part of its efforts to cost climate change damage. It remains the responsibility of each WG to gather and evaluate 'sound', i.e. trusted, knowledge and base advice upon it. The next report was due in 2000, and work schedules have already been defined. Science is to be 'discovered' according to politically defined schedules and to policy targets that have not only already been set, but if not supported by science, can be excused as 'no regrets' or 'win-win' policies. Biologists and ecologists have so far contributed little to these global scientific efforts, just as the softer, less quantitative social sciences have stayed away.

For example, WGI gathers and assesses available scientific information on climate change science. Its co-chairman remains Sir John Houghton, a British meteorologist with a long career as a government and ICSU advisor. Only his group was able to base itself, in 1987, on a well-established research network and had close links with large climate research institutions in North America, the then USSR and Germany (Max Planck Institutes), as well as national meteorological offices. This group is still working effectively from a small secretariat within the Hadley Centre for Climate Prediction and Research in the UK, partly funded by the UK Department of the Environment (DoE) and using the MET Office research facilities and observational data funded by the DoE (Boehmer-Christiansen 1995a,b; Courtney 1999). The Russians have been badly treated inside the IPCC, in part because of methodological differences (weak modelling capacity), in part because they have remained sceptics (on scientific grounds, e.g. Kirill Kondratyev who believes that the IPCC, by ignoring the complex effects of aerosols and weak observational inputs, has greatly exaggerated the 'danger' of warming), or because they have supported warming on the grounds that this would benefit mankind.

The IPCC organizational structure reveals a highly linear model of policy process in which 'science' thinks and recommends in isolation from society, while politicians accept and implement the facts and uncertainties. This model is unrealistic because it ignores politics and does not admit to the uncertainties of science. The assumption (known as the 'Enlightenment fallacy') that knowledge automatically produces response is accepted without question (Berry 1993: 314). This may serve the interests of science (or at least certain sections of the scientific establishment, which benefit from increased funding). The political class also benefits from its support of the IPCC because of the opportunities this offers to pursue low-cost UN environmental diplomacy and to transfer information, capacity and technology to the 'South' (with all the attendant commercial and electoral appeal).

Policy advice from science and the politicization of the IPCC

Warming predictions made in the UK have declined significantly since the mid-1980s (Mason 1996), though one would hardly think so on the basis of political announcements, press reports, even IPCC statements as reported by the UN when negotiations for the Kyoto Protocol began in earnest (DOE 1996, press statements June 1997) and in association with the Hague conference in December 2000. In any case there has been a significant decline in the predicted average warming, and large uncertainties surrounding aerosols and other factors remain. The average warming range predicted per decade in 1990 was 0.3°C for a doubling of carbon dioxide, today it is 0.2°C per decade. A former British Secretary of State for the Environment, John Gummer, pressed for urgent precautionary action by exaggerating the IPCC conclusions and by linking an IPCC phrase to his own, i.e. 'discernible human influence on the climate mainly as a result of greenhouse gas emissions from burning fossil fuels . . .'. Many politicians since have repeated his claims. The IPCC carefully avoids making such a link. How much of this calculated increase is linked to human activity is, of course, the essence of the policy response problem. The Blair government appears to have adopted the Gummer position to the full, apparently for reasons linked less to diplomacy than to revenue raising.

Few have dared to challenge the science underlying the prediction of even moderate warming (Emsley 1996), although the famous scientific consensus has been described as 'limited' in a scientific journal (Houghton 1996: 572). As mentioned, solar scientists have staged yet more serious challenges. Nevertheless, global warming believers continue to argue that the IPCC confirmed the need for mitigation action. Closer examination confirms that previous ambiguities continue.

In March 1995 it was agreed that:

> . . . This qualitative evidence does not prove conclusively that a cause and effect relationship exists between anthropogenic activities and the response of the climate system, nor does it allow us to quantify the magnitude of the effect. However, the best evidence that we have at present, drawn together from quantitative studies and qualitative sources, indicates that human activities have had an identifiable effect on climate. (Callender 1996: Appendix)

Note that the claim is not 'climate change' or even warming, but simply climate. By November 1995, after intense debate and several other formulations, this became the better-known statement:

> Our ability to quantify the human influence on global climate is cur-
> rently limited because the expected signal is still emerging from the noise
> of natural variability, and because there are uncertainties in key factors.
> These include the magnitude and patterns of long term natural variabil-
> ity and the time-evolving pattern of forcing by, and response to, changes
> in the concentrations of greenhouse gases and aerosols, and land surface
> changes. Nevertheless, the balance of evidence suggests that there is a
> discernible human influence on global climate. (ibid.)

By what criteria was the evidence 'balanced'? Advocates may still choose
between two positions, which may be contrasted with the 1990 IPCC
Science Report. This predicted 'with certainty' a rate of increase of global mean temperature during the next century 'of about 0.3°C per
decade . . . this will result in a likely increase in global mean temperature
of about 1°C above the present value by *2025*' (Houghton *et al.* 1990:
11). Certainty apparently created by the use of 'will', is taken away by
the use of 'likely'. TAR does not alter this picture.

As far as 'closed' science understanding is concerned, a view made in
1987 that '. . . the range of scientific uncertainty is currently so large that
neither "do nothing" nor "prevent emissions" can be excluded from
consideration' (Warrick and Jones 1988: 62), surely still holds today.
These two authors pleaded in the late 1980s that it was imperative to
extend full support to a two-pronged research effort that would narrow
the range of scientific uncertainty regarding the greenhouse effect, while
identifying and defining ways in which socio-economic and environmental
systems were likely to be affected. This early plea ignored two research
areas, solutions and adaptation, which have since been battling for inclusion in the research agenda.

The impact of research agendas on policy raises major questions about
the nature and funding of government-funded science with its short-term
contracts and the myopia inherent to political decision making. Natural
science appears to promote itself increasingly with reference to environmental threats, which it promises to be able to measure and predict. The
reduction of uncertainties is then linked to cost savings and competitiveness, the mantra of contemporary politics and business. If 'big' research
plays its cards right, it may be allowed to research climate change for
another decade or so. It will continue to predict warming by relatively
simple mathematical computations that are easily manipulated. The activities of the scientific community and the IPCC as suppliers of information
to environmental administrations are therefore of significance in any
explanation of climate policy; but they are certainly not sufficient – for
this, one has to look at the non-environmental benefits arising in the
short term to other lobbies, and to government itself.

Conclusions

Scientific advice derived at the frontiers of research is by nature inconclusive, and climate is too complex a phenomenon to be left to the physical sciences alone. The big issue, even if we knew everything, is whether humanity can control itself sufficiently to prevent climatic change without destroying its fragile political structures in the process. If emission behaviour cannot be changed 'enough', then adaptation should be the research and policy agenda. This may not suit the global research enterprise, for adaptation must be local, culture-bound and may not promise lucrative trade and investment flows. Research thrives on debate and controversy, not on consensus. The political weakness of science that is funded to be consensual and policy-relevant tends to result in advice that is ambiguous because it strives to serve all parties. Scientists may become unwilling to involve themselves in policy debates (Proctor 1991), or seek to keep their knowledge to themselves.

Climate change politics so far have largely created proposals, data and vague emission reduction targets that have transformed the noble goal of protecting the global atmosphere into a utilitarian game about regulatory intervention. This is likely to be of particular danger to the weakest countries, who may lose more of whatever independence they may have had in their 'development' policies. Yet global warming is also a justification for bureaucratic planning everywhere at a time when governments are trying to escape their responsibilities and return difficult social decisions to 'the market'. The structures dominating climate change science make reasoned opposition difficult, but on the positive side the process has enhanced the political attractiveness of the 'environment' and 'ecology'.

Climate policy cannot be understood, and certainly cannot be implemented effectively from an environmental perspective, without a more sophisticated view of the role and limitations of the natural sciences and their need for controversy and public funding. In addition, a better understanding of the coalition of non-environmental interests is needed, especially those of a commercial and bureaucratic nature, which now drive the issue internationally. This coalition drives the climate change debate irrespective of the scientific and ecological uncertainties facing society–nature relationships.

Note

1. An earlier, pre-Kyoto version of this paper was published in 1997 as 'A winning coalition of advocacy: climate research, bureaucracy and "alternative" fuels', *Energy Policy* 25(4): 439–444. For this volume, it has been updated and substantially amended.

References

Adams, D. (1996) *Joint Implementation: Opportunities for Business under the UN Framework Convention on Climate Change.* London: FT Energy Publishing.

Boehmer-Christiansen, S.A. (1993) Scientific consensus and climate change: the codification of a global research agenda. *Energy and Environment* 4(4): 362–406. See also in IPCC-related papers in *Environmental Politics* (1995) 4 (1 and 2) and in *Global Environmental Change* (1994) 4 (2 and 3).

Boehmer-Christiansen, S.A. (1994) Global climate protection policy: the limits of scientific advice – Parts I and II. *Global Environmental Change* 4 (2 and 3).

Boehmer-Christiansen, S.A. (1995a) Britain and the Intergovernmental Panel on Climate Change: The impacts of scientific advice on global warming: Integrated policy analysis and the global dimension. *Environmental Politics* 4(1) Spring.

Boehmer-Christiansen, S.A. (1995b) Britain and the Intergovernmental Panel on Climate Change: The impacts of scientific advice on global warming. Part II: The domestic story of the British response to climate change. *Environmental Politics* 4(2) Summer.

Boehmer-Christiansen, S.A. (1996) Political pressures in the formation of scientific consensus. In: Emsley, J. (ed.) *The Global Warming Debate.* London: European Science and Environment Forum.

Boehmer-Christiansen, S.A. (1997) A winning coalition of advocacy: climate research, bureaucracy and 'alternative' fuels. *Energy Policy* 25(4): 439–44.

Berry, S. (1993) Green religion and green science. *RSA Journal* CXLI(5438), April.

Calder, N. (1997) *The Manic Sun: Weather Theories Confounded.* London: Pilkington Press.

Callender, B. (1996) Global climatic change – the latest scientific understanding. Draft paper presented to the 28th International Geographical Congress, the Hague, Netherlands, 4–10 August: mimeo.

Churchill, S. and Freestone, D. (1992) (eds) *International Law and Climate Change.* London: Graham and Trotman/Nijhoff.

Courtney, R. (1999) An assessment of validation experiments conducted on computer models of global climate at the UK Hadley Centre. *Energy and Environment* 10(5): 491–502.

Department of the Environment (UK) (1996) *Climate Change Briefing*. October. London: Department of the Environment.

Emsley, J. (1996) (ed.) *The Global Warming Debate*. London: European Science and Environment Forum.

ENDS (Environmental Data Services) (1998) *Report 278. EC Climate Policy Takes Shape*. March. London: ENDS.

Green, O. and Skea, J. (1997) After Kyoto: making climate policy work. *Global Environmental Change Briefing, Special Briefing No.1*. University of Sussex: ESRC.

Hasselmann, Klaus, MPI (1997) From an an open email to the climate modelling community. June.

Houghton, J. *et al.* (1990) (eds) *Climate Change, The IPCC Scientific Assessment*, London: Cambridge University Press for UNEP/WMO. (First and second assessment reports.)

Houghton, J. (1996) Letter to *Nature 384*. November.

Jepma, C. (1998) Kyoto protocol and compatibility. *Quarterly Magazine on Joint Implementation*. Vol. 4(1), April. Groningen, Netherlands.

Kellogg, W.W. and Schware, R. (1981) (eds) *Climate Change and Society: Consequences of Increasing Carbon Dioxide*. Boulder: Westview.

Mason, B.J. (1996) Predictions of climate changes caused by man-made emission of greenhouse gases: a critical assessment. *Contemporary Physics 36(5)*: 299–319.

OECD (1991) *Greenhouse Emissions. The Energy Dimension*. Paris: OECD.

O'Neill, J. (1993) *Ecology, Policy, Politics*. Routledge: London.

Pearce, F. (1998) Playing dirty in Kyoto. *New Scientist*. 17 January: 48.

Proctor, R. (1991) *Value Free Science? Purity and Power in Modern Knowledge*, Cambridge, Mass: Harvard University Press.

Shackley, S. and Wynne B. (1997) Global warming potentials: ambiguity or precision as an aid to policy? *Climate Research* 8 (May): 89–107

Stiglitz, J. (1997) Stepping towards balance: addressing global climate change, Paper read at the *Fifth World Bank Conference on Environmentally and Socially Sustainable Development*. World Bank: Washington DC. mimeo. October 6.

UN, Framework Convention on Climate Change (1997) Conference of the Parties, 1–10 December 1997, Kyoto Protocol to the UN FCCC, FCCC/CP/1997/L.7/Add.1, 10 December 1997, 24 pages (subject to technical revisions).

US, National Research Council (1990) *Research Strategies for the US Global Change Research Program*. Washington: National Academy Press.

Wade, Robert (1997) Greening the Bank. In: Kapur, D., Lewis, J.P. and Webb, R. (eds) *The World Bank: Its First Half Century*. Washington, DC: Brooking Institution Press.

Warrick, R.A. and Jones, P.D. (1988) The greenhouse effect: impacts and policies. *Forum for Applied Research and Public Policy*. Autumn: 44–62.

World Bank (1997) *Guidelines for Climate Change Overlays*. ESD, Washington: The World Bank. February.

World Energy Council (WEC) (1993) *Journal*, July.

World Energy Council (WEC) (1998) *Bulletin*, Issue 5, April.

World Meteorological Organisation (1986) *Report of the International Conference on the Assessment of Carbon Dioxide and other Greenhouse Gases in Climate Variations and Associated Impacts*. Villach 9–15 October 1985, WMO Nn. 661, Geneva.

Young, Z. and Boehmer-Christiansen S.A. (1998) Green energy facilitated? The uncertain function of the global environment facility. *Energy and Environment* 9(1): 35–60.

7 | Sustainability, uncertainty and global fisheries

Rosemary Rayfuse and Martijn Wilder

Introduction

The exploitation of the world's oceans and their resources has provided a source of wealth and food security to the international community for centuries. However, as international demand for marine products has grown so, too, have the pressures placed upon existing stocks and marine ecosystems. In the last few decades there has been an unprecedented exploitation of global fishery stocks, to the extent that landings from fisheries are now in serious decline. Fisheries in 13 out of 17 of the world's major fishing regions are now either depleted or in serious decline and 70 per cent of the world's commercially important fish stocks are either fully fished, overexploited, depleted or very slowly recovering from overfishing (FAO 1998). The total closure of certain North American and European fisheries, with their attendant political, economic and social costs and the fishing wars between fishers, both domestically and internationally, are indicators of the underlying problem – the unsustainable conduct of global fisheries.

The causes of the crisis in global fisheries have been well documented. With respect to high seas fisheries, the United Nations Programme of Action for Sustainable Development, Agenda 21, lists problems of inadequate adoption, monitoring and enforcement of effective conservation measures in many areas, and overutilization of resources (UNCED 1992). Unregulated fishing, overcapitalization, excessive fleet size, vessel reflagging to escape controls, insufficiently selective gear, unreliable databases and lack of co-operation between states are also listed as contributing factors.

The problems plaguing fisheries within national jurisdictions are similar. Agenda 21 refers to local overfishing, unauthorized incursions by foreign fleets, ecosystem degradation, overcapitalization and excessive fleet sizes, underevaluation of catch, insufficiently selective gear, unreliable databases and increasing competition between artisanal and large-scale fishing and between fishing and other types of activity.

The adverse impacts of these practices on marine biodiversity are also well known. Modern fishing methods, in particular, large-scale trawl and seine (net) gear have been found to trap large amounts of marine wildlife including birds (particularly the wandering albatross), sea turtles, dolphins and other marine mammals as well as commercial and non-commercial fish.

These problems are further exacerbated by practices such as discarding (throwing overboard unwanted accidental by-catch including birds, mammals and fish) and high grading (throwing back smaller fish caught of a particular stock being fished in order to maximize size and quality of catch). The FAO has officially estimated that up to 20 per cent of the global fish catch is thrown back into the oceans (FAO 1991: 33). Unofficially, the estimates are as high as 32 per cent (Doulman 1996: 21). The number of marine mammals such as turtles, dugongs, dolphins and 'trash fish' caught as part of the annual by-catch often outweighs the catch of target species, while the number of birds killed globally is in the tens of thousands in the Southern Ocean alone.

As a renewable resource, fisheries can be exploited indefinitely, both on the high seas and within national jurisdictions – if managed sustainably. However, the challenge of such sustainable management is great. Internationally, it requires an unprecedented level of co-operation and goodwill to deal with resources about which little is known. Within national jurisdictions sustainable management of fisheries is no less complex.

Australia, for example, is responsible for the management of one of the world's largest ocean expanses, with an exclusive economic zone covering approximately 9 million square kilometres. Yet although Australia is widely regarded as at the forefront of fisheries management, managing such an expanse in a sustainable fashion has proved difficult, with many species either over- or fully-exploited and with general stock levels in decline. Furthermore, for many other species stock levels are simply unknown (McLoughlin *et al.* 1994, 1997; BRS 1998). It is clear that many of the problems facing Australia (and other coastal states) are similar to those plaguing international fisheries management. It is a matter of scale. The experience of Australia is thus a useful indicator from which to study sustainable management in international fisheries as well. In this chapter we look at the nature of global fisheries, the concept of sustainability and the nature and role of fisheries management. We examine these issues first within the context of international fisheries: those which occur between two or more exclusive economic zones, between exclusive economic zones and the high seas, or wholly on the high seas. Then we examine these issues within the context of fisheries within national jurisdiction, using the experience of Australia as a case study. The two sections of the chapter are brought together in our conclusion on the approaches to be followed for the sustainable management of global fisheries.

The nature of fisheries

Ninety per cent of the world's fish are found within the 200-mile exclusive economic zones (EEZ) of coastal states. The remaining 10 per cent are located on the high seas. Fish stocks located wholly within EEZs are subject to the jurisdiction of, and management by, coastal states. Thus the vast majority of global fisheries resources are controlled by the coastal states in whose EEZ they are found.

However, fish do not respect this ecologically arbitrary 200-mile limit. Anadromous species, such as salmon, spawn in fresh water within the jurisdiction of a coastal state but spend most of their life cycle at sea – both within and beyond 200-mile limits – before returning to their spawning grounds. Effective management therefore requires co-operation between states of origin, range states and fishing states. Other species of fish – or stocks of those species – may be found straddling two or more EEZs or an EEZ and the high seas. The former are generally referred to as transboundary stocks, and their effective management requires co-operation between the states in whose EEZs they are located. The latter are known as straddling fish stocks. Effective management of straddling fish stocks requires co-operation between any coastal states in whose waters the stocks are found and any fishing states who fish for those stocks on the high seas. Highly migratory species, such as tuna, migrate through any number of EEZs and high seas areas. As with anadromous species, their management requires the co-operation of states of origin, all range states and all fishing states.

It is axiomatic that fish resources are biologically renewable if conserved and managed appropriately and destructible if not. The difficulty is to manage 'appropriately' a resource which is under increasing global demand, is open to exploitation by so many potential competing interests, and about which – in global terms – little scientific knowledge exists. It is in this context that the issue of 'sustainable' management of fishery resources arises.

The concept of sustainability in the fisheries context

The concept of 'sustainable development' first appeared in the early conservation agreements designed to protect fisheries, fauna and flora (Sands 1995: 305). However, it gained popular definition and political currency with *Our Common Future*, the 1987 Report of the World Commission on Environment and Development (see Chapter 1, this volume.)

Sustainable development is an anthropocentric concept. Human beings are at the centre of concerns for sustainable development (UNCED Rio Declaration: Principle 1) – its goal is not conservation of resources for conservation's sake, but rather conservation and management of resources to ensure their continuing availability to meet the needs of both present and future generations. It is in this context that we talk about 'sustainability' and 'sustainable management' of fisheries.

Of course, to talk of 'fisheries management' is itself something of an oxymoron. What has occurred in the past has been neither management, nor has it been of fish. It has been mis-management or non-management of people. As recognized by many commentators, 'sustainable fisheries management' requires effective control over the human element of the exploitation equation (Maguire, Neis and Sinclair 1995: 142).

Furthermore, truly sustainable management requires an understanding of the larger environmental picture – including the interrelations of species and marine ecosystems, the effect of pollution from land- and sea-based sources, the effects of habitat loss, the effects of potential future deep sea bed mining and the contribution of fish to food security (Kyoto Declaration 1995). While there may be agreement on the need for an interdisciplinary, all-encompassing approach, the manner in which this can best be accomplished is the subject of much debate.

International fisheries management

Recognition of the need to manage and conserve international fisheries, through international agreements, is now accepted as fundamental. However, managing the resource and those who exploit it involves far more than agreement on the legal obligations of doing so. Translating the obligations that exist into practice has been difficult. Agreement on what to manage, how to manage, who is to manage and how the legal obligations are to be enforced has been rare. While conflicting interests make agreement difficult, uncertainty regarding what is to be managed has been the real hurdle to achieving effective sustainability.

The difficulties of international fisheries management: lack of certainty

With the possible exception of the question of enforcement, the greatest difficulty plaguing international fisheries management has been that of uncertainty: scientific, practical and legal.

Scientific uncertainty

Scientific uncertainty relates to a lack of reliable data and understanding, and includes uncertainty as to quantity, location and range, genetic origin, reproductive capacities and interdependence of species; uncertainty as to the relationship of fish species with aspects of the larger marine ecosystems and the global ecosystem as a whole; and uncertainty as to what is being taken and by whom. It is uncertainty as to the impact of fishing itself and those techniques best suited to managing fisheries (Burke, Freeberg and Miles 1994) and it is uncertainty over the impacts of pollution, habitat degradation and broader issues such as climate change. In political terms this scientific uncertainty is open to manipulation and (mis-)use in management decisions (Maguire, Neis and Sinclair 1995: 144). Reasons for the uncertainty involve both straightforward considerations of time and cost, and the more complex issues of human interactions and the limits of knowledge.

Even though considerable scientific research effort has been expended on global fisheries, available biological information has been limited (FAO 1992b: 1) In part, this is a function of the practical and political realities of the nature of scientific research (Johnston 1995: 157) and of the vast spaces involved. More importantly, most research has been conducted on stocks and species that have already been decimated (FAO 1992a). Lack of historical reference points for many species means that research has often been conducted in a vacuum. Discovery of new fisheries often means that fishing is taking place on stocks for which there is no scientific information.

To reduce this uncertainty, data and information on catches has been sought directly from those exploiting the resource – fishers. The effectiveness of this has, however, been limited due to lack of co-operation, both intentional and inadvertent, inadequate monitoring, control and surveillance, and inadequate funding of regional fisheries organizations (Marashi 1996).

In addition, until relatively recently, data were only required relating to the targeted species, not to by-catch. It was thus not possible to evaluate the effects of the fishing effort on untargeted stocks and species. These uncertainties, when combined with overfishing, have made assessing stock abundance notoriously difficult, and largely a failure. They have also given rise to significant conflicts between states over the viability of particular fisheries.

Practical uncertainty

Practical uncertainty relates to the question of appropriate strategies, reference points and management techniques. It is complicated by difficulties in defining the appropriate area of management – local, national, regional or international – and authority over them (Borgese and Saigal 1995: 1).

Although numerous management strategies, each with different reference points, are possible, the one most widely adopted has been to set catch levels on the basis of maximum sustainable yields (MSY) of a single target species. In theory, MSY aims to maintain the productivity of the oceans by allowing fishers to take only that number of fish from a stock that is replaced by the annual rate of new recruits (young fish of harvestable size) entering the stock. Once the MSY is determined, total allowable catches (TACs) are set for individual fisheries.

However, MSY has been criticized for failing to take account of other factors impacting on sustainable resource exploitation, such as the economic value of the catch, the cost of catching, the natural instability of some stocks and the interrelations between the targeted species and other species in the ecosystem. Its determination, and hence that of TACs, which is based on the incomplete, suspect, and often out-of-date data received, is therefore fraught with scientific uncertainty. The use of the MSY approach is therefore now generally considered to be incompatible with the sustainable development of multispecies resources in an ecosystem context (FAO 1992b).

In recent years the notion of an optimum sustainable yield (OSY) incorporating all of the above absent factors, and set at a level lower than the MSY, has been adopted, as seen in the EEZ and high seas fishing regimes established under the 1982 United Nations Convention on the Law of the Sea (UNCLOS), where similar economic and environmental factors must also be considered in determining the MSY when deciding the TAC (UNCLOS: Articles 61 and 119).

Perhaps the most fundamental practical uncertainty arises from conflicting notions of why fisheries are being managed: for the economic benefit of present fishers or for the long-term economic and nutritional benefit of the world's present and future population as a whole (Paul 1996: 26). Even when management strategies are agreed, considerable uncertainty has arisen as to which techniques should be used to best implement them. As Johnston notes, the techniques preferred have depended on the 'mind-set' of the proponent (Johnston 1995: 1). Whatever approach is adopted, the ability to make an accurate judgement as to the level of sustainability is dependent upon the ability to accumulate accurate knowledge of the fishery.

Legal uncertainty

Legal uncertainty is essentially the question of who manages, for whom, and what legal obligations are imposed on those who manage and those who exploit. It relates both to the vagueness of diplomatically negotiated and legally drafted instruments and to their non-existence in some cases. It includes uncertainty as to who has jurisdiction to regulate and how to regulate and manage effectively in the absence of enforcement powers.

The extension of coastal state jurisdiction over 200-mile exclusive economic zones as provided for in the 1982 United Nations Convention on the Law of the Sea (UNCLOS) did not lead to the sustainable management of fish resources within EEZs. Coastal states saw the acquisition of control over vast ocean areas as a bonanza and treated it as such. Open unregulated access led to unbridled competition, which in turn led to technological developments to increase catch size and profit, which led to overfishing and the decline and collapse of stocks and the communities dependent on their exploitation.

On the high seas the situation was, if anything, exacerbated. Vessels from foreign nations which had fished in waters now within the 200-mile limits were pushed out onto the high seas, where freedom to fish and open access still existed. This led to a free for all which placed increased pressure on the small percentage of the world's fish population located in the high seas. Additionally, many of these vessels continued to fish illegally within the new 200-mile zones, giving rise to increased tensions and conflict between coastal states and distant water fishing states.

Conflicts over exploitation of transboundary stocks, straddling fish stocks (SFS) and highly migratory fish stocks (HMFS) also increased. Despite provisions in UNCLOS directed at managing such stocks, and the emergence of regional and subregional fisheries organizations, the effectiveness of such provisions and bodies has proved to be limited, especially given the absence of any real management mandate. With conflict as to who should actually manage certain areas, many high seas areas were simply left unregulated.

Uncertainty also surrounded the obligations of those involved in fisheries exploitation. This, and human ingenuity, led to such problems as reflagging of vessels to avoid compliance with existing regimes, and to irresponsible fishing practices (such as dumping, high grading and strip mining). The issue of legal uncertainty is, thus, closely linked with the other major problem of international and national fisheries management – lack of enforcement.

The difficulties of international fisheries management: lack of enforcement

Enforcement in international fisheries is a difficult task. In the absence of express consent to the content of, and agreement to submit to, enforcement mechanisms – and states cannot be forced to give their consent – no enforcement is possible.

States have, of course, consented to a number of obligations. However, consent to accept obligations is not consent to their enforcement. Perhaps the most important obligation, in the fisheries context, is that of flag state responsibility. That obligation, spelled out in Article 117 of UNCLOS, requires states to ensure that vessels flying their flags comply with all agreed conservation and management measures and, where they do not, to take prompt and effective action to deter and prevent any further violations.

An obvious problem has been that of flags of convenience. Some states allow registration of vessels under their flags as essentially a revenue-raising activity, taking no interest in the activities of these vessels (Doulman 1996: 22). A related issue is that of reflagging of vessels to avoid flag state control. In this practice (also referred to as flags of convenience) the idea is to reflag in a state which is not a member of a particular regional fisheries organization or party to particular multilateral treaties. FAO figures show that more than 2000 fishing vessels reflagged between 1991 and 1994 (Doulman 1996: 23). The trend has continued.

Flag states, even those not generally considered 'flag of convenience' states, have been less than diligent in living up to their obligations. They often fail to enact the legislation required to bring their international obligations into force. Where legislation exists, domestic political considerations often influence states' decisions not to prosecute or take other action against their vessels and/or nationals. Similarly, penalties may be so slight as to be ineffective deterrents. Economic factors such as cost of enforcement and prosecution also intrude. This is particularly true for developing states, whose economic priorities may be focused elsewhere. Geographical factors also play a role. Often vessels are fishing in waters far distant from their state of registry and seldom, if ever, return to port there. The exercise of flag state jurisdiction in such circumstances may be impossible.

Effective enforcement also requires effective monitoring, control and surveillance (MCS). Cost and geographical factors play a significant role here as well. Adequate monitoring of vessel activity by flag states, particularly when they are far from port or when large numbers of vessels are concerned, can be prohibitively expensive and logistically impossible. Fisheries patrol vessels are expensive and limited in range of operation. Monitoring by observer is possible. However, this presupposes availability of sufficient numbers of willing and adequately trained observers to be placed on each vessel flying a state's flag. In addition, observers may be subject to abuse or corruption. MCS technology such as 'black box' type transponders to monitor vessel movements and operations is available – however, they are expensive and subject to human intervention.

Effective enforcement of conservation and management measures in international fisheries also requires that states accept to be bound by measures decided upon by the regional fisheries organizations of which they are members. The ability of states to 'opt out' of measures, such as established quotas, with which they disagree can simply render nugatory the activities of these organizations. Fishing by non-member states in areas under the jurisdiction of these regional organizations can have the same result.

The responses of international fisheries management to uncertainty

The difficulties in the management of international fisheries are not new. Nor are many of the solutions which have been invoked. However, with the post-UNCED emphasis on sustainable development, some new and wide-ranging solutions have been sought and are being introduced.

Scientific uncertainty

The precautionary principle

By far the most important response to scientific uncertainty, and indeed to all uncertainty in the fisheries context, has been the acceptance of the application of the precautionary principle to fisheries conservation and management (Garcia 1994; Cameron and Aboucher 1991).

According to the precautionary principle, 'where there are threats of serious or irreversible damage to the environment, lack of full scientific certainty shall not be used as a reason for postponing cost-effective measures to prevent environmental degradation' (UNCED, Rio Declaration, Principle 15 1992).

The precautionary approach is now mandated by article 7 of the FAO Code of Conduct for Responsible Fishing (FAO 1995a) and by article 6 of the UN Agreement on Straddling Fish Stocks and Highly Migratory Fish Stocks (the UN Fish Stocks Agreement) (UN 1995) and its application is being incorporated into national fisheries legislation and into the activities of regional fisheries organizations. It applies at all levels of the fishery system: to development planning, management, research, technology development and transfer, legal and institutional frameworks, fish capture and processing, fisheries enhancement and aquaculture (FAO 1995a).

A significant aspect of the precautionary approach is that it reverses the burden of proof. Those wishing to undertake a particular activity are required to show that significant harm is not likely to occur (FAO 1995a). In practice, this requires the preparation of such things as environmental impact assessments and detailed management plans before starting to harvest new fish stocks.

Data acquisition

To successfully implement the precautionary approach and to overcome its burden of proof requires evidence in the form of scientific data. Agenda 21 calls for states to co-operate, with the support of appropriate international organizations, to promote enhanced collection of data necessary for the conservation and sustainable use of the marine living resources of the high seas, to exchange data and information adequate for fisheries assessment, and to develop databases on the high seas marine living resources and fisheries. These data can then be made available to scientists and decision makers to assist in determining TACs and other conservation and management measures.

Increasingly, regional organizations are requiring more extensive, detailed, complex and finer-scale data from fishers. Efforts are under way to rationalize and standardize reporting requirements (FAO 1993a). To ensure accuracy, data on landings and transshipped fish are also sought from port states and transshipment states. On-board scientific observers are increasingly being used to assist in verification as well. Data can, however, only be acquired from member states. Fishing by non-member states and entities therefore continues to go largely unmonitored. The effect of this fishing must now – given the precautionary approach – be taken into account.

Practical uncertainty

Ecosystem approach

Another significant step in international fisheries management has been the development and gradual adoption by a number of instruments and organizations of the ecosystem approach to fisheries conservation and management. Although the exact parameters of ecosystem management depend very much on the 'ecosystem' in question (Maguire, Neis and Sinclair 1995), in the fisheries context it generally refers to management of the marine ecosystem. Ecosystem management does not necessarily involve 'managing' multiple species. Rather, it entails managing exploited species in such a way as to minimize risks of ecological linkages failing. For very practical reasons, this may often require managing the exploited species alone, but in such a way that allowance is made for 'escarpment' to meet ecosystem needs.

The ecosystem approach recognizes ecosystem relationships such as those between birds, marine mammals, and fish, as well as between various species of fish. It also recognizes the inherent limitations of lack of scientific knowledge about marine ecosystems and inadequate and finite financial resources. Working to overcome these limitations, it aims to reduce or avoid detrimental impacts on associated and dependent species and ecosystems as a whole. In doing so it squarely addresses the by-catch and incidental mortality issues while seeking to ensure sustainability of all species.

The first, and for many years, only international agreement that required the ecosystem approach was the 1980 Convention on the Conservation of Antarctic and Marine Living Resources. It has now been introduced, in varying degrees, into a number of other international agreements and the operations of regional fishery organizations such as the 1992 Convention for the Conservation of Anadromous Stocks in the North Pacific Ocean and the 1995 UN Fish Stocks Agreement.

Biodiversity approach

A relatively recent development, the ramifications of which for international fisheries conservation and management have yet to be fully explored, stems from the 1992 Convention on Biological Diversity. The Biodiversity Convention defines biological diversity as 'the variability among living organisms from all sources including, *inter alia*, terrestrial, marine and other aquatic ecosystems, and the ecological complexes of which they are a part; this includes diversity within species, between species and of ecosystems' (United Nations 1992a: article 2). It calls upon states to take appropriate conservation measures, including the protection and management of biological resources important for the conservation of biological diversity and the rehabilitation and restoration of degraded ecosystems (Paul 1996). The implications of these obligations are far-reaching.

Although little in the Law of the Sea Convention directly addresses the issue of conservation of biodiversity, recent fisheries agreements have sought to ameliorate the dire consequences of modern commercial fishing techniques for biodiversity. Article 5(g) of the UN Fish Stocks Agreement specifically mandates protection of biodiversity in the marine environment. The global moratorium on large-scale pelagic driftnet fishing and agreements aimed at reduction of by-catch of unwanted fish, marine turtles, sea birds and marine mammals are also manifestations of attempts to regulate conservation of biodiversity. The larger implications of co-ordination between fisheries conservation and management and biodiversity conservation are, however, still little understood and little explored. How the biodiversity approach will be implemented in practice thus remains uncertain (Rose 1999).

Management techniques

Many varied techniques have been adopted by international fisheries organizations in their attempts to conserve and manage the stocks and species under their jurisdiction. These include the imposition of quotas, area restrictions (e.g. to protect spawning grounds), the closure of fishing grounds and gear restrictions.

Licensing is another mechanism whose use is growing. Technically most licensing of fishing vessels is done by their flag state. International fisheries organizations are, however, increasingly requiring member states

to submit lists of their licensed vessels which will be fishing in the regulated area. Vessels which are found to violate conservation measures can
thus be more easily identified and their flag states requested to revoke
their licenses. Related to this is the possibility of linking licensing to fish
within an EEZ with the requirement to provide information about high
seas fishing (Bergin and Michaelis 1996: 295).

Vessels can also be required to announce their entry into and exit
from a regulated area, to provide information on their location within
the area and to identify when they are engaged in fishing activities. Hail
systems are in use in a number of areas, although, given vast expanses of
ocean spaces, they are clearly open to abuse. Many organizations are
now considering and introducing compulsory on board vessel monitoring systems (VMS) and satellite tracking. Also being discussed in some
organizations is the use of real-time data reporting, which clearly allows
more prompt assessment of whether allowable catch limits are being
reached, so closures can be introduced quickly and effectively.

Legal uncertainty – clarification of approaches, rights and obligations

The UNCED process and its products, the Rio Declaration, Agenda 21
and the Biodiversity Convention, set the stage for the approaches to be
taken to international fisheries conservation and management. An important step towards legal certainty in international fisheries has been
the solution of the issue of jurisdiction over straddling fish stocks and
highly migratory fish stocks. The 1995 Fish Stocks Agreement puts an
end to claims by coastal states for extended fisheries jurisdiction beyond
their 200-mile zones. It mandates co-operation of coastal states and fishing
states either directly or through regional or sub-regional international
fishery organizations to ensure the long-term conservation and sustainable use of straddling stocks and highly migratory fish stocks. It provides
enforcement mechanisms to ensure the attainment of that objective. It
sets out more clearly the rights and obligations of flag, coastal and port
states and provides for the special requirements of developing states.

Also adopted in 1995 was the FAO Code of Conduct for Responsible
Fisheries (the Code of Conduct). The code sets out, in a non-mandatory
way, the principles and standards that are applicable to the conservation, management and development of all fisheries. Its objectives are
multiple but it is essentially aimed at establishing a 'framework for
national and international efforts to ensure sustainable exploitation of
aquatic living resources in harmony with the environment' (FAO 1995a).

An integral part of the Code of Conduct is the 1993 Agreement to Promote Compliance with International Conservation and Management Measures by Fishing Vessels on the High Seas (the Compliance Agreement) (FAO 1993b) which is aimed at reducing the incidence of reflagging of fishing vessels for the purposes of avoiding international conservation and management measures.

Importantly, each of these instruments provides for an enhanced role for international fishery organizations. In response to these calls, the FAO has been considering how to strengthen the regional fisheries organizations under its auspices (FAO 1997). In addition, new multilateral agreements are being negotiated which establish new commissions to regulate fishing activity in previously unregulated areas of the high seas. A draft convention on the conservation and management of highly migratory fish stocks in the western and central Pacific Ocean and an agreement on straddling and highly migratory fisheries in the south east Atlantic are also currently under negotiation.

Many legal issues such as the relationship between the Biodiversity Convention and fisheries conventions are yet to be resolved. Additionally, many new ones, such as the interrelationship between the new global and regional fisheries instruments, have been created. However, the legal developments, particularly since UNCED, have given new clarity and certainty to much of international fisheries regulation.

Responses of international fisheries to enforcement problems

A plethora of interesting approaches and mechanisms is being introduced on the international level to improve enforcement of conservation and management measures.

Monitoring of vessel movement and activity is being enhanced by air reconnaissance, VMS systems and satellite tracking. Nevertheless, difficulties still exist with respect to standardization of systems and hardware. Observers are being increasingly used to assist in data gathering and incidentally in ensuring respect for conservation and management measures. Additional verification of data is accomplished through port state inspection and verification upon landing or during at-sea transshipment. Attempts have been made to shore up port state verification by prohibiting unlicensed fishing and transshipment within EEZs. This, however, may not be possible in respect of vessels that fish and transship on the high seas. Maintaining and exchanging registers of violators between members of organizations and pursuant to the Compliance Agreement and Code of Conduct will increase the efficacy of these moves (Bergin and Michaelis 1996).

Flag state enforcement, with all its limitations, remains the mainstay of international law. However, flag states are increasingly entering into agreements to assign some of their rights to others better suited to enforce them. In some cases, such as the Canadian arrest on the high seas of the Spanish fishing vessel the *Estai*, coastal states are simply taking the matter into their own hands. A less dramatic approach is, of course, diplomatic protest and the possibility of trade or other sanctions.

Scientific research which will assist enforcement is also under way. Techniques similar to DNA testing in humans can establish the genetic identity of species and stocks and may assist in determining whether specimens caught were from a legally fished stock or otherwise. Economic incentives and disincentives are also being utilized to aid enforcement (Sutton 1996: 22).

Having examined the international fisheries scene we turn now to our case study of fisheries management in Australia.

Case study: Australian fisheries management

Australia has been hailed by many as being at the forefront of fisheries management on the world scene (House of Representatives 1996–1997). Nonetheless, all is not well. A number of its national fisheries are overfished, while the status of many others is uncertain, thereby making accurate and effective management difficult. The latest national Fishery Status Reports indicate that of the species covered four are overfished, 12 are fully fished (i.e. sustainable) and the status of 13 is uncertain or unknown. Only one remains underfished (Bureau of Resource Sciences 1998). In recent years the total fishery production has largely levelled off, although the value has grown. A performance audit, carried out by the Australian National Audit Office (ANAO) in 1996, on the Australian Fisheries Management Agency (AFMA), was highly critical of the agency and its management of Australia's fisheries (ANAO 1996). In addition, a 1997 Parliamentary review of Commonwealth fisheries management designed to further investigate the ANAO Report, put forward a further range of recommendations for improving national fisheries management (House of Representatives Standing Committee 1997).

The problems facing Australia's fisheries and their management are not unique to Australia. On the contrary, they largely mirror those problems experienced in international fisheries. Nevertheless, the difficulties faced by one of the world's recognised leaders in fisheries policy and management highlight the challenges that sustainable fisheries management poses.

The nature of Australian fisheries

Australia is responsible for the management of one of the world's largest exclusive economic zones covering approximately 9 million square kilometres of ocean and diverse marine ecosystems. These waters belong to three large oceanic basins: the Indian, Southern and Pacific Oceans. The waters enclosed within this area are distributed within almost 60 degrees of latitude, from Torres Strait in the north to Antarctica in the south. The longitudinal range is 72 degrees, from the Cocos Islands in the west to Norfolk Island in the east. If the Australian Antarctic Territory is included, the span extends from 40°E to 175°E.

Despite its ocean expanses, Australia ranks only fifteenth in world fisheries production in terms of tonnes of fish landed, largely because Australia's waters are low in nutrients and because much of the surrounding continental shelf is narrow (McLoughlin, Wallner and Staples 1997). Nonetheless, many of the fisheries target high-priced species such as prawns, lobsters, abalone and tuna.

Until the early 1950s, Australian fisheries were largely of a cottage type, serving local domestic markets. However, by the early 1980s overcapitalised fleets threatened the commercial viability of a number of stocks, while by the late 1980s real concerns over the future of Australia's fisheries management were being publicly voiced and substantial changes to national fisheries management called for (Industry Commission 1992; Senate Standing Committee 1993).

At the same time the jurisdictional division of federal (i.e. Commonwealth) and state/territory fisheries, combined with the lack of an effective management structure, further compromised the sustainability of the nation's fisheries. Not until 1983 with the Offshore Constitutional Settlement (OCS) was there a shift towards a co-ordinated national approach to oceans management, with the states and territories being given title to their 'coastal waters', i.e. all waters landward of the three-nautical-mile limit. All waters beyond that three-nautical-mile limit remained, and still remain, under the control of the Commonwealth. In addition, states and territories were given concurrent legislative power over coastal waters, i.e. the same power to legislate over coastal waters as exists over their own land territories.

The OCS arrangements led to the production in 1989 of the first Commonwealth government fisheries statement, followed by a suite of fisheries legislation in 1991, including the *Fisheries Management Act 1991*, under which AFMA was established. AFMA was given responsibility for ensuring the sustainable use of Commonwealth fisheries resources. The Act also provided for the Commonwealth and the states/territories to agree on arrangements to allow management of a fishery both within and outside state coastal waters to be undertaken by one authority (state or Commonwealth) and under one law (state or Commonwealth). However, many fisheries continue to require collaborative management between these governments, with the result being that management of Australia's fisheries resources is now a complex mix of Commonwealth and state/territory responsibility.

Despite substantial changes to the policy, legal and management framework of Australian fisheries in the last decade, and concerted attempts by authorities to do so, many of Australia's fisheries are not managed sustainably or with any certainty. In the 1996 Australian National Audit Office audit (ANAO 1996) (which was later endorsed by the House of Representatives Standing Committee on Managing Commonwealth Fisheries that followed (House of Representatives Standing Committee 1997)), it was stated that the existing management framework is failing to deliver. The report highlighted the problems of uncertainty and enforcement based on:

- the need for a greater understanding of the nature of Australia's fisheries, both in terms of accurate stock assessment and in terms of the environmental impact that fishing places on individual stocks and the ecosystems in which they live;
- the need for adoption of conservative decision making in developing fisheries policy, including greater use of the precautionary approach; and
- the need for enhanced management, regulatory control and enforcement.

Since the ANAO and House of Representatives reports, significant developments have occurred, many of which are directed at improving Australia's oceans management and will, in particular, address some of the concerns raised by the ANAO.

In particular, in late 1998 Australia launched a national Oceans Policy which sets out a framework for the integrated and ecosystem planning and management of all of Australia's marine jurisdictions based on large regional marine ecosystems, with the principles of ecologically sustainable development and the precautionary approach being applied in the use of oceans. The policy binds all Commonwealth agencies (including AFMA), although state participation is optional.

The policy (Commonwealth of Australia 1998: Vol 2, 10–12) recognises that to ensure a sustainable national fishery issues of overcapacity, management of the marine environment, uncertainty and enforcement must be dealt with. It specifically calls for:

- improvements in fisheries management, including a review of all fisheries laws, regulations and management arrangements and increasing and improving the collection of scientific information;
- ensuring ecologically sustainable fisheries practices that focus on protection of the wider ecosystem;
- fostering stewardship through encouraging codes of responsible fishing behaviour;
- adopting regulatory approaches that improve sustainability and efficiency such as Individual Transferable Quotas (ITQs);
- undertaking structural adjustment of the fishing industry with the specific intention of reducing excess capacity;
- supporting and enhancing research and development capacity with a view to improving scientific knowledge; and
- addressing issues of illegal fishing and compliance through increased legal certainty and enforcement.

Furthermore, the policy also identifies key policies and principles for ecologically sustainable ocean use, including the precautionary principle, the maintenance of ecosystem integrity and the promotion of ecologically sustainable marine-based industries. In doing so, it specifically identifies the need to 'manage for uncertainty' by accommodating uncertainty in resource assessment and being capable of rapid response, such as a cessation of fishing, in the case of adverse impacts on the stock (Commonwealth of Australia 1998: Vol 1, 38).

In mid-1999 the federal government overhauled its existing environmental legislation with the passing of the *Environment Protection and Biodiversity Act 1999* which identifies the marine environment as one of a range of matters of national environmental significance. All actions, activities and decisions which may have a significant impact on Commonwealth marine areas or which take place within Commonwealth marine areas and may have a significant effect on the environment will require environmental assessment and approval. This will include the approval of new plans of management for fisheries and the establishment of new fisheries within Commonwealth waters.

Sustainability in Australian fisheries

Agenda 21 lays out guidelines for the sustainable use and conservation of marine living resources under national jurisdiction (UNCED 1992: Chapter 17). Recognising the mounting problems facing national fisheries, coastal states are obliged to commit themselves to the conservation and sustainable use of marine living resources under national jurisdiction in line with the principles set out in Agenda 21.

In response to this, Australia released its own National Strategy for Ecologically Sustainable Development in December 1992 (Commonwealth of Australia 1992). In relation to fisheries, the strategy recognised the need for fisheries management agencies throughout Australia to adopt a fisheries ecosystem management framework which would provide a more holistic and sustainable approach to the management of aquatic resources. Such an approach was to be consistent with the principles of ecologically sustainable development (ESD) – in particular, as they relate to Australia's needs.

Furthermore, in the case of AFMA the *Fisheries Management Act 1991* provides (as does its Corporate Plan 1996–2001) that it must pursue the objective of:

> Ensuring that the exploitation of fisheries resources and the carrying on of any related activities are conducted in a manner consistent with the principles of ecologically sustainable development and the exercise of the precautionary principle, in particular the need to have regard to the impact of fishing activities on non-target species and the long term sustainability of the marine environment.

However, AFMA was heavily criticised in 1996 by the ANAO for failing to pursue its obligations with respect to sustainability (ESD). In particular, it was noted that at the policy level, AFMA's interpretation of the general ESD obligation had obscured its own primary role, which is to conserve fish stocks and the associated environment. Instead, AFMA policy has been directed towards managing stocks rather than regulating fish catches for their long-term productive potential. It also identified a lack of policy direction with respect to assessing the broader environmental impacts from fishing particular stocks (ANAO 1996). Furthermore, the actual management practices adopted, including the use of input controls and allowing fishing in the absence of adequate stock assessments or environmental impact assessments, were said to be further negating the ability to achieve sustainability.

Interestingly, the House of Representatives Standing Committee was not as critical of AFMA in its efforts to ensure sustainable fisheries, noting that in some fisheries AFMA's performance had in fact improved (House of Representatives Standing Committee 1997: 7.21). It has since been recognised in the latest Australian Fisheries Status Reports that the focus of management and research has progressively moved from considering mainly the fishing of target species to one that includes consideration of the non-target species and the marine environment (BRS 1998). In this regard there is evidence that many of the criticisms levelled against AFMA are being addressed.

Difficulties for Australian fisheries management

As has been the case for international fisheries, achieving sustainable and effective management of Australia's fisheries has been difficult. Aside from the practical and enforcement difficulties of dealing with one of the world's largest national fishing areas, Australian fisheries management must also deal with the difficulties presented by scientific, practical and legal uncertainty.

Australia's new Oceans Policy clearly recognises such difficulties and, as indicated above, lays out a range of broad objectives to be pursued and specific measures to directly overcome the problems that exist. The way in which such objectives will actually materialise and the extent to which the measures outlined are actually implemented remains to be seen. One of the major difficulties that still needs to be resolved is the legal uncertainty that persists given that the policy only binds the Commonwealth and not the states. Nonetheless, it is a significant step forward in improving the framework for sustainable fisheries management in Australia.

Having addressed causes of uncertainty and responses to them in the context of international fisheries, they are now examined in the context of Australian fisheries.

Lack of certainty in Australian fisheries

Scientific uncertainty

Australia's understanding of its own fisheries, both in terms of the size and nature of stocks, has historically been limited and even today remains very much in the developmental stage. Until the establishment of AFMA no significant attempt had been made to assess national fishery stocks. Available data simply came from routine logbook (catch) figures provided by commercial fishers, supplemented only by ad hoc research. It was not until 1992, with the establishment of AFMA, that a formalised stock assessment process was introduced, together with improved monitoring and research programmes. In addition, at this time the Bureau of Resource Sciences was appointed to undertake ongoing assessments of national stock levels (BRS 1998).

Australia's national fishery status reports released in 1992 and 1993 and as recently as the latest, in 1998, continue to find that the status of many species is simply unknown (McLoughlin, Wallner and Staples 1994; BRS 1998). As in so many fisheries, the desire of numerous fishers to maximise catches by circumventing management controls has resulted in the recording of incorrect or misleading catch data, not to mention the problems of high-grading and discarding (ANAO 1996). Furthermore, the historical absence of a formal stock assessment policy has resulted in a lack of direction in what is actually being assessed and for what purpose.

The adequacy of assessments has also been compromised by the fact that there is often a significant lack of knowledge about the actual stocks being assessed and about the wider ecosystem impacts that occur as a result of their exploitation. Environmental impact assessments of national fisheries has been almost nonexistent, despite the Australian and New Zealand Fisheries and Aquaculture Council having called, in 1992, for the Commonwealth and state fishery management agencies to undertake fisheries ecosystems management assessments (ANAO 1996). Both the new Oceans Policy and the *Biodiversity and Conservation Act* will see a change in such an approach, with recognition of the need to carry out environmental assessments, particularly in relation to fisheries management plans.

Adequate knowledge of the nature of the fishery, stock levels and interrelationships with other species is an essential precondition to determining the most effective practical management techniques. For Australia, as for others, this continued scientific uncertainty makes achieving sustainability extremely difficult.

Policy uncertainty

As part of its attempt to establish a more comprehensive fisheries management framework, in 1989 the Commonwealth government released Australia's first national fisheries policy, *New Directions for Commonwealth Fisheries Management in the 1990s: A Government Policy Statement*. Regarded at the time as extremely visionary, it recognised the importance of achieving ecologically sustainable fisheries.

By the mid-1990s, however, the policy was criticised as being outdated, though it remains the guiding policy document for fisheries management in Australia. There has, therefore, been a general failure in the development of national fisheries policies, the manner in which they are guided and in the determination of the criteria on which they are based. In addition, as recognised in the 1996 ANAO Report, there has been a clear absence of policy with respect to day-to-day operational issues (*ANAO Report 1996*: 32–33).

Of even greater concern is the fact that historically decisions have been made that are both contrary to the policy itself and to achieving ESD. There have been occasions where AFMA has decided to maintain existing levels of commercial fishing activity despite stock assessment reports acknowledging that the data were insufficient or unreliable and/or such reports recommending reductions in allowable catches (ANAO 1996: 29). On other occasions, where AFMA has imposed a strategy to achieve a lower catch, it has been through either additional input controls or lowering the TAC, but still to levels above those believed to be too high.

The use of management advisory committees (MACs), made up from representatives with an interest in a particular fishery, to advise on the management of that fishery and to provide advice on research, compliance and financial issues based on five-year plans has also proved problematic. While the MAC framework for effective management exists, the actual implementation of management plans and strategies has been limited.

In the absence of a specific, current and implemented national fisheries policy the uncertainties of what it is that we are managing, and the manner in which it is to be done, remain. There is little doubt that the new Oceans Policy recognises and attempts to address many of these issues within a broader oceans ecosystem context. However, fisheries are but one part of that policy framework and it remains to be seen how the specific issues of policy uncertainty are to be addressed in practice.

Finally, the practical implementation of the national fisheries policy has been further complicated by the division of Australian waters between the Commonwealth and state governments. There remains some uncertainty as to who is to manage what stocks, particularly for those fish that migrate between jurisdictional areas. It is this final problem of multi-jurisdictional control that has resulted in the problems of legal uncertainty.

Legal uncertainty

Uncertainty regarding control over fisheries management in Australian waters and the division in control that still exists have been a major impediment to achieving an efficient and co-ordinated approach to fisheries management. The existence of different management authorities and the absence in many cases of a single jurisdiction for species, irrespective of where and how caught, does little to optimise effective management.

The division of legal control over Australia's fisheries has historically resulted in ineffective management owing to the inability to act across jurisdictional boundaries. It has also allowed controls to be avoided and conservation strategies impeded. The further ability to circumvent fishing controls has not only led to increased pressure on species, but it has also allowed for greater uncertainty with respect to determining accurate stock levels and appropriate management tools.

While the OCS arrangements have clarified many of the legal uncertainties that existed by allowing for the Commonwealth and the states/territories to agree on arrangements to allow single management of a fishery both within and outside state coastal waters, many fisheries still require collaborative management among these governments. The result is that management of Australia's fisheries resources remains a complex mix of Commonwealth and state/territory responsibility. While the Oceans Policy, like many other statements before it, calls for such jurisdictional issues to be resolved and OCS arrangements to be finalised it has ruled out any legislative revision, at this stage, of the current arrangements. In the light of this and of the reluctance by states to hand control over coastal fisheries to the Commonwealth, it is difficult to see how such legal uncertainties will be removed.

Lack of enforcement in Australian fisheries

The effectiveness of any management regime ultimately relies upon the ability to implement and enforce those measures adopted to achieve the desired outcomes. In Australia, substantial efforts have been made towards improving surveillance levels and towards ensuring compliance with and enforcement of the national fisheries management programme. In particular, in the last eight years surveillance-compliance has been subject to significant review and assessment (Commonwealth of Australia 1989; AFMA 1993; Senate Standing Committee 1993). As a consequence, AFMA, whose Operations Branch is responsible for the development and co-ordination of licensing, surveillance and compliance programmes for all its fisheries, has worked closely with state agencies, MACs and industry to address the difficulties that have previously been identified. In many areas self-regulation has delivered positive results.

Nevertheless, while progress has been made, Australian fisheries continue to face enforcement difficulties. There are occasions where MACs, which are often made up largely of industry members, will recommend compliance plans that are more suitable to their own short-term self-interest than to the sustainability of the fishery. While MACs clearly have a valuable role to play in the areas of compliance and enforcement, it is very often the MACs and their members that need to be managed to ensure appropriate outcomes. Furthermore, previous reviews have shown that the powers of fisheries enforcement officers and the levels of penalties that can be imposed are not sufficient to encourage compliance (AFMA 1993; Senate Standing Committee 1993).

Effective enforcement is also compromised by the domestic legal uncertainty that exists. With different management agencies at different levels there has been a lack of harmonisation and co-operation in enforcement methods adopted between states. In view of this, and the lack of single jurisdiction over all fish within the Australian fisheries zone, the opportunities to evade fishing controls have clearly been open to exploitation.

Perhaps Australia's greatest hurdle is the need to police vast areas of exclusive economic zone with limited resources. Not only is there a need to control the national fishing fleet within that zone and to manage catches of foreign fleets in national waters under regional arrangements, but there is also a need to stop continued illegal incursions of foreign fishing fleets. In October 1997, a Royal Australian Navy frigate, following an extensive surveillance operation, caught and seized two foreign vessels poaching icefish and Patagonian toothfish within the exclusive economic zone of Australia's sub-Antarctic waters. A further vessel was apprehended in February 1998. While these arrests were a significant success, the task and cost of policing the entire EEZ of Australia's Southern Ocean is enormous (Bateman and Rothwell 1998).

This is further complicated by the difficulties faced in managing straddling and migratory stocks, such as orange roughy and southern bluefin tuna (SBT), that move between Australia's EEZ and the high seas. In the case of orange roughy the problem was highlighted in July 1999 when two South African vessels were caught fishing spawning stock on the Tasman rise, an area on the immediate edge of Australia's EEZ. Although not in breach of international law, the catch threatened the sustainability of the species, which is managed under arrangements between New Zealand and Australia, the two states bordering the high seas area. After diplomatic protests by Australia to South Africa, the vessels left the area, but not before some weeks of sustained fishing effort causing potentially serious damage to the biomass. In the case of SBT, the recent unsuccessful appeal under the Convention for the Conservation of Southern Bluefin Tuna (CCSBT) to the International Tribunal for the Law of the Sea has only further highlighted the difficulties that exist.

These examples clearly illustrate that Australia has had some success in enforcing the principles of sustainability beyond national boundaries for species that move through national waters. However, they also demonstrate the tremendous difficulties and uncertainties that exist. In particular, unless Australia can control all fishers of a particular stock then sustainability of that stock will not be possible. The challenge is great in respect of fisheries wholly within Australia's EEZ. For fisheries which straddle into the open access area of the high seas the challenges seem almost insurmountable. For example, in the case of SBT, catches by non-CCSBT states now exceed the CCSBT-TAC by 30 per cent, while the taking of orange roughy by third states in the area managed by Australia and New Zealand makes any targets set simply too high (BRS 1998).

Ensuring the sustainability of Australia's fisheries

Many of the problems and difficulties faced in Australian fisheries management have not yet been adequately dealt with, despite the fact that achieving sustainability should, in both theory and practice, be far more readily attainable at the domestic level. Only through addressing those issues of uncertainty identified will it be possible to achieve sustainable fisheries. This includes adopting appropriate management policies, tools and techniques as well as enforcing them. To this extent there are valuable lessons to be learned from international experience.

Scientific uncertainty

Comprehensive assessments of all Australian fisheries, state and Commonwealth, are required in order to determine the current status of all Australian stocks, especially those whose levels are unknown. The current annual Fisheries Status Reports, prepared by the BRS, play a key role in this regard, but significant gaps in Australia's understanding of its stocks remain. Management authorities obviously have an important role to play here.

Environmental impact assessments should also be conducted to determine the impact of fishing not only on particular stocks, but also on their dependent and associated ecosystems. The provisions of the new *Biodiversity and Conservation Act* will assist in this, although Commonwealth fisheries already subject to management plans and state fisheries are exempt from such assessment under the Act. In the absence of accurate data, adopting a precautionary approach to fisheries management, as envisaged by the Oceans Policy, is essential.

The Oceans Policy also commits the government to supporting and enhancing research and development capacity with a view to improving stock assessments, increasing knowledge of the nature of stocks themselves (especially in terms of life cycles), and understanding the impacts of fishing and how best to mitigate them.

Policy uncertainty

In the absence of an accurate understanding of the fishery and the ecosystem in which it is found, management policies and practices that account for this uncertainty are required. While the Oceans Policy outlines new directions in fisheries management policy, a dedicated national fisheries policy, which fits into the Oceans Policy, remains fundamental. The now dated 1989 Fisheries Policy Statement needs to be reviewed and revised. As with the Oceans Policy, the central focus of such a policy should remain ensuring sustainability by regulating resource usage through restricting and/or reducing the fishing effort where it is excessive; exercising precaution where there is uncertainty; and promoting fishing where underutilised resources are thought to exist.

The establishment, under the Oceans Policy, of regional marine plans, based on large marine ecosystems, will for the first time see a shift to a holistic ecosystem- and biodiversity-based management approach. At the practical level, this includes implementing measures to reduce the impact of fishing on seabirds, dugongs, turtles and marine habitats. While historically such measures have tended to be ad hoc and research-focused and have not been part of a general overall ecosystem approach (McLoughlin, Wallner and Staples 1997) some important steps have recently been taken, including a decision requiring the compulsory use of turtle exclusion devices in the Northern Prawn Fishery and finalisation of a by-catch action plan for the fishery. In addition, pelagic longlining was listed in 1996 as a key threatening process under the *Endangered Species Protection Act* because of the impact on seabirds. The listing made obligatory the subsequent introduction of a threat abatement plan, released in August 1998 (BRS 1998), to minimise seabird by-catch.

Furthermore, the Oceans Policy will see the Commonwealth develop and implement a Commonwealth and National Fisheries By-catch Policy as well as other measures to reduce by-catch and implement recovery and threat abatement plans for endangered species. The states/territories and the Commonwealth have already made progress on the development of by-catch policies (BRS 1998).

Practical uncertainty will also be minimised through ensuring that all management decisions with respect to policy direction, policy implementation, research priorities and enforcement requirements are made on an independent basis. While the role played by MACs is a cornerstone of national fisheries management, as no doubt will be the role to be played by Regional Marine Plan Steering Committees under the Oceans Policy, it is necessary that all views are represented. Managing the potential for conflicts of interest, especially in the case of industry, and the potential for MACs to put short-term interests above long-term objectives is fundamental.

Finally, realistic mechanisms and controls which can be implemented, enforced and are capable of achieving sustainability are required. Importantly, the Oceans Policy endorses the use of the more effective output controls, including individual transferable quotas (ITQs), and the reduction of excess capacity.

Legal uncertainty

The achievement of single jurisdiction control over fish species to be managed remains Australia's greatest legal obstacle to providing successful fisheries management. With little real progress having been made in this area, it is significant that the Oceans Policy will seek to ensure greater cross-jurisdictional management of fisheries through initiating a review of all fisheries laws and regulations and through further review and revision of the OCS arrangements.

Improving enforcement in Australian fisheries

As indicated previously, while there has been substantial improvement in the enforcement of Australia's national fisheries management programme, difficulties remain.

At a practical level Australia is committed to improving surveillance and enforcement, both through legislative reform and increased resources (Commonwealth of Australia 1998: Vol 2, 10–11). However, the specifics of such commitment remain to be determined. Improved catch data recording and logbook reporting, together with landing, transport and fish receiver records, are crucial if accurate stock assessments and catch levels are to be determined (ANAO 1996: 110–115). It is also necessary for the federal government to make the necessary legislative amendments to increase the powers of fisheries officers and to improve the enforcement capabilities of AFMA. In this respect finalisation of the OCS arrangements is required to enable a single enforcement agency to operate.

Importantly, the Oceans Policy specifically states that the government will both increase fishing patrols in the sub-Antarctic and the waters of northern Australia and increase the number of fisheries officers to control illegal fishing. It will also amend the fisheries laws to make surveillance of foreign fishing more effective and to strengthen and better enforce its international commitments in a way that will ensure stock sustainability in high sea areas (Commonwealth of Australia 1998: Vol 2, 11).

Complete enforcement is unrealistic. Effective enforcement for sustainability is, however, achievable. Ultimately, the level of enforcement will depend upon the extent to which the Oceans Policy objectives of sustainability and ecosystem management are sought. Actual enforcement will depend upon the mechanisms adopted and the funding levels and cost recovery measures that exist. Nonetheless, improvements in domestic measures such as satellite tracking and the adoption of certified catch schemes and the actions taken against illegal fishers in Australian waters demonstrate a clear commitment to protect the sustainability of our national resources. The key is ensuring that such commitment continues to be implemented.

Conclusion

For international fisheries, long-term sustainability requires both an integrated holistic approach and the wider participation of all fishing and coastal nations in relevant fisheries organizations and arrangements to ensure that approach. In addition, the mandates of these organizations and arrangements must be strengthened to ensure their ability to meet the objectives of sustainability. Nevertheless, without adequate enforcement of conservation and management measures, both on the high seas and within areas under national jurisdiction, the decimation of global fisheries resources will continue, with its attendant adverse consequences for political, economic and food security. Long-term sustainability of international fisheries thus also requires the development of new and effective enforcement mechanisms, both legal and practical, beyond the reach of which fishing vessels and their flag states cannot hide.

For Australia, as for all other coastal nations, effective, sustainable, long-term fisheries management will only be achievable if policy objectives, and the way in which those policies are implemented, deliver ecologically sustainable outcomes. Policies and practices that provide some certainty as to stock levels, and that present an understanding of the broader environmental impacts and relationships within and between fisheries and ecosystems, are fundamental. Only then will it be possible to manage fisheries along safe exploitable levels upon which sustainability can be realized. The new Oceans Policy is clearly a step in the right direction.

Most important of all for all global fisheries, both international and national, sustainability requires the political will and fortitude to fully implement the precautionary approach and to make the difficult decisions required.

References

Australian Bureau of Agriculture and Resource Economics (1996) *Australian Fisheries Survey Reports 1995*. Canberra: Australian Bureau of Agriculture and Resource Economics.

Australian Fisheries Management Authority (AFMA) (1993) *Review of Compliance and Observer Programs (Internal Report)*. Canberra: Australian Government Publishing Service.

Australian Fisheries Management Authority (AFMA) (1996) *Annual Report 1995–96*. Canberra: Australian Fisheries Management Authority.

Australian National Audit Office (ANAO) (1996) *Audit Report No. 32 1995–96: Performance Audit on Commonwealth Fisheries Management & Australian Fisheries Management Authority*. Canberra: Australian Government Publishing Service.

Bateman, S. and Rothwell, D. (1998) *Southern Ocean Fishing: Policy Challenges for Australia*, University of Wollongong.

Bergin, A. and Michaelis, F.B. (1996) Australia, the South Pacific and UNCED's oceans agenda. In: Krikowen, L.K. *et al.* (eds) *Oceans Law and Policy in the post-UNCED Era: Australian and Canadian Perspectives*, 289–311.

Borghese, E.M. and Saigal, K. (1995) Managerial implications of sustainable development in the oceans. In: *Ocean Governance and the United Nations*. Halifax: Centre for Foreign Policy Studies, Dathousie University.

Bureau of Resource Sciences (BRS) (1998) *Fishery Status Reports 1997: Resource Assessments of Australian Commonwealth Fisheries*. Canberra, Bureau of Resource Sciences. (December).

Burke, W., Freeberg, M. and Miles, E. (1994) United Nations Resolutions on Driftnet Fishing: An Unsustainable Precedent for High Seas and Coastal Fisheries Management in 25. *Ocean Development and International Law* 127.

Cameron, J. and Aboucher, J. (1991) The precautionary principle: a fundamental principle of law and policy for the protection of the global environment. *Boston College International and Comparative Law Review*. XIV (1).

Commission on Conservation of Antarctic Marine Living Resources (CCAMLR) (1995) Report of the Fourteenth Meeting of the Commission. Hobart: CCAMLR.

Commonwealth Government (1989) *New Directions for Commonwealth Fisheries Management in the 1990s: A Government Policy Statement*. Canberra: Australian Government Publishing Service.

Commonwealth of Australia (1989) *Surveillance and Compliance in the Fishing Industry*. Canberra: Australian Government Publishing Service.

Commonwealth of Australia (1992) *National Strategy for Ecologically Sustainable Development*. Canberra: Australian Government Publishing Service.

Commonwealth of Australia (1998) *Australia's Ocean Policy*. Canberra: Australian Government Publishing Service.

Convention on the Conservation and Management of Pollock Resources in the Central Bering Sea (1994) *International Legal Materials* 67: 34.

Doulman, D. (1996) An overview of world fisheries: challenges and prospects for achieving sustainable resource use. *Law of the Sea Institute Proceedings*.

Food and Agriculture Organisation (1991) *Fish for Food and Development*. Rome: FAO.

Food and Agriculture Organisation (1992a) *The State of Our Knowledge on High Seas Living Resources and their Controlled Exploitation*. Doc FI/HSF/TC/92/3. Rome: FAO.

Food and Agriculture Organisation (1992b) *High Seas Management: New Concepts and Techniques*. Doc. FI/HSF/TC/92/5. Rome: FAO.

Food and Agriculture Organisation (1993a) *Conservation and Rational Utilisation of Living Marine Resources with Special Reference to Responsible Fishing*. Doc. COFI/93/5. Rome: FAO.

Food and Agriculture Organisation (1993b) Agreement to Promote Compliance with International Conservation and Management Measures by Fishing Vessels on the High Seas. Rome: FAO.

Food and Agriculture Organisation (1995a) Code of Conduct for Responsible Fisheries. Rome: FAO.

Food and Agriculture Organisation (1995b) *Rome Consensus on World Fisheries*. Rome: FAO.

Food and Agriculture Organisation (1997) *Strengthening FAO Regional Fishery Bodies*. Doc. COFI/97/4. Rome: FAO.

Food and Agriculture Organisation (1998) *The State of the World Fisheries and Aquaculture (SOFIA)*. Rome: FAO.

Garcia, S.M. (1994) The precautionary approach to fisheries with reference to straddling fish stocks and highly migratory fish stocks, FAO Fisheries, No. 871, FIRM/C87, Rome: FAO.

House of Representatives Standing Committee (1996a) House of Representatives Standing Committee on Primary Industry, Resources and Rural and Regional Affairs, of 1996–1997: *Submissions to the Inquiry into the Management of Commonwealth Fisheries*. Canberra: Commonwealth of Australia.

House of Representatives Standing Committee (1996b) House of Representatives Standing Committee on Primary Industry, Resources and Rural and Regional Affairs, of 1996–1997: *Transcripts of Evidence presented to the Committee's Inquiry into the Management of Commonwealth Fisheries*. Canberra: Commonwealth of Australia.

House of Representatives Standing Committee (1997) House of Representatives Standing Committee on Primary Industry, Resources and Rural and Regional Affairs of June 1997: *Managing Commonwealth Fisheries: The Last Frontier*. Canberra: Commonwealth of Australia.

Industry Commission (1992) *Report on Cost Recovery for Managing Fisheries*. Canberra: Australian Government Publishing Service.

Johnston, D.M. (1995) Stresses and mind-sets in fishery management. *Dalhousie Law Journal* 18(1): 154–67.

Kyoto Declaration and Plan of Action on the Sustainable Contribution of fisheries to Food Security (1995) London: World Wildlife Fund.

Maguire, J.-J., Neis, B. and Sinclair, P.R. (1995) What are we managing anyway? The need for an interdisciplinary approach to managing fisheries ecosystems. *Dalhousie Law Journal* 18(1): 141–53.

Marashi, S.H. (1996) Regional fisheries organisations: mandates, operations and issues. *Law of the Sea Institute Proceedings.*

McLoughlin, K., Wallner, B. and Staples, D. (1994) (eds) *Fishery Status Reports 1993: Resource Assessments of Australian Commonwealth Fisheries.* Canberra: Bureau of Resource Sciences.

McLoughlin, K., Wallner, B. and Staples, D. (1997) (eds) *Fishery Status Reports 1996: Resource Assessments of Australian Commonwealth Fisheries.* Canberra: Bureau of Resource Sciences.

Paul, L. (1996) Expanding awareness: emerging approaches to fisheries management. *Law of the Sea Institute Proceedings.*

Rose, G. (1999) Marine biodiversity protection through fisheries management – legal developments. *Review of European Community and International Environmental Law* 8(3).

Sands, P. (1995) International law in the field of sustainable development. In: *The British Yearbook of International Law 1995.* Oxford: Clarendon Press.

Senate Standing Committee (1993) *Fisheries Reviewed.* Inquiry by the Senate Standing Committee on Industry, Science, Technology, Transport, Communications and Infrastructure on the adequacy of the Commonwealth's fisheries legislation. Canberra: Australian Government Publishing Service.

Sutton, M. (1996) Reversing the crisis in marine fisheries: the role of non-governmental organisations. *Law of the Sea Institute Proceedings.*

Treaty on Fisheries between the Government of Certain Pacific Island States and the Government of the United States of America (1987) Reproduced in *International Journal of Marine and Coastal Law* 3(1): 60.

UNCED (1992) *Agenda 21: Programme of Action for Sustainable Development, Rio Declaration on Environment and Development, Statement of Forest Principles.* The final texts of agreements negotiated by governments at the United Nations Conference on Sustainable Development (UNCED) 3–14 June 1992, Rio de Janiero, Brazil. New York: United Nations.

United Nations (1979) Convention on Conservation of Migratory Species of Wild Animals. New York: United Nations Publishing.

United Nations (1982) Convention on the Law of the Sea. New York: United Nations Publications.

United Nations (1992a) *Convention on Biodiversity.* New York: United Nations Publications.

United Nations (1992b) Convention on the Conservation of Anadromous Fish Stocks in the North Pacific. New York: United Nations Publishing.

United Nations (1995) UN Agreement for the Implementation of the Provisions of the Law of the Sea Convention of 10 December 1982, Relating to the Conservation and Management of Straddling Fish Stocks and Highly Migratory Fish Stocks.

Wilder, M. (1995) Quota systems in international wildlife and fisheries regimes. *Journal of Environment and Development* 4(2) (Summer).

8 | Sustainability, uncertainty and environmental policy: lessons from New Zealand's pastoral high country

Michael Harte and Janet Gough

Introduction

The management of the pastoral high country of New Zealand's South Island has reached a turning point. Increasing bare land, declining fertility, invasion by weeds and pests, and reduced floral and faunal diversity over much of the high country are symptoms of unsustainable land uses. It is generally recognized that fundamental changes are required to prevent further environmental degradation and to ensure that resources are preserved for future generations. Actions now being taken by diverse groups such as landholders, local government, research agencies, conservation groups, and central government departments aim to secure a more sustainable future.

There is no one policy path to achieving the goal of sustainability in the South Island pastoral high country. Developing appropriate policies requires an understanding of the many dimensions of sustainability, including high country environmental processes, land use practices, and the varied social and institutional factors that help shape environmental outcomes.

Addressing these different dimensions of sustainability would be a relatively straightforward task if policy makers had a reasonable degree of certainty about the likely outcomes of potential policy actions. Unfortunately, the high country policy problem is complicated by diverse ecological structures and processes, differing land use practices, and varied social and institutional factors. As a consequence, it is fair to say that uncertainty – rather than certainty – is the norm in all but the most trivial of high country policy issues. Making effective policy decisions in the face of uncertainty requires that policy makers focus not just on the biophysical environment, but also on the institutions and processes that form the policy environment in which the sustainability issue is debated and initiatives developed.

In the remainder of this chapter we present a case for recasting existing institutional arrangements and policy processes so that the implications of uncertainty are given greater prominence in the development of policies for the sustainable management of the high country. We begin by describing the major biophysical, social, and institutional influences shaping the high country since its settlement by humans. This 'contextualization' of the sustainability problem sets the scene for discussion of the implications of uncertainty for sustainable management and the policy adjustments we believe necessary to better promote the sustainable management of the high country.

The South Island pastoral high country: an evolving social, institutional and ecological mosaic

In this section we attempt to unravel the complex web of ecology, social forces and institutional factors that have shaped the landscape of the South Island high country since its settlement by humans. This description, divided into three time periods – pre-1948, 1948–1990, and 1990 to the present – sets the scene for our subsequent exploration of the policy issues associated with the sustainable management of the pastoral high country.

Pre-1948

The first humans to settle the South Island high country were the Maori around AD 1200 (Davison 1986). The earliest recorded South Island Maori were the Kahui Tipua. The land that greeted these early inhabitants was heavily vegetated, with plant associations dominated by tussock and herb fields at the highest elevations, by southern beech at mid-elevations, and podocarp species at the lowest elevations. The diverse flora and fauna of the high country's basins provided rich areas for hunting and gathering. Parts of the high country were also a source of pounamu (also called greenstone or nephrite jade). This was used to make highly prized taonga (treasures) (Park 1996).

By the time Europeans first visited New Zealand, the Ngai Tahu occupied most of the South Island. The landscape utilized by the Ngai Tahu was markedly different from that first settled by the Kahui Tipua 600 years earlier. Human-induced fires had destroyed much of the forest, leaving only remnant stands, and tall tussock grass had taken the place of trees. Many species of large flightless birds were extinct and the Maori had turned to predominantly coastal sources of staple foods.

Prior to 1840 European settlers in New Zealand were chiefly involved in sealing, whaling and logging. After 1840 there was a concerted attempt to turn New Zealand into an agrarian economy by the colonial administration in London. While early provincial governments encouraged new settlers to purchase fertile low country land, they adopted a cautious approach to the alienation of 'waste land'. This meant that most of the pastoral land of the South Island high country remained under the jurisdiction of the Crown.

The first high country pastoral runholders (ranchers) came from England and Scotland and were a mixture of adventurers, innovators and speculators. To become runholders they needed sufficient money to buy livestock, because the government required the runs to be stocked within a few months of being taken up. Sheep were grazed on the sunny hillsides in summer and in the valleys in winter. O'Connor (1982, 1983) describes the method of pastoralism used at this time as exploitative because the land resources were used without any thought for replenishment or maintenance of nutrients. Fire was an essential management tool used for both access and for controlling feed supply. As well as fire, the early settlers introduced animals, all of which would have an impact on the environment, and some of which would later become major uncontrollable pests: rabbits, deer, goats, sheep and possums.

Early high country settlement also had a speculative nature, partly because of the high price of wool in the mid-1800s. Runs (ranches) therefore changed hands frequently, as investors bought and sold land. A number of runs were also taken up and lost because of bad conditions, inexperience, or because early settlers were lured to the gold fields of Australia by the promise of easier and greater wealth. The runholders who stayed through the 1850s and 1860s became wealthy and gained considerable political power.

Government attempts to rationalize land tenure arrangements in the late 1800s caused instability in the management of the high country. The *Land Act* of 1877 established a national policy for pastoral leases, with licences being granted for a maximum of 11 years with no right of renewal. At the end of the term they were auctioned to the highest bidder. An amendment in 1882 increased the maximum term to 21 years but had little practical effect. The *Land Act* of 1882 also imposed an obligation on licensees to control rabbits and weeds, with severe financial penalties for failing to do so. In 1889 when the first runs came up for auction in accordance with the provisions of the 1877 Act, economic and ecological conditions meant that existing runholders faced little competition at auction. Coupled with the destabilizing influence of tenure reform, other economic, environmental, and social factors increased difficulties for high country pastoralists in the last decades of the century (Eldred-Grigg 1980). Wool prices fell, heavy snow storms caused stock losses, and the development of refrigerated shipping to Europe reduced the attractiveness of high country pastoralism relative to other forms of agriculture. The pastoral runholders would have also viewed the land reforming policies of government during the 1890s with a sense of concern and insecurity (Centre for Resource Management 1983).

In the early 1900s social conditions began to change in the high country. Higher incomes for farm workers and the decreasing attractiveness of high country farming meant that shepherds and other high country workers were able to become runholders and managers. Their knowledge of the high country and improving farm practices meant that, while never highly profitable, high country pastoralism slowly became more stable and settled.

Severe ecological degradation resulting in the virtual eradication of some tussock grassland species and increased flooding was not widely recognized until the 1920s. By then, deer, rabbits and overgrazing of sheep had reduced vegetation and soil fertility to the extent that a number of the major runs, or stations, had become unprofitable. Loss of protection meant that water, wind and temperature extremes seriously depleted the topsoil, causing slips and making farming impossible. Recognition of these problems led to a major change of direction in the management of the high country in the 1940s and 1950s.

Before the 1940s the dominance of extensive pastoralism in the South Island high country was unquestioned. Tourism and recreation, although long associated with the high country, were never on a sufficient scale to compete or conflict with pastoralism. Recreational hunters assisted runholders in their statutory responsibilities to control deer and other pests. Forestry was recognized as an alternative land use for the lower areas of the high country as early as the turn of the century, but social and political pressure meant pastoralism remained the dominant land use.

1948–1990

A variety of legislative moves in the 1940s bought about a substantial institutional change in the management of the South Island pastoral high country. For the first time government policy began to acknowledge the multidimensionality of ecosystem degradation in the high country. This was demonstrated in the enactment of major pieces of legislation such as the *Soil Conservation and Rivers Control Act* in 1941, and the *Land Act* in 1948.

Although considerable concern about the state of the high country and its impact on lower areas through increased flooding and decreased water quality had been expressed for a number of years, little was done to alleviate the situation before the 1940s. There were a number of reasons for this, including the effects of the Depression and war, a lack of capital, and insufficient knowledge about the ecosystems involved.

During the 1930s and 1940s soil conservation, or management of the land to prevent soil erosion, reduce runoff and protect water quality, became recognized as an important aspect of land management in the South Island high country. In 1941 the passage of the *Soil Conservation and Rivers Control Act* linked the administration of soil conservation with river control and drainage activities. These initiatives marked some of the first attempts to underpin land management systematically with scientific principles.

The *1948 Land Act* was the first major reform of the land tenure system since 1882. The concept of the pastoral lease was applied to land suitable or adaptable for pastoral purposes only. Leases were granted for 33 years (with perpetual right of renewal) and initially a rental was fixed for the term of the lease. The most significant aspect of this Act was the removal of the right to freehold pastoral land. The removal of the right to freehold went against the general intent of the Act, which was to provide much greater freedom to own other agricultural land outright. The exception of pastoral land was motivated by doubts about its long-term future (or sustainability in current terms). The Minister of Lands is recorded as saying during the Land Bill's second reading in Parliament:

> If there is any doubt as to the suitability of the land for permanent aliena-
> tion, obviously the Crown must retain some control over it. That is why
> there is no right of purchase in these hill country leases called pastoral
> licences (New Zealand Parliamentary Debates, 1948, Vol 284, p. 3999).

The development of a land use capability mapping system during the 1950s gave land managers new information about high country soils. In

1959, a new *Soil Conservation and Rivers Control Act* gave catchment authorities the right to limit or prohibit land occupiers from grazing at certain times, and from the early 1960s destocking of susceptible land identified by the land classification scheme was introduced. The 1970s saw extensive retirement of the poorest land from pastoralism. By 1979, 300 000 hectares had been retired and a further 400 000 hectares had grazing restrictions in place (Whitehouse 1985).

At the same time as the most degraded high country soils were being removed from pastoral use, technological and scientific advances provided for the intensification of land use on soils of higher quality. Developments such as aerial topdressing and oversowing resulted in dramatic pasture improvement. Newly trained soil conservators demonstrated the benefits of tree planting, and the removal of sheep and rabbits. They also undertook experiments to determine the amount of sediment and debris carried by mountain streams, and the subsequent effect on lower reaches of the main rivers. While these original soil conservators were trying to enhance the productivity of the high country in order to maximize pastoral production, many were also ecologists and recognized and recorded the damage that had been done to native plants and ecosystems.

The adoption of a science-based land management regime initially proved successful and high country productivity increased. The impact of rabbits and other pests and weeds became less evident, and by 1979 it was claimed that rabbits were under control in parts of the high country (Whitby 1979). Recent experience with RCD (rabbit calicivirus disease) in New Zealand has followed a similar path with gains from initial dramatic reductions in rabbit numbers being eroded by lack of appropriate follow-up management strategies.

Countervailing economic and social forces subsequently eroded the gains made through improved land management practices. New Zealand's high standard of living throughout the 1950s and 1960s had been based on the export of primary commodities, especially wool and dairy products. However, following the oil price shocks of the early 1970s, farm commodity prices fell and, in order to generate much needed foreign exchange earnings, the government of the time put in place subsidies, incentives and loan schemes intended to maintain agricultural exports. High country pastoralists responded to these measures by increasing stock numbers.

From the early 1980s major economic reforms and institutional changes in New Zealand had dramatic impacts on farming and land use management. Following the recognition that the continued subsidization of agriculture was a major and unjustifiable drain on public funds, farm subsidies were removed. This caused considerable hardship for many farmers. As farming income and profits dropped, the cost of improvement of land became prohibitive. Farmers seeking to reduce farm costs and boost incomes kept stocking rates high while reducing investment in pasture improvement. Almost overnight, ecological degradation accelerated in the pastoral high country. The latest in a series of pests, *Hieracium*, commonly known as hawkweed, began to invade the high country.

1990 to the present: the *Resource Management Act*

Today, pastoralism is no longer the land use of choice in the South Island high country. Changes in relative commodity prices mean that forestry, tourism, recreation, and conservation are now viable, profitable uses for the high country. Each has differing social, ecological, and economic dimensions and the present debate over the future of the high country reflects this multidimensionality. The institutional context in which this debate takes place has also changed significantly since 1990.

The most significant piece of environmental legislation in New Zealand's history was passed in 1991. The *Resource Management Act 1991* replaced over 60 pieces of disparate legislation and changed the institutional environment in which environmental management is practised. Significant resource management responsibilities were devolved from central government to local government. This was done with the expectation that local government would have a greater understanding of local environmental and social conditions than central government and would therefore make more informed resource management decisions.

Unlike previous environmental management legislation, the *Resource Management Act* neither requires nor sanctions an approach whereby a single activity such as pastoralism is singled out for special treatment, unless that approach can be justified in terms of effects on the environment. The Act requires all persons exercising powers and functions under it to achieve the purpose of the Act. The purpose is set out in Section 5 which states:

> **5. Purpose-**
> (1) The purpose of this Act is to promote the sustainable management of natural and physical resources.
> (2) In this Act 'sustainable management' means managing the use, development and protection of natural and physical resources in a way, or at a rate, which enables people and communities to provide for their

social, economic, and cultural wellbeing and for their health and safety
while –

(a) Sustaining the potential of natural and physical resources (excluding
minerals) to meet the reasonably foreseeable needs of future generations;
and

(b) Safeguarding the life-supporting capacity of air, water, soil and ecosystems; and

(c) Avoiding, remedying or mitigating any adverse effects of activities on
the environment.

Achieving the purpose of the Act requires that people and communities
should be allowed to use the resources of the high country in ways that
achieve their own ends, provided they operate within the constraints set
by subsections 5(2)(a)–(c). Identifying what constitutes the environmental
constraints therefore becomes critical in determining how the high country
is managed. Unfortunately, determining these environmental constraints
is not always straightforward. Environmental problems in the South Island
high country are characterized by a diversity of ecological structures and
processes and conflicting social, economic, and ecological objectives. Problems, their context, and their solution may change over time and in
response to each other.

Although the *Resource Management Act* has changed the way in which
the environmental effects of high country land use are considered, it is
deliberately indifferent to economic and land tenure considerations. Recognizing the limitations of the present tenure system, especially pastoral
leases, the government is currently reviewing the *Land Act 1948*. The
present pastoral lease arrangements result in neither the Crown nor the
lessee having full responsibility for their land management practices. As
a consequence, there is a diminished responsibility for the sustainable
management of the high country – regardless of the *Resource Management Act*.

The basic idea behind the proposed land tenure reforms is to make some
leasehold land freehold, thereby partitioning pastoral land into publicly
owned conservation land and privately owned commercial land. By assigning private property rights to occupiers of land capable of sustained
commercial production, the government intends to improve their economic incentives to manage the land well and to make them accountable
for the management of their land. It is expected that the pastoral lease
land suited to privatization will be land of high quality on which farming, forestry and other commercial land use practices can be sustained.
The reform of land tenure is supported by high country runholders and

has the backing of influential lobby groups including Federated Farmers and the Business Roundtable.

The move by the government to reduce its involvement in the high country has met with substantial opposition from other public interest groups including conservation organizations such as the Royal Forest and Bird Protection Society of New Zealand and recreational groups such as Federated Mountain Clubs (Sage 1995). There is considerable concern about the future health of the high country ecosystems and alpine vegetation if they are allowed to be held in private hands. There is also doubt that an underfunded Department of Conservation will be able to manage large amounts of additional conservation land. Despite government assurances that there will be little change to the status quo, there is widespread public belief that access to the high country will be more restricted and that future generations will lose the resource. In addition to concerns about private owners of high country land restricting access, there is a concern that if the Department of Conservation is left with large areas of degraded land and insufficient funds to manage that land it will also restrict access.

Another important aspect of policy development for the sustainable management of the high country is the role and status of tangata whenua (people of the land). Ngai Tahu have had a land claim over much of the South Island high country settled recently, giving the tribe a much stronger role in its management. Maori must also be consulted as a matter of course by the government over issues related to the Treaty of Waitangi. In the absence of any written constitution, the Treaty of Waitangi, signed in 1840, is one of New Zealand's most important constitutional documents. Under the English version of the treaty, Maori ceded sovereignty to the Crown in return for the full rights of ownership of their lands, forests, fisheries and other prized possessions, as well as the rights and privileges of British subjects (Orange 1987). In terms of the Maori version, the Crown was accorded *Kawanatanga*, or governorship. However, this right is constrained, in that the right to rule over Maori interests in natural resources is permitted only to the extent necessary for the common good.

It is still too early to tell whether recent environmental management and proposed land tenure reforms will reverse or are capable of reversing 1000 years of ecosystem degradation. It is clear from past experiences, however, that if sustainable management of the South Island high country is to be achieved, the sustainable management problem needs to be placed in a broader policy framework; one that recognizes the difficulty we have in acknowledging the damage we do to ecosystems before that damage becomes irreversible. The next part of this chapter will argue that short-sighted policy arises from a failure to recognize the indeterminate nature of the human–environment relationship.

Despite the pervasive nature of uncertainty and indeterminacy in the high country policy debate, policy makers have considered its implications for the development of sustainable land management policies at a fairly superficial level. It is usually assumed that careful and considered use of adequate scientific knowledge will give rise to sustainable land management practices. In the development of policy, this means giving primacy to the quantifiable findings of science and secondary consideration to other policy dimensions such as the links between uncertainty and sustainability. But because of their potential influence on sustainable land management, such 'secondary' considerations deserve considerably more attention than they currently receive.

Uncertainty, sustainable management and policy development

It is difficult to provide a precise characterization of uncertainty. Uncertainty is not so much an entity in its own right, but a description of instances where there is doubt about some dimension, or dimensions, of a suspected outcome. It commonly refers to an absence of information regarding: (1) the effects of a particular action, event or circumstance; and/or (2) the likelihood of an event, circumstance or action occurring. Recently, a large amount of work has been carried out exploring the concept of uncertainty and classifying its different manifestations (see Dovers and Handmer 1995; Faber *et al.* 1992; Funtowicz and Ravetz 1990, 1992, 1993; Smithson, 1989; and Wynne 1992a). Wynne (1992a), for example, has developed the following taxonomy of uncertainty:

- *Risk*: system behaviour is essentially known, and outcomes can be assigned a probabilistic value.
- *Uncertainty (or scientific uncertainty)*: significant systems parameters are known, but not the probability distributions.
- *Ignorance*: what is not known is not known. Ignorance increases when the degree of action or commitment based on what we think we know increases.
- *Indeterminacy*: causal links, networks and/or processes are open, and therefore defy prediction.

Recognition that certainty in environmental policy development is rarely possible is an important first step in dealing with uncertainty. However, it is just as important that we also acknowledge potential sources of uncertainty as well as its existence. Without exploring the origins of uncertainty it is difficult to develop workable policy responses.

The generally well developed and robust theories available to applied science practitioners are largely absent in research related to environmental policy. Policy analysts and decision makers have few metatheories to guide practice. Policy research is also typically interdisciplinary. Many environmental policy issues involve human health, ecological, and economic issues. Different disciplines have very different theoretical, empirical and social aspects. Researchers and policy makers must often borrow from unfamiliar fields of study, meaning they cannot make the same judgements about information quality that they would in their own areas of specialization. In many environmental policy contexts the result is considerable uncertainty. Moreover, although policies for sustainable management of the environment are often assumed to be underpinned by robust ecological principles and evidence, ecology is often unable to provide the degree of guidance sought by policy makers. Many of ecology's theoretical foundations have been challenged in recent times (see, for related discussions, Sagoff 1985; McIntosh 1987; Botkin 1990; Peters 1992; Pickett et al. 1992; Caldwell and Shrader-Frechette 1993; Demeritt 1994; Shrader-Frechette and McCoy 1994; Zimmerer 1994). Ecology has neither unambiguous definitions of basic concepts such as species, niche, or ecosystems, nor a testable general theory capable of providing predictive or prescriptive guidance to environmental managers and analysts. Emerging ecological debates, both theoretical and empirical, also tend to undermine the ecological rationale for management practices based around preconceived notions of sustainability. It is often not possible to identify a single organizational state of an ecosystem that corresponds to sustainability. Instead, there may be a range of states that can be sustained with different levels or types of management effort.

The problems created by such uncertainties are heightened by the need for policy makers to be accountable to the public at large. Policy is judged by its performance in the 'real world'. Few people are willing to accept uncertainty as an excuse for failed policy. It is expected that analysts and decision makers will already have factored in relevant considerations. But eventual policy outcomes are influenced by a multitude of factors, many of them uncertain or indeterminate.

Policy issues associated with the sustainable management of the high country are vivid examples of uncertainty at work. Policy conclusions

Table 8.1 Some key sources of uncertainty.

Economic[1]	Ecological[2]	Scientific[3]
■ Uncertainties often exist about the ownership of, and access to, resources. ■ There are potential spillovers from other activities. ■ Natural resource investments tend to be long-term investments. ■ Natural resource commodity prices fluctuate more and are more difficult to forecast than other commodity prices. ■ Most resource commodities are at greater risk of being replaced by cheaper substitutes developed by cheaper but unpredictable technological change.	■ Ecosystems exhibit a high degree of complexity in structure and function. ■ Ecosystems frequently exhibit both hidden resilience and threshold responses. ■ There are often significant time lags between an event and its effects. ■ Effects are often cumulative and synergistic. ■ Ecological relationships are spatially and temporally contingent and not homogeneous.	■ An inability to conduct well-designed field experiments and long-term studies at all scales of ecological study, but especially at the ecosystem level. ■ Uncertainties in the statistical analyses used in ecological studies. ■ Complications arising from the extrapolation of laboratory studies to the real world. ■ Limited predictive ability of ecology because of (1) the extreme complexity of natural systems, and (2) an inability to verify and validate predictive models

[1] See Panayotou (1993).
[2] See Peters (1992); Jorgensen (1990); Harte (1995).
[3] See, for example: M'Gongle et al. (1994); MacGarvan (1993).

reached in the past have been notable for their superficial assumptions about complex ecological, social, and economic processes. These assumptions have received little critical review, despite considerable uncertainty and often indeterminacy and ignorance. Some of the possible causes of uncertainty in the high country policy development process are presented in Table 8.1.

It is not possible to make absolute statements about many of the ecological processes in the high country because of their considerable spatial and temporal variation. Tussock grasslands are inherently unstable dynamic systems. The present composition and structure of these grasslands are the products of disturbances such as grazing, burning and droughts.

Vegetation changes and invasions are to be expected. It is likely that, were it not for the pressures of burning and grazing, some of the modified lands would revert to a woody species-dominant vegetation. The recent rapid spread of *Hieracium* species in other areas may indicate a shift from one system to another.

It is also unclear that there is any one ecological state that can be defined as sustainable, or identified as the goal of sustainable land management. Demands by some environmental groups to return the high country to some sort of pristine state are simplistic and unrealistic. The evidence for ecosystem recovery in the high country when grazing pressure is removed is mixed. Kerr and Treskanova (1993) point to the return of indigenous (and other exotic) vegetation species at the expense of *Hieracium* on one high country run once sheep were removed and rabbits controlled in 1991. (In the 1950s *Hieracium* was controlled on Molesworth Station by applying large amounts of fertilizer and grazing cattle in the area.) O'Connor (1983), however, suggests that biological conservation by exclusion of livestock and fire may be interesting ecologically but is often futile in conserving native flora and its dependent fauna. O'Connor points out that exotic shrub and tree species may invade grasslands under low grazing intensity in many localities and that other exotic elements already present may end up dominating.

The indeterminate nature of ecological processes in the high country makes the development of generic land management policies difficult. At one level, environmental analysts and policy makers should ideally be aware of (1) controlling system processes; (2) the context of system processes; (3) the historical range of system perturbations; (4) the evolutionary and physiological limits of system components; and (5) the characteristics and effects of short- and long-term biophysical phenomena. At another level it may be appropriate to devolve many land management responsibilities to individual land managers and to encourage the practice of adaptive management. This involves, for example, initiatives designed to promote self-monitoring programmes by land managers that have the potential to help individuals responsible for short-term localized land management to establish the links between different management techniques and the health of ecosystems. The rural community, aided by central and local government, is already taking initiatives to ensure more sustainable land management practices. The self-monitoring programmes discussed above are being developed and promoted by organizations like the Rural Futures Trust (Rural Futures Trust 1996), which are also promoting other community-based initiatives to ensure sustainable land management practices.

Improving the policy development process

To prevent repeating past environmental policy failures in the South Island high country we believe that two metapolicy adjustments are necessary. First, policy makers, land users, owners, and researchers need to acknowledge that sustainability is fundamentally a social concept. Second, sustainable land management policy initiatives must recognize and explicitly take into account the existence of uncertainty and indeterminacy.

Sustainability as a social concept

Although sustainable management is the guiding concept for land management in the high country, its translation into practical policies is proving difficult. It has been argued that it is usually not possible to conclusively identify any overall ecological state as sustainable because high country ecosystems are dynamic and their structure, function and composition are constantly changing in time and space, both in the short term and in an evolutionary sense. This does not imply that sustainability in the high country is an illusion, nor is it suggested that there are not very real ecological limits to the use of the high country. Nevertheless, there is a need to look beyond the biophysical dimension of sustainability to make real progress in developing effective policies for sustainability. The present resources accorded ecological research in the high country need to be redirected, and their role needs to change from being the driving force behind policy development to one of informing it.

To regain some of its usefulness as a practical policy objective, sustainability in the high country needs to be recast as a social concept. Past high country policies and associated social and institutional arrangements reflected a preference for a high country landscape dominated by extensive pastoralism and the economic, environmental and amenity services it offered. People were willing to accept the degradation of indigenous ecosystems to allow the expansion of this preferred landscape. Recent changes in collective preferences, together with changing economic incentives and a better understanding of ecological and economic realities, have forced a rethink of expectations for the high country, and we are now in the process of creating a new mix of high country land uses.

The role of the policy maker becomes one of developing policies and institutions that facilitate those land use choices that fit with the ecological realities of the high country and discourage those activities that have long-term adverse impacts on the life-supporting capacity of ecosystems. The legislative and institutional reforms, research, creation of community groups and other policy measures used to guide choices should not, however, be seen as ends in themselves. They are simply potential aids to achieving sustainable management of the high country. For these techniques and others to be successful in reversing the decline of the natural values of the high country they must operate within a framework where the overall guiding principle is the sustainability of resources.

Without an understanding of the social dimension of sustainable management it is unlikely that government policy will be structured in a way that promotes solutions that are ecologically sound, socially acceptable and politically supported. For example, unconvinced by the economic and environmental arguments presented, many individuals and organizations still distrust the government's motives for the current reform of land tenure in the high country and are reluctant to be involved in the policy development process. Thus, paradoxically, an attempt to remedy one aspect of an environmental problem may reduce the likelihood of achieving sustainable management, due to an increased polarization of society's values and attitudes toward the environment, and the government's role in its management.

A number of recent initiatives may improve future policy analysis of the social dimension of sustainable land management. Recently, for example, it has been recommended that the government initiate a national science strategy for sustainable land management. One of the major recommendations of the Strategic Group on Sustainable Land Management Research is that social science be a major focus of a national science strategy. The group also recognized the need for:

> consultation, participation and partnerships among policy makers, land users, owners, scientists and educators at all stages of the science process. This partnership is essential to attain sustainable land management. It must be based on shared responsibilities, shared commitment, shared benefits and shared control. (Strategic Consultative Group on Sustainable Land Management Research 1995: 7)

Many new opportunities, such as the *Resource Management Act*, land tenure reforms, and sustainable land management strategies, exist for government, industry and communities to adopt both a wider range of policy options and more effective tools for promoting sustainable land management. They represent a chance to rethink attitudes and approaches to uncertainty, science and sustainability, and adopt innovative problem solving and policy development processes that are informed rather than restricted by the multidimensionality of sustainable management in the high country.

It is also clear, however, that previous generations of policy makers, land users, owners, and researchers believed that policy changes in their time afforded similar opportunities to promote the 'sustainability' of the high country. The passage of time has shown this optimism to be misplaced. The question remains then, given the inherent uncertainty and indeterminacy associated with the sustainable management of the South Island high country, what can current policy makers do to bring about sustainable management?

Accommodating uncertainty and indeterminacy via the precautionary principle and stakeholder participation

There are many potential approaches to managing uncertainty in the context of sustainability (see, for related discussions, Dovers 1995; Dovers, Norton and Handmer 1996). In the case of the sustainable management of the South Island pastoral high country, we believe policy makers can improve the policy development process in two important areas. First, they can make better use of the precautionary principle. Second, they can adopt a more participatory approach to development of sustainable land management policies.

Building on the precautionary principle

The precautionary principle is a relatively recent and widely debated concept in environmental policy development (see, for related discussions, Bodansky 1991; O'Riordan 1993, 1994; Dovers and Handmer 1995; Myers 1993). The precautionary principle is important in dealing with uncertainty. It suggests that, rather than await certainty, policy makers should act in anticipation of any adverse effect to prevent it. Most current applications of the precautionary principle require policy makers to act *judiciously* when there is *adequate* evidence and where action can be *justified* on the grounds of *cost-effectiveness* and where inaction could lead to potentially *irreversible effects* or *serious harm* to future generations.

In New Zealand, the precautionary principle is articulated most fully in the government's *Environment 2010 Strategy* where it is given prominence as one of 11 key principles for integrating environment, society and economy. The *Strategy* states that the '. . . precautionary principle should be applied to resource management practice, where there is limited knowledge or understanding about the potential for adverse environmental effects or the risk of serious or irreversible environmental damage'.

Perhaps more importantly, many people view the *Resource Management Act* as a precautionary statute (Board of Inquiry 1995). Although the precautionary principle is not explicitly mentioned, the Act's articulation of sustainable management clearly legitimates a precautionary approach to environmental problems. The definition of adverse effects in the Act includes any potential effect of high probability and any potential effect of low probability that has a high potential impact. Dealing with such effects, while meeting the reasonably foreseeable needs of future generations, and safeguarding the life-supporting capacity of the environment clearly requires adoption of the precautionary principle.

Dovers and Handmer (1995: 94) believe that the precautionary principle is 'a composite of value-laden notions and loose, qualitative descriptors that are themselves subject to intense debate'. This makes it extremely difficult for the policy maker to implement the precautionary principle in many specific contexts. They must somehow construct a policy solution that anticipates the unknown and indeterminate. Little, if any, guidance is given as to what precautionary measures should be taken. A simple yes or no type policy decision will rarely be adequate in these circumstances.

These concerns or reservations are borne out by experience with the *Hazardous Substances and New Organisms Act* (HSNO), which came into law in 1996 but as of March 2001 is still only partially implemented. The purpose of the HSNO Act is 'to protect the environment, and the health and safety of people and communities, by preventing or managing the adverse effects of hazardous substances and new organisms'. Under the Act decision makers are required to adopt a *precautionary approach* (Section 7).

> All persons exercising functions, powers, and duties under this Act, including but not limited to, functions, powers, and duties under sections 29, 32, 38, 45, and 48 of this Act, shall take into account the need for caution in managing adverse effects where there is scientific and technical uncertainty about those effects.

While it is still in its infancy, there is little evidence to date that decision makers have used this part of the Act.

Despite these limitations, we believe policy makers can make better use of the precautionary principle during the policy development process. For example, in an attempt to keep the costs of government regulation to a minimum, increasing emphasis is placed by policy makers on avoiding inappropriate and often costly policy interventions. In terms of policy research this is translated into a desire to avoid type I statistical errors. These errors occur when a statistical study concludes that there is an effect or relationship between variables when one does not exist. However, the greater the emphasis placed on avoiding type I errors, the greater the chance of making a type II statistical error. A type II statistical error concludes that there is no effect or relationship when one actually exists. This potentially leads to a situation of no policy intervention when intervention is required.

Different parties often incur the costs of too much or too little policy intervention, and this too affects the environmental policy development process. The cost of unnecessary intervention in the high country is borne directly by runholders and government. Agricultural lobby groups have a specific interest in informing policy makers of the specific costs of a proposed policy. In contrast, the costs of policy inaction are often manifested as externalities, borne by the environment and future generations. Promoting an awareness of these costs is often left to disparate groups representing sometimes intangible public and environmental interests. The political influence of these groups is often less than that of the pastoral lobby. Even when these groups are given a hearing, they may only be able to point to uncertain, long-term costs imposed disproportionately on wide and underrepresented future interests.

Greater internalization of the precautionary principle by policy makers could help strike an appropriate balance between inappropriate policy action and no policy action. Its key contributions in this regard include:

- *Encouraging proactive policy development*: This entails a willingness to take action in advance of conclusive evidence of the need for the proposed action. Further delay may prove costly to the environment and society, and in the longer term discriminate against future generations.
- *Better safeguarding of the assimilative and regenerative capacities of natural systems*: This requires holding back from possible but undesirable uses of resources that may exceed the assimilative and/or regenerative capacity of the high country
- *Placing the onus of proof on those who propose actions or policies with potentially adverse environmental effects*: This aspect of the precautionary principle could hasten the switch to less environmentally damaging technologies or management regimes that lead to the internalization of many social and environmental externalities, and effectively promote sustainable management of the environment.

Enhancing stakeholder participation

Emerging in recent literature is a persuasive argument that suggests that the existence of indeterminacy requires a fundamental reconsideration of the role of science and institutional decision making in environmental policy development (Funtowicz and Ravetz 1992, 1993; Wynne 1992a, 1992b). It calls for a more open, 'democratized' approach to environmental policy – an approach in which scientific and technical evidence loses or gains credibility in accordance with the degree of uncertainty engendered and the decision stakes. In many instances, it is argued, the existence of indeterminacy will mean that the expert should have no more privilege or standing than the lay person in the policy development process.

Environmental decisions in the South Island high country depend on evaluations of future states of ecosystems, economy and society which are unknown and perhaps unknowable. Wynne (1992a: 120) writes:

> Thus uncertainties in the scientific knowledge for environmental protection decisions cannot be properly described as objective shortfalls of knowledge, as most treatments suppose. The extent of uncertainty seen in the scientific knowledge base is itself a subjective function of complex social and cultural factors. Scientific uncertainty can be enlarged by social uncertainties in the context of its practical interpretation, and it can be reduced by opposite social forces.

Acceptance of this perspective leads to the conclusion that environmental science and related problem-solving processes, while rapidly expanding our knowledge of natural systems and their interaction with human systems, can only go so far in their contribution to environmental policy development. When faced with uncertainty and indeterminacy, science by itself can no longer guide policy makers. The challenge facing environmental policy makers is to broaden a reliance on science and experts by tapping into a wider pool of social knowledge and skills. The increasingly rapid dissemination of formal and informal knowledge means that it is possible to broaden involvement in policy making by developing institutions and problem-solving methodologies that enhance participatory policy development.

When dealing with environmental problems policy makers often introduce both particular social meanings of sustainability (such as Section 5 of the *Resource Management Act*) and also particular policy models with which to analyse sustainable management problems (such as property rights analysis of land tenure problems). These policy models often abstract the problem analysis from the social context in which it occurs, but policy makers rarely examine the dependency of the analysis on the original social context. This means that policy recommendations are often divorced from the social and institutional dimensions of the original problem.

The imposition of meanings and models on the high country by experts and related institutions can, in turn, create a social threat to those high country stakeholders, and thus intensify land management problems. Such a position tends to avoid the constructive, self-critical development of processes that foster broad-based discourse over meanings and approaches relevant to sustainable land management. For example, many of the social anxieties and tensions associated with the proposed high country land tenure reforms tend to be represented by policy makers as symptoms of ignorance, irrationality, or naive expectations of sustainable management in the high country. This approach denigrates and marginalizes many individuals and groups with an interest in the sustainable management of the high country. It may further enhance their feelings of being threatened by institutions that do not respect their identity, rationalities, and standing with respect to the issue in question. Without the support of all high country stakeholders, it is unlikely that sustainable management can be achieved.

Acceptance by policy makers of a *social learning* approach to sustainable land management in the high country may break a long tradition of issue polarization there. It would encourage a greater awareness of the need to understand the social context in which the problem of sustainable management is debated. Social learning at its simplest means that policy makers: (1) actively seek to understand the assumptions tacitly shaping their own understanding of sustainable management; (2) recognize other stakeholders' assumptions and values; and (3) engage in the collective negotiation of meanings, definitions and assessments that form the basis for subsequent policy development.

In practice, this means adopting a more open participatory process for deciding appropriate policy action. The knowledge held by all stakeholders in the South Island high country needs to be acknowledged as an essential element of sustainable management. Most pressing environmental problems cannot be solved in the laboratory or in government offices – only through real-world trial and error. The process of social learning necessarily involves many diverse participants, whose status as stakeholders is derived from having knowledge relevant to the problem being studied, being part of the problem, and/or as an individual or community affected by wider implications of sustainable land management.

Conclusion

Resolving the complex issue of sustainable management in the South Island high country in a way that achieves widely acceptable economic, ecological, and social outcomes is a process characterized by uncertainty. In the past, judgements about policy approaches have been based on a scientific approach to the management of the high country without examining the assumptions behind the way that knowledge is used. The knowledge and values of 'non-privileged' stakeholders tended to become relatively unimportant and were sometimes seen as impediments to policy development. Thus, policy institutions often inadvertently worked against the very outcomes they sought because they imposed unshared definitions of and assumptions about high country environmental issues.

Consequently, policy horizons have been limited to maintaining the status quo and established approaches to environmental management. To counter this situation environmental policy debates need to be carried out in a wider social context, accepting of different philosophical stances, types of knowledge and expectations of outcome. Only when the proper scope and basis of environmental knowledge and action are debated and negotiated in public, and the outcomes subject to extended peer review, will environmental policies provide for sustainable economic, ecological and social processes in the South Island high country. Present policy initiatives may help create the potential for a significant shift in the way in which we conceive and implement environmental policy. But this potential will remain substantially untapped unless policy institutions learn to engage as participants in, rather than arbitrators of, wider policy debates. They must also reorganize the process of policy development to accord uncertainty, informal knowledge, and subjective values the same policy privilege as scientific knowledge.

Viewed in a broader context, the sustainable management of the South Island high country becomes not so much a policy problem as an opportunity to engage fully with the complex and indeterminate nature of environmental policy. Once it is realized that policy certainty is largely an illusion, policy makers may let go of a fear of being 'wrong' and focus more on achieving outcomes that have broad-based stakeholder backing. Concomitantly, equal participation of many stakeholders – including central, regional and territorial government, sectoral interest groups, research institutions, public interest groups, communities and individuals – will do much to improve the credibility of environmental policy institutions and their objective of sustainable management.

References

Board of Inquiry (1995) *Proposed Taranaki Power Station – Air Discharge Effects*. Report and Recommendation of the Board of Enquiry pursuant to Section 148 of the *Resource Management Act 1991*.

Bodansky, D. (1991) Scientific uncertainty and the precautionary principle. *Environment* 33: 4–5; 43–4.

Botkin, D.B. (1990) *Discordant Harmonies: A New Ecology for the Twenty-first Century*. New York: Oxford University Press.

Caldwell, L.K. and Shrader-Frechette, K. (1993) *Policy for Land: Law and Ethics*. Savage, Md: Rowan and Littlefield.

Centre for Resource Management (1983) *Pastoral High Country: Proposed Tenure Changes and the Public Interest*. Lincoln Papers in Resource Management No. 11. Centre for Resource Management. Lincoln University.

Davison, J.J. (1986) Ohau: a study of the evolution of New Zealand mountain land recreation. *New Zealand Man* (sic) *and the Biosphere Report*, No 10. Lincoln College, Canterbury, New Zealand.

Demeritt, D. (1994) Ecological objectivity and critique in writings on nature and human societies. *Journal of Historical Geography* 20: 22–37.

Dovers, S. (1995) *Risk, Uncertainty and Ignorance: Policy Process and Institutional Issues*. Proceedings of the 1995 Fenner Conference on the Environment: *Risk and Uncertainty in Environmental Management*, Canberra, 13–16 November.

Dovers, S.R. and Handmer, J.W. (1995) Ignorance, the precautionary principle and sustainability. *Ambio* 24: 92–7.

Dovers, S.R., Norton, T.W. and Handmer, J.W. (1996) Uncertainty, ecology, sustainability and policy. *Biodiversity and Conservation* 5: 1143–67.

Eldred-Grigg, S. (1980) *A Southern Gentry*. Auckland: Heinemann Reed.

Faber, M., Manstetten, R. and Proops, J. (1992) Toward an open future: ignorance, novelty and evolution. In: Costanza, R., Norton, B.G. and Haskel, B.D. (eds) *Ecosystem Health: New Goals for Environmental Management*. Washington DC: Island Press.

Funtowicz, S.O. and Ravetz, J.R. (1990) *Uncertainty and Quality in Science for Policy*, Dordrecht: Kluwer Academic Press.

Funtowicz, S.O. and Ravetz, J.R. (1992) Three types of risk assessment and the emergence of post normal science. In: Golding, D. and Krimsky, S. (eds) *Social Theories of Risk*. New York: Greenwood Press.

Funtowicz, S.O. and Ravetz, J.R. (1993) Science for the post-normal age. *Futures* 25: 735–55.

Harte, M.J. (1995) Ecology, sustainability, and environment as capital. *Ecological Economics* 15: 157–64.

Jorgensen, S.E. (1990) Ecosystem theory, ecological buffer capacity, uncertainty and complexity. *Ecological Modelling* 52: 125–33.

Kerr, I.G.C. and Treskanova, M. (1993) Evaluation of the grasslands of Mount John for sustainable management. Unpublished report to New Zealand Defence Forces.

MacGarvan, M. (1993) The Precautionary Principle and the Limits of Science. Paper presented at the Precautionary Principle Conference, Institute of Environmental Studies, University of New South Wales, September.

M'Gongle, R.M., Jamieson, T.L., McAllister, M.K. and Peterman, R.M. (1994) Taking uncertainty seriously: from permissive regulation to preventative design in environmental decision making. *Osgoode Hall Law Journal* 32: 99–169.

McIntosh, R.P. (1987) Pluralism in ecology. *Annual Review of Ecology and Systematics* 18: 321–41.

Myers, N. (1993) Biodiversity and the precautionary principle. *Ambio* 22: 74–9.

New Zealand Parliamentary Debates (1948) Second Reading of the Land Bill, Vol 284, Wellington: Queens Printer.

O'Connor, K.F. (1982) Implications of past exploitation and current development in the conservation of South Island tussock grasslands. *New Zealand Journal of Ecology* 5: 97–107.

O'Connor, K.F. (1983) Land use in the hill and high country. In: Bedford, R.D. and Sturman, A.P. (eds) *Canterbury at the Cross-roads*. Miscellaneous Series No 8. Christchurch: New Zealand Geographical Society Special Publication.

Orange, C. (1987) *The Treaty of Waitangi*. Wellington: Allen & Unwin.

O'Riordan, T. (1993) *Interpreting the Precautionary Principle*. CSERGE PA working paper 93–03. Centre for Social and Economic Research on the Global Environment, University of East Anglia, Norwich.

O'Riordan, T. (1994) *The Precautionary Principle in UK Environmental Law and Policy*. CSERGE GEC working paper 94–11. Centre for Social and Economic Research on the Global Environment, University of East Anglia, Norwich.

Park, G. (1996) *Nga Uruora: The Groves of Life*. Wellington: Victoria University Press.

Panayotou, T. (1993) *Green Markets: The Economics of Sustainable Development*. San Francisco: ICS Press.

Peters, R.H. (1992) *A Critique for Ecology*. Cambridge: Cambridge University Press.

Pickett, S.T.A., Parker, V.T. and Fiedler, P.L. (1992) The new paradigm in ecology: implications for conservation biology above the species level. In: Fiedler, P.L. and Jain, S.K. (eds) *Conservation Biology: The Theory and Practice of Nature Conservation Preservation and Management*. London: Chapman and Hall, 66–88.

Rural Futures Trust (1996) Project report: Improving Farmon and linking it with a GIS. *Rural Futures Trust*, No. 3, March: 2–3.

Sage, E. (1995) The big steal. *Forest and Bird*. No. 277: 18–19.

Sagoff, M. (1985) Fact and value in environmental science, *Environmental Ethics* 7: 99–116.

Shrader-Frechette, K. and McCoy, E.D. (1994) What ecology can do for environmental management. *Journal of Environmental Management* 41: 293–307.

Smithson, M. (1989) *Ignorance and Uncertainty: Emerging Paradigms.* New York: Springer-Verlag.

Strategic Consultative Group on Sustainable Land Management Research (1995) *Science for Sustainable Land Management: Towards a New Agenda and Partnership.* Wellington, NZ: Ministry of Research, Science and Technology.

Whitby, M.C. (1979) *Economics of Pastoral Development in the Upper Waitaki. New Zealand Man and Biosphere Report No. 3.* Christchurch, NZ: Lincoln University.

Whitehouse, I. (1985) South Island high country erosion: Is anyone to blame? *Soil and Water,* No. 4: 29–39. Wellington: NWASCA.

Wynne, B. (1992a) Uncertainty and environmental learning: reconceiving science and policy in the preventive paradigm. *Global Environmental Change* 2: 111–127.

Wynne, B. (1992b) Risk and social learning: reification to engagement. In: Golding, D. and Krimsky, S. (eds) *Social Theories of Risk.* New York: Greenwood Press.

Zimmerer, K.S. (1994) Human geography and the 'new ecology': the prospect and promise of integration. *Annals of the Association of American Geographers* 84: 108–25.

9 | Fire and biodiversity:
understanding and managing the impacts of fire on forest biodiversity in south eastern Australia

Jann E. Williams

Introduction

Fire is used as a management tool in many of the world's terrestrial ecosystems. In forest ecosystems of North America and Australia, for example, it is often employed to reduce plant biomass, with the aim of making wildfires more manageable and reducing the perceived threat to human life and property. However, the effectiveness of 'prescribed' fire on a wide scale is often debated and remains controversial. While there is no doubt that the careful use of fire may promote some management objectives, there is also no doubt that the cumulative effects of prescribed and natural fires can threaten the long-term biological productivity of ecosystems and the biological diversity that they support. Fire can be a threatening process even where it is considered a 'natural' part of ecosystem ecology (e.g. Russell-Smith *et al.* 1997; Williams *et al.* 1999).

Here I review scientific understanding of the way that fire may affect the structure, function and biodiversity of forests and discuss the implications of this knowledge for the sustainable management of such ecosystems. The forested region of south eastern Australia is used as a case study. Prescribed fires are used extensively and regularly in many forested landscapes in the region, and the uncertainties raised by imperfect knowledge and ignorance of the ecological significance of cumulative impacts have important implications for forest policy, the formulation of management prescriptions and the mechanisms that are exercised to evaluate the sustainability of forest management. Many of these implications hold as true for forest ecosystems worldwide as they do for south east Australia (Williams and Woinarski 1997).

Fire and the structure, function and biodiversity of forests in south eastern Australia

Ecological impacts of different fire regimes

While fire is a natural component of the environment, its effects may vary according to the fire regime and the properties of the ecosystem (Gill 1986). The term 'fire regime' refers to the type, frequency, season and intensity of a fire (the components of the regime) experienced at a specified location in the forest. Vegetation may be adapted to a particular regime of fires (Gill 1975) that may change through time and differ from place to place.

Bushfires are commonly described in terms such as 'Byram's intensity' or 'rate of spread' (Luke and McArthur 1978) which reflect suppression difficulty or general damage potential. However, meaningful descriptions of individual bushfires for interpreting ecological effects of the fire itself may be those that reflect the amount and rate of heat energy released and its distribution (Tolhurst 1993). By systematically identifying the critical components of the physical effects of fire, a more thorough understanding of environmental impacts of fires may be deduced.

The distribution of heat through vegetation during a fire is critical for the survival of plants and plant tissues (Gill 1995; Williams *et al.* 1999). This 'temperature exposure' has a direct impact on the extent of leaf scorch, cambial death and seed survival in shrubs and trees (Mercer *et al.* 1994). In addition, the susceptibility of a plant to injury or death depends on the nature and distribution of critical tissues, such as the cambial layer in trees, and the nature and extent of its insulating material. Heat is required to stimulate the *en masse* release of seeds from within woody fruits of some serotinous species (Gill 1976), but heat from combustion may also kill seeds in woody fruits if temperatures exceed critical thresholds (e.g. Ashton 1986; Judd 1993; Bradstock *et al.* 1994).

Heat transfer down the soil profile will have implications for survival of soil flora and fauna, seeds and subterranean organs of plants. The main heating effect is within the top few centimetres of the soil. Factors that modify heat transfer into the soil are initial soil temperature, moisture content and thermal diffusivity. Marked spatial heterogeneity in soil temperatures during fires has been repeatedly detected. The importance of heat in breaking seed dormancy has been illustrated in plant families such as *Fabaceae*, *Lamiaceae* and *Rhamnaceae* in different fire-prone environments throughout the world (e.g. Keeley and Keeley 1989; Bradstock and Auld 1995). Recent overseas work has implicated other fire-related cues, such as smoke and aqueous extracts of smoke and charred wood, in the breaking of seed dormancy (see Dixon *et al*. 1995; Brown and van Staden 1997). Knowledge of the mechanism underlying seed dormancy may be important, in particular, for the management of rare plant species. Understanding what triggers the germination of seed in these species could improve management intended to maintain viable populations of species.

The microclimate/environment of recently burnt areas can be profoundly different from that of unburnt areas, although the degree of difference will depend on both fire characteristics and ecosystem properties (Russell-Smith *et al*. 1997; Williams *et al*. 1999). Biologically derived legacies from the pre-disturbance ecosystem can have important influences on fine-scale 'patterning' and on patterns of ecosystem recovery after disturbance (Franklin 1990). Pattern legacies include those created in soil properties – chemical, physical and microbiological – through actions of plants and their litter and patterns in understorey vegetation associated with variations in canopy light conditions. Large woody structures are important in terms of animal habitat. Spatial and temporal variation in post-fire environments leads to variation in the relative success of species over the landscape and from fire to fire.

There have been a number of general reviews on changes in the environment in relation to fire, and several recent reviews focusing on nutrient dynamics in the post-fire environment (e.g. Williams and Gill 1995; Keith 1997; Gill 1997). However, few published studies address the effects of fire on the micro-environment in temperate eucalypt forests (Bell and Williams 1997). This applies in particular to the fine-scale heterogeneity in soil moisture, air temperature, nutrient and radiation regimes that is likely to have a major impact on post-fire dynamics. Fires may significantly influence the mosaic of biological patterns and ecological processes within their sphere of influence. Understanding the effects of fire regimes on biotic patterns and landscape processes requires a consideration of the landscape context within which the fires occur, and the scale of the impact. While a single fire may significantly influence the response of biota, consideration of the fire regime is necessary in order to understand better the responses of species and assemblages of species to such disturbance. Further, the effects of fires on biota often need to be considered in light of the fire regime experienced by a region as much as any individual forest site (Gill 1997).

The responses of the biota to fire regimes may be strongly influenced by their life-history characteristics (e.g. habitat requirements, mode of reproduction). Additional factors affecting response include the conditions at the time of the fire (e.g. life-history stage, plant condition, fire edge to habitat area ratio) and the post-fire environment. Tolhurst *et al.* (1992), for example, studied the impacts of fire on the structure of plant communities in central Victoria. They noted that frequent, repeated fires would be likely to alter the vegetation structure significantly, and suggested that there was a need for long-term studies to understand the effects of repeated burning at various frequencies. In particular, the potential simplification of vegetation structure, through the impact of fires on shrubs and small trees is of concern, and was identified for further investigation. In a study of floristic variation in dry sclerophyll forest communities, Cary (1992) investigated several aspects of fire regimes on plant species composition. He found that the time since the most recent fire and the length of the inter-fire interval in the recent fire history (less than 30 years) are equally important determinants of above-ground floristic variation, and that their effects are strongly interactive.

The effect of litter-reduction burning on plant species composition will depend largely on the life-history characteristics of species at the site. The season, frequency and intensity of these fires may differ from natural (non-human) fires and hence their impact on plant species composition may also differ. For example, litter-reduction fires often occur in autumn/ spring whereas wildfires usually occur in summer. The intensity of litter fires is typically low compared to wildfires, and their frequency is typically regular and frequent. These factors may act independently, in combination, or synergistically, to influence biotic response.

Post-fire recovery of animal species is reliant to a large extent on the recovery patterns of vascular plants and vegetation. Currently, knowledge of the potential impacts of different fire regimes on forest fauna is limited (e.g. Catling 1991; Tolhurst *et al.* 1992; Gill 1997). For example, knowledge of the impacts of fire on a whole suite of small organisms such as invertebrates and soil fauna is poor, although a recent review by York (1999) at least sets the scene for studying the impact of prescribed burning on forest invertebrates. The understanding of the effects of variable fire regimes on these fauna and their contribution to various energy flows, and decomposition and nutrient cycling is even more limited.

Nutrients

Nutrient losses from an ecosystem are generally believed to be highest after high-intensity fire. Raison (1980, 1981) argued that nutrient losses after slash-burning in forests may be greater than those following wildfire. Raison (1980) estimated that 10–20 times more nitrogen was lost after high-intensity slash fires in wet sclerophyll/mixed forest than in harvested wood, and that it would take in excess of 150 years for phosphorous to be replaced in the soil. Turner and Lambert (1980) suggested that replacement times for phosphorus after slash fires may be much shorter in other forest types, ranging from 19 years in *Eucalyptus muellerana* forest to 88 years in *Eucalyptus signata* forest on Stradbroke Island. Raison (1981) noted that for soils with an organically enriched surface, losses of phosphorus through leaching and ash blow may be very high. The effect of fire on nutrients other than nitrogen and phosphorus may also be important, but does not appear to have received as much attention (Keith 1997). Adams and Attiwill (1991) noted that calcium may be the nutrient at greatest risk of depletion after severe perturbation (logging, fire and nutrient addition).

Overall, much remains to be learnt about 'nutrient budgets' after fire and, in particular, the effects of repeated fires on these (Bell and Williams 1997; Gill 1997). Many studies have been published on the ecological effects of a single fire, but few are concerned with the response of the biota to different fire regimes, particularly cumulative effects arising from more recent human-induced changes in fire regimes. As a result of this limited understanding, it is often difficult to predict adequately the likely response of an ecosystem or its biota to fire (Gill 1997).

Modelling approaches

In order to address the long-term effects of repeated fires on plant species a range of modelling approaches have been suggested. All of these approaches have limitations that need to be kept in mind if the outputs are used for operational management. For example, the behaviour of forest fires can be studied experimentally at low fire intensity, but for extreme conditions fire behaviour can only be modelled using extrapolated data of uncertain reliability. In the case of the effects of fire on plant demography in south east Australia, several studies have attempted to use quantitative demographic data to predict the fate of obligate-seeding species experiencing repeated fires; only one comparable study does so for a resprouting species (Williams and Gill 1995; see also Gill 1997). Serotinous obligate seeders have been shown to be particularly prone to extinction under frequent fire regimes (Morrison *et al.* 1996; Bradstock *et al.* 1997). Overall, the task of modelling demographic responses of long-lived species to fire is complex. Morrison and Cary (1994) highlighted some of the potential problems with this and emphasized the need for long-term studies of replicate populations to characterize the underlying variability in plant responses to fires. Because most of the available data is on the response of vascular plants to single fires, Gill (1999) has recently developed a series of graphical models using this information. By trying to improve the understanding of these dynamics, there is a greater chance to interpret the impact of repeated fires.

At the community level, Noble and Slatyer (1980) modelled a series of 'vital attributes' of plant species to depict relative performances of species experiencing various fire regimes. Fire frequency is considered the major or sole determining factor of a fire-regime in this model. The ability of a species to regenerate during the inter-fire period is incorporated into the model, but only the presence/absence of a plant species is considered. Moore and Noble (1990), building on this earlier work, developed the FATE model of vegetation dynamics. Life-history characteristics, response to local environmental conditions and response to disturbance are utilized to model the response of functional groups of plants to disturbance regimes. Unlike the vital attributes approach, FATE provides a discrete quantitative basis for modelling vegetation, in that it uses ranked scales for abundance and for plant attributes. These models allow an examination of scenarios, but users should be aware of the underlying assumptions and simplifications in these models. More recently, Keith (1996) has promoted the use of functional similarities between plant species to help reduce the complexity of managing biodiversity.

Techniques such as population viability analysis have been developed to model the response of biota to disturbances such as fire. As a means to avoid some of the limitations, Gill and Bradstock (1995) suggested that simple models of biodiversity conservation and conditions appropriate to fire control can be used to clarify issues and identify solutions. For example, they proposed that 'critical' and 'threshold' fuel loadings may be identified to characterize the risk of ignition, guide fire control and inform the use of prescribed fire. They also outlined an initial rationale for attempting to avoid plant species extinctions and to identify appropriate intervals between fires to maintain certain conservation values. More recently, risk assessment is being applied to examine the merits and limitations of alternative fire management scenarios, such as the work by Bradstock *et al.* (1997) on *Banksia ericifolia*. Currently, however, the extent to which fire management prescriptions and simple models can be reliably transferred between different forest sites is often unclear (Williams *et al.* 1994; Gill 1997).

Fire protection versus fire management

The issue of fire protection versus fire management is controversial and has been debated for several decades in Australia. The main issues associated with fuel-reduction burning are the efficacy of the practice for fuel reduction and the lowering of potential fire intensities, on the one hand, and possible negative impacts on biodiversity, on the other (Gill and Bradstock 1995; Morrison *et al.* 1996). In particular, two key areas of attention in the debate are the effect of fuel-reduction burning on 'wildfire' behaviour and control, and the ecological effects of both 'wildfires' and 'prescribed' fires (Gill *et al.* 1987).

The volume of fuel that can be burnt directly affects the intensity of a fire (Byram 1959). Hence, if fuel load can be reduced, then fire intensity is reduced. Given this, it follows that lower fuel weights will significantly reduce the period of time when high-intensity fires can occur (Raison *et al.* 1983) and improve the chances of wildfire control (Underwood *et al.* 1985). The amount of fuel reduced, and its rate of recovery to pre-fire levels, is of special interest to managers because the primary basis of fire protection is to keep potential fuels below specified levels (Tolhurst 1996). It has been argued that the potential effectiveness of fuel-reduction burning may be limited by rapid fuel-accumulation rates. A number of studies have reported rapid accumulation of fine fuel in the first few years post-fire for several forest types (Fox *et al.* 1979; Birk and Simpson 1980; Attiwill 1985; Birk and Bridges 1989; Tolhurst *et al.* 1992), but the extent to which this may undermine fire control is often unclear.

The broad benefits of prescribed burning in modifying fire behaviour and assisting fire suppression are generally accepted but more research is required to quantify the changes in fire behaviour as fuels re-accumulate after burning (Cheney 1996). Few quantitative data are available on the effect of fuel-reduction burning on the subsequent rate of spread of high-intensity fires (Cheney 1996). Hence, experience and qualitative judgements, rather than empirical data, are often used to assess the value of fuel-reduction burning as a management technique (Birk and Bridges 1989; Resource Assessment Commission 1992).

Concerns about the effectiveness of broad-area burning in ecosystems to reduce the risk of wildfire and its impact on environmental values are often raised (e.g. Raison *et al.* 1983; Australian Biological Resources Group 1984; Good 1989; Resource Assessment Commission 1992). Such practices increase the potential risk of wildfire due to unplanned 'escapes' from controlled burning exercises. While the number of studies specifically addressing the impact of fuel-reduction burning are increasing (Clark 1988; Hamilton *et al.* 1991; Tolhurst *et al.* 1992; York 1999), the Forest and Timber Inquiry of the Resource Assessment Commission concluded that scientific information was inadequate to assess the nature and extent of past and current impacts of this practice on the environmental values of forests (Resource Assessment Commission 1992). Monitoring of the effects of fire on forest biota is exceptionally limited across most of the forest estate.

There is no question that fuel-reduction burning affects fire behaviour. It also affects forest biota and soil nutrients. While there have been many studies published on the effects of a single fire, few studies have appeared on the response of the biota to fire regimes. Depending on the primary land use in an area, frequent low-intensity fires may or may not satisfy management objectives; in particular, the maintenance of species diversity. Clearly, greater effort is required to characterize and track the effects of fuel-reduction burning to determine to what extent it satisfies all management objectives.

There are relatively few days per year during which the use of prescribed burning is practicable (Gill *et al.* 1987). It is only safe to burn, from a fire-control perspective, under a narrow range of climatic and environmental conditions (McArthur 1962; Cheney 1978). Low-intensity fires are normally lit outside the driest season, usually in autumn and spring. Factors to be taken into account before deciding to burn include air temperature, wind strength and direction, fuel loads and types, and forest type, age and condition (Forestry Commission of New South Wales 1993). In addition, economic, staffing and equipment supply factors may further constrain options for using fire. All of these variables require adequate consideration if fire is to be used effectively and safely.

Some implications for resource managers and policy

From the above discussion it is evident that many uncertainties exist as to the use of fire. These relate to issues such as the true effectiveness of prescribed fire in reducing the potential risk to life and human property, the potential impacts of fire on ecosystems (structure, function, biodiversity), the most appropriate ways to use fire, and the best ways to control the spread of fire. These issues and uncertainties have many implications for resource managers responsible for operational management, and for policy concerned with ecosystem and environmental management.

It is sometimes suggested that contemporary fire regimes should attempt to replicate those occurring prior to European occupation of the Australian continent. However, the merits of widely adopting this approach in south eastern Australia are unclear. The nature of fire regimes in the forests of this region, prior to the arrival of Europeans, is poorly known. Few reliable scientific data are available and historical accounts are difficult to interpret precisely (Benson and Redpath 1997). Furthermore, given the altered nature of the contemporary landscape (e.g. landscape fragmentation, presence of feral species), proposals to replicate the pre-European fire regime often appear inappropriate (Williams and Gill 1995). Fire policy and operational management may be more effective if based on clear management objectives that are developed with specific regions, ecosystems, biota and ecological processes in mind, rather than close adherance to the 'pre-European' situation. For example, when trying to conserve biodiversity, attempts to mimic inferred historical fire regimes are likely to fail because many landscapes have been transformed in various ways in the last 200 years. However, where available and appropriate, the incorporation of Aboriginal knowledge of landscapes and the role of fire is recommended when it helps meet management objectives. Due to the history of European occupation, this knowledge is principally found in northern and central Australia (Bowman 1998).

Fire policy is likely to be most effective if it is conceived and implemented in a manner that accommodates the uncertainties arising from limited knowledge and the constraints generated from existing institutional and funding arrangements. The development of management objectives for different forest regions and systems should be informed by scientific knowledge that is subject to evaluation and refinement over time; a concept often referred to as 'adaptive management'. Such an approach requires flexibility, adequate infrastructure to evaluate the efficacy of management regimes in different forests, and means to critically assess their transferability or portability between forests.

Currently, the extent to which fire management prescriptions and simple models can be reliably transferred between different forest sites is often unclear. For example, a fire model developed for litter fuels may not apply in other forests when the local fuels are a mixture of grass, shrubs and litter. In a similar way, the findings of an appropriate fire frequency for low-intensity fires may have been ascertained in relation to risk of uncontrollable fires and plant species persistence in one area, but the plant assemblages and environmental context may be quite different in another area. Gill and Bradstock (1995) recognized the difficulties raised by imperfect knowledge when using models of fire behaviour and biotic response to fire regimes. These difficulties could be minimized with the adoption of a more integrated, landscape-based approach to fire management that enables, inter alia, evaluation of the assumptions underlying management approaches.

The determination of a fire management regime will depend to some extent on the management objectives in the area being considered. A different regime, for example, may be implemented close to urban areas than in more remote locations (Simmons and Adams 1986). In areas where the maintenance of species diversity is of concern, a number of considerations are necessary. Gill (1977) indicated that plant species are adapted to a particular fire regime. Fire frequency becomes especially significant for species requiring a long period of time (relative to the interval between fires) to grow from seedling to reproductive age. Thus, species regenerating from seed are especially vulnerable to frequent fire, although species regenerating from root-stock may also be adversely affected. To ascertain the response of the flora to fires at a particular location, certain life-history characteristics of individual species should be systematically recorded (Gill 1986). Post-fire weather will also affect the establishment and growth of plant species. In the case of rare species intervention, such as watering seedlings during dry periods, may be required (for examples, see Williams and Gill 1995).

Clear objectives need to be set whenever a forest area is deliberately burnt. Fires with an intensity other than that prescribed for a particular management objective may result in variation in the floristic composition of the subsequent regrowth, which will have a significant effect on conservation values of plant communities in the long term. Kendall (1993) suggested that calculating the 'ecologically safe fire window' for fire management compartments would allow managers to determine when an area could be safely burnt in order to minimize the likelihood of species extinction. This method uses the time between the primary juvenile period of an indicator 'fire-sensitive species' (vulnerable to extinction through recurrent fire) and the longevity of an indicator 'fire-dependent' species (vulnerable to extinction in the absence of fire) to determine the necessary fire frequency. It has been suggested this information could be incorporated into a register similar to the one discussed by Gill and Bradstock (1992).

Scientifically based methods for predicting the types and abundances of fauna from floristic and/or structural data in fire-affected plant communities also require further development. While it appears necessary to ensure a diversity of age classes in appropriate vegetation types, further investigations are needed into the habitat requirements of different animal species. Other areas that require investigation are how important it is to avoid:

- peak breeding/nesting periods of at least flora and fauna which are perceived as significant
- flowering and seed development periods (potential reduction of food for fauna).

Variability in plant response to fire has important implications when attempting to generalize patterns from a single location and should be considered in the context of both research and management. Therefore, caution is advised when considering the transfer of fire management prescriptions from one area to another. One avenue for identifying species that exhibit ecotypic variation is through the national fire response register (Gill and Bradstock 1992). It has been stated that wildfires are incompatible with the objectives of forest management (Forestry Commission of New South Wales 1993). However, it is likely that at least some biota require high-intensity fire for their persistence (Whelan and Muston 1991; Catling 1991). While it is difficult to study these fires directly, the use of models, post hoc observations or experiments in controlled conditions may increase our understanding of this phenomenon.

Monitoring is the major component of a scientific management programme that is lacking in many areas of Australian natural resource management. Because of limited understanding and the variable context of forest sites, systematic survey of plant species in relocatable sites throughout the forest estate is required to help provide the base for effective monitoring of the resource and for the further documentation of the effects of fires on forest biodiversity. This will require the monitoring of biotic responses in relation to various characteristics of the fire and the post-fire environment. For plant species, at least, an apparently practical monitoring system has been devised to assist managers in their selection of suitable fire regimes (Gill and Nicholls 1989). This would entail systematically monitoring selected sites before and after an 'event'. As well as the cumulative effects of fires at a site, exogenous factors such as climatic variation need to be considered.

This or a similar system needs to be developed for areas where logging and/or grazing are also involved. While a minimal-set approach may not capture all of the variation in behaviour within and between populations of plant species, it can provide important baseline information that is currently not available. However, the transfer of information to other locations for some species at least must be cautioned (Williams *et al.* 1994). A similar approach could be developed for monitoring animal species within the overall biotic assemblage at a site. Collection of environmental information at the monitoring site may allow the process responsible for the changes to be identified (Wardell-Johnson *et al.* 1989).

Effective adaptive management is also reliant on good communication between researchers and operational managers. Co-operation at the operational level is required to ensure that prescribed burns can be employed to test the many assumptions upon which management regimes are based. Similarly, enhanced communication is required when evaluating the portability of techniques. The development of electronical databases such as FireNet (Green *et al.* 1993) could facilitate greater communication of research results, on the one hand, and conveyance of management problems and experiences, on the other. FireNet enables communication among students, managers and researchers and provides courses, aids and research materials to users (Gill 1997).

Conclusions

The effects of fire regimes on ecosystem components and on landscape-scale processes are complex and poorly understood. And technical assessment of the extent to which current fire management regimes achieve their desired effects is often not possible because of the lack of monitoring related to management aims. A more sophisticated and integrated approach to landscape planning and management is required if fire policy is to be based on scientific information that is amenable to evaluation, adaptation and revision. In particular, more attention needs to be given to monitoring the effects of fire in relation to specific management goals, and using operational management activities as opportunities for scientific experimentation and learning. Situations will arise when it will not be possible to meet all management objectives while employing fuel-reduction burning; judgements between fire protection and conservation objectives will need to be made. Until existing knowledge, management infrastructure and community awareness are much further developed, it would appear prudent that such judgements and the use of fire as a management tool in native forests be undertaken conservatively. Science has an important role to play in informing these judgements, but more effective processes are required to achieve this.

Acknowledgements

This work was based in part on work undertaken with Dr A.M. Gill from the Centre for Plant Biodiversity Research, Division of Plant Industry, CSIRO, Canberra. This work was principally prepared while I was visiting research scientists based in the Environment Department at the University of York, UK. I thank the Department and Professor Charles Perrings for providing access to resources.

References

Adams, M.A. and Attiwill, P.M. (1991) Nutrient balance in forests of northern Tasmania. 2. Alteration of nutrient availability and soil-water chemistry as a result of logging, slash-burning and fertilizer application. *Forest Ecology and Management* 44: 115–31.

Ashton, D.H. (1986) Viability of seeds of *Eucalyptus obliqua* and *Leptospermum juniperinum* from capsules subjected to a crown fire. *Australian Forestry* 49: 28–35.

Attiwill, P.M. (1985) Effects of fire on forest ecosystems. In: Landsberg, J.J. and Parsons, W. (eds) *Research for Forest Management*. Melbourne: CSIRO, 249–68.

Auld, T.D. and O'Connell, M.A. (1991) Predicting patterns of post-fire germination in 35 eastern Australian *Fabaceae*. *Australian Journal of Ecology* 16: 53–70.

Australian Biological Resources Group (1984) The impacts of timber production and harvesting on native flora and fauna. Report by the Australian Biological Research Group. In: Ferguson, I.S. *Report of the Board of Inquiry into the Timber Industry Vol II.* (1985). Victoria: Victorian Government Publishing Service.

Bell, D.T. and Williams, J.E. (1997) Eucalypt ecophysiology. Chapter 8 in Williams, J.E. and Woinarski, J.C.Z. (eds) *Ecology of Eucalypts: Individuals to Ecosystems*. Cambridge: Cambridge University Press, 167–96.

Benson, J.S. and Redpath, P.A. (1997) The nature of pre-European native vegetation in south-eastern Australia: a critique of Ryan, D.G., Ryan, J.S. and Starr, B.J. (1995) *The Australian Landscape – Observations of Explorers and Early Settlers. Cunninghamia* 5: 285–328.

Birk, E.M. and Bridges, R.G. (1989) Recurrent fires and fuel accumulation in even aged blackbutt (*Eucalyptus pilularis*) forests. *Forest Ecology and Management* 29: 59–79.

Birk, E.M. and Simpson, R.W. (1980) Steady state and the continuous input model of litter accumulation and decomposition in Australian eucalypt forests. *Ecology* 61: 481–85.

Bowman, D.M.J.S. (1998) Tansley Review No. 101: The impact of Aboriginal landscape burning on the Australian biota. *New Phytologist* 140: 385–410.

Bradstock, R.A. and Auld, T.D. (1995) Soil temperatures during experimental bushfires in relation to fire intensity: consequences for legume germination and fire management in south-eastern Australia. *Journal of Applied Ecology* 32: 76–84.

Bradstock, R.A., Auld, T.D., Ellis, M.E. and Cohn, J.S. (1992) Soil temperatures during bushfires in semi-arid, mallee shrublands. *Australian Journal of Ecology* 17: 433–40.

Bradstock, R.A., Bedward, M., Scott, J., and Keith, D.A. (1997) Simulation of the effect of temporal and spatial variation in fire regimes on the population viability of a *Banksia* species. *Conservation Biology* 10: 776–84.

Bradstock, R.A., Gill, A.M., Hastings, S. and Moore, P.H. (1994) Survival of serotinous seedbanks during bushfires: comparative studies of *Hakea* species from southeastern Australia. *Australian Journal of Ecology* 19: 276–82.

Braithwaite, R.W. (1987) Effects of fire regimes on lizards in the wet-dry tropics of Australia. *Journal of Tropical Ecology* 3: 265–75.

Brown, N.A.C. (1993) Seed germination in the Fynbos fire ephemeral, *Syncarpha vestita* (L.) B.Noed. is promoted by smoke, aqueous extracts of smoke and charred wood derived from burning the ericoid-leaved shrub, *Passerina vulgaris* Thoday. *International Journal of Wildland Fire* 3: 203–6.

Brown, N.A.C. and van Staden, J. (1997) Smoke as a germination cue: a review. *Plant Growth Regulation* 22: 115–24.

Byram, G.M. (1959) Combustion of forest fuels. In: Davis, K.P. (ed.) *Forest Fire: Control and Use*. New York: McGraw-Hill, 61–89.

Cary, G. (1992) Fire Frequency and Floristic Variation in Dry Sclerophyll Communities. Hons. Thesis, Department of Applied Biology, University of Technology, Sydney.

Catling, P. (1991) Ecological effects of prescribed burning practices on the mammals of southeastern Australia. In: Lunney, D. (ed.) *Conservation of Australia's Forest Fauna*. Mosman: Royal Zoological Society of NSW, 353–63.

Cheney, N.P. (1976) Fire disasters in Australia 1945–75. *Australian Forestry* 39: 245–68.

Cheney, N.P. (1978) *Guidelines for Fire Management on Forested Watersheds, Based on Australian Experience*. FAO Conservation Guide No. 4. Special Readings in Conservation. Rome: FAO.

Cheney, P. (1996) The effectiveness of fuel reduction burning for fire management. In: *Fire and Biodiversity: The Effects and Effectiveness of Fire Management*, 9–16. Biodiversity Series, Paper No. 8. Canberra: Biodiversity Unit, Department of the Environment, Sport and Territories.

Clark, S.C. (1988) Effects of hazard-reduction burning on populations of understorey plant species on Hawkesbury sandstone. *Australian Journal of Ecology* 13: 473–84.

Dixon, K.W., Roche, S. and Pate, J.S.P. (1995) The promotive effect of smoke derived from burnt native vegetation on seed germination of Western Australian plants. *Oecologia* 101: 185–92.

Forestry Commission of New South Wales (1993) *Environmental Impact Statement. Kempsey/Wauchope Management Areas.* Vol. 1. Prepared by Truyard Pty. Ltd. Sydney: Forestry Commission of New South Wales.

Fox, B.J., Fox, M.D. and McKay, G.M. (1979) Litter accumulation after fire in a eucalypt forest. *Australian Journal of Botany* 27: 157–65.

Franklin, J.F. (1990) Biological legacies: a critical management concept from Mt St Helens. *Transactions of the 55th North American Wildlife and Natural Resources Conference*: 216–19.

Friend, G.R. (1993) Impact of fire on small vertebrates in mallee woodlands and heathlands of temperate Australia: a review. *Biological Conservation* 65: 99–114.

Gill, A.M. (1975) Fire and the Australian flora: a review. *Australian Forestry* 38: 4–25.

Gill, A.M. (1976) Fire and the opening of *Banksia ornata* F.Muell. follicles. *Australian Journal of Botany* 24: 329–35.

Gill, A.M. (1977) Management of fire-prone vegetation for plant species conservation in Australia. *Search* 8(12).

Gill, A.M. (1986) *Research for the Fire Management of Western Australian State Forests and Conservation Reserves.* Western Australian Department of Conservation and Land Management Technical Bulletin 12.

Gill, A.M. (1995) Stems and fires. In: Gartner, B.L. (ed.) *Plant stems. Physiological and Functional Morphology.* San Diego: Academic Press, 323–42.

Gill, A.M. (1997) Eucalypts and fire: interdependent or independent? Chapter 7 in Williams, J.E. and Woinarski, J.C.Z. (eds) *Ecology of Eucalypts: Individuals to Ecosystems.* Cambridge: Cambridge University Press, 151–67.

Gill, A.M. (1999) Biodiversity and bushfires: an Australia-wide perspective on plant-species changes after a fire event. In: *Australia's Biodiversity – Responses to Fire. Plants, Birds and Invertebrates.* Biodiversity Technical Paper, No. 1. Canberra: Environment Australia, 9–54.

Gill, A.M. and Bradstock, R.A. (1992) A national register for the fire responses of plant species. *Cunninghamia* 2: 635–60.

Gill, A.M. and Bradstock, R.A. (1995) Extinction of biota by fires. In: Bradstock, R.A., Auld, T.D., Keith, R.T., Kingsford, R.T., Lunney, D. and Sivertsen, D.P. (eds) *Conserving Biodiversity: Threats and Solutions.* Sydney: Surrey Beatty and Sons, 309–22.

Gill, A.M. and Nicholls, A.O. (1989) Monitoring fire-prone flora in reserves for nature conservation. In: Burrows, N., McCaw, L. and Friend, G. (eds) *Fire Management on Nature Conservation Lands.* Western Australian Department of Conservation and Land Management Occasional Paper 1/89, 137–151.

Gill, A.M., Christian, K.R., Moore, P.H.R. and Forrester, R.I. (1987) Bushfire incidence, fire hazard and fuel-reduction burning. *Australian Journal of Ecology* 12: 299–306.

Good, R.B. (1989) The planned use of fire on conservation lands – lessons from eastern states. In: Ford, J.R. (ed.) *Fire Ecology and Management of Western*

Australian Ecosystems. WAIT Environmental Studies Group Report No. 14, 147–52.

Green, D.G., Gill, A.M. and Trevitt, A.C.F. (1993) FireNet: an international network for landscape fire information. *Wildfire* 2(4): 22–30.

Hamilton, S.D., Lawrie, A.C., Hopmans, P. and Leonard, B.V. (1991) Effects of fuel-reduction burning on a *Eucalyptus obliqua* forest ecosystem in Victoria. *Australian Journal of Botany* 39: 203–17.

Judd, T.S. (1993) Seed survival in small myrtaceous capsules subjected to experimental heating. *Oecologia* 93: 576–81.

Keeley, J.E. and Keeley, S.C. (1989) Allelopathy and the fire-induced herb cycle. In: Keeley, S.C. (ed.) *The California Chapparal: Paradigms Reexamined.* No. 34 Science Series. Natural History Museum, Los Angeles County.

Keith, D.A. (1996) Fire-driven extinction of plant populations: a synthesis of theory and review of evidence from Australian vegetation. *Proceedings of the Linnean Society of New South Wales* 116: 37–78.

Keith, H. (1997) Nutrient cycling in eucalypt ecosystems. Chapter 9 in Williams, J.E. and Woinarski, J.C.Z. (eds) *Ecology of Eucalypts: Individuals to Ecosystems.* Cambridge: Cambridge University Press, 197–226.

Kendall, D. (1993) Solving the Fire Manager's Paradox? Decision Support for Australian Fire Effects Management: The Manager's Perspective. Hons. Thesis. Department of Forestry, The Australian National University.

Luke, R.H. and McArthur, A.G. (1978) *Bustifires in Australia.* Australian Government Publishing Service. Canberra.

McArthur, A.G. (1962) *Control Burning in Eucalypt Forests.* Commonwealth of Australia Forest Timber Bureau Leaflet No. 80.

Mercer, G.N., Gill, A.M. and Weber, R.O. (1994) A time-dependent model of fire impact on seed survival in woody fruits. *Australian Journal of Botany* 41: 71–81.

Moore, A.D. and Noble, I. (1990) Individualistic nature of vegetation stand dynamics. *Journal of Environmental Management* 31: 61–81.

Morrison, D.A. and Cary, G.J. (1994) Robustness of demographic estimates in studies of plant responses to fire. *Australian Journal of Ecology* 19: 110–14.

Morrison, D.A., Buckney, R.T., and Bewick, B.J. (1996) Conservation conflicts over burning bush in south-eastern Australia. *Conservation Biology* 76: 167–75.

Noble, I.R. (1984) Mortality of lignotuberous seedlings of *Eucalyptus* species after an intense fire in montane forest. *Australian Journal of Ecology* 9: 47–50.

Noble, I.R. and Slatyer, R.O. (1980) The use of vital attributes to predict successional changes in plant communities subject to recurrent disturbance. *Vegetatio* 43: 5–21.

Norton, T.W. and Possingham, H.P. (1993) Wildlife modelling for biodiversity conservation. In: Jakeman, A.J., Beck, M.B. and McAleer, M.J. (eds) *Modelling Change in Environmental Systems.* Melbourne: John Wiley and Sons, 243–66.

Raison, R.J. (1980) Possible site deterioration associated with slash-burning. *Search* 11: 68–72.

Raison, R.J. (1981) More on the effects of intense fires on the long-term productivity of forest sites: Reply to comments. *Search* 12: 10–14.

Raison, R.J., Woods, P.V. and Khanna, P.K. (1983) Dynamics of fine fuels in recurrently burnt eucalypt forests. *Australian Forestry* 46: 294–302.

Raison, R.J., Woods, P.V., Jakobsen, B.F. and Bary, G.A.V. (1986) Soil temperatures during and following low-intensity prescribed burning in a *Eucalyptus pauciflora* forest. *Australian Journal of Soil Research* 24: 33–47.

Resource Assessment Commission (1992) Final Report. Volume 1. Forest and Timber Inquiry. Canberra: Australian Government Publishing Service.

Russell-Smith, J., Ryan, P.G. and Durieu, R. (1997) MSS-derived fire history of Kakadu National Park, monsoonal northern Australia, 1980–94: seasonal extent, frequency and patchiness. *Journal of Applied Ecology* 34: 748–66.

Simmons, D. and Adams, R. (1986) Fuel dynamics in an urban fringe dry sclerophyll forest in Victoria. *Australian Forestry* 49: 149–154.

Tolhurst, K.G. (1990) Response of bracken to low intensity prescribed fire in open eucalypt forest in west-central Victoria. In: Thomson, J.A. and Smith, R.T. (eds) *Bracken Biology and Management*. Australian Institute of Agricultural Scientists Occasional Publication No. 40, 53–62.

Tolhurst, K.G. (1993) Fire from a plant, animal and soil perspective. In Abstracts of Landscape Fires Conference, Perth, Western Australia.

Tolhurst, K.G. (1996) The effects of fuel reduction burning on fuel loads in a dry sclerophyll forest. In: *Fire and Biodiversity: The Effects and Effectiveness of Fire Management*. Biodiversity Series, Paper No. 8. Canberra: Biodiversity Unit, Department of the Environment, Sport and Territories, 17–20.

Tolhurst, K.G., Flinn, D.W., Loyn, R.H., Wilson, A.A.G. and Foletta, I.J. (1992) *Ecological Effects of Fuel Reduction Burning in a Dry Sclerophyll Forest: a Summary of Principal Research Findings and their Management Implications*. Victoria: Department of Conservation and Environment.

Turner, J.S. and Lambert, M.J. (1980) Slash burning on forest sites. *Search* 11: 316–317.

Underwood, R.J., Sneeuwjagt, R.J. and Styles, H.G. (1985) The contribution of prescribed burning to forest fire control in Western Australia: case studies. In: Ford, J.R. (ed.) *Fire Ecology and Management of Western Australian Ecosystems*. WAIT Environmental Studies Group Report No. 14, 153–70.

Wardell-Johnson, G., McCaw, W.L. and Maisey, K.G. (1989) Critical data requirements for the effective management of fire on nature conservation lands in south western Australia. In: Burrows, N., McCaw, L. and Friend, G. (eds) *Fire Management on Nature Conservation Lands*. Western Australian Department of Conservation and Land Management Occasional Paper 1/89, 59–73.

Whelan, R.J. and Muston, R.M. (1991) Fire regimes and management in south-eastern Australia. *Proceedings of the Tall Timbers Fire Ecology Conference* 15: 235–58.

Williams, J.E. and Gill, A.M. (1995) *The Impact of Fire Regimes on Native Forests in Eastern New South Wales*. Forest Issues Monograph No. 1. Sydney: New South Wales National Parks and Wildlife Service.

Williams, J.E. and Woinarski, J.C.Z. (1997) (eds) *Ecology of Eucalypts: Individuals to Ecosystems*. Cambridge: Cambridge University Press.

Williams, J.E., Whelan, R.J. and Gill, A.M. (1994) Fire and environmental heterogeneity in southern temperate forest ecosystems: implications for management. *Australian Journal of Botany* 42: 125–37.

Williams, R.J., Cook, G.D., Gill, A.M. and Moore, P.H.R. (1999) Fire regime, fire intensity and tree survival in a tropical savanna in northern Australia. *Australian Journal of Ecology* 24: 50–9.

York, A. (1999) Long-term effects of repeated prescribed burning on forest invertebrates: management implications for the conservation of biodiversity. In: *Australia's Biodiversity – Responses to Fire. Plants, Birds and Invertebrates*. Biodiversity Technical Paper, No. 1. Canberra: Environment Australia, 181–266.

10 | Wetlands: policy ahead of knowledge?

Paul Adam

Introduction

Over the past few decades there has been a significant change in the public appreciation of wetlands. From the view that wetlands are wastelands of no benefit to humanity and whose loss is a positive benefit, we have moved to an era in which wetlands are amongst the jewels in the environmental crown, to be protected at almost any cost. This change in opinion is reflected in policies, legislation, regulation and administrative practice in many countries (see among others, Government of Canada 1991; Davidson and Gauthier 1993; Lynch-Stewart *et al.* 1993; Nakashima and Khan 1994; ANCA 1996; DLWC 1996; Environment Australia 1997). In part this greater appreciation of wetlands reflects more general concerns about the environment, but more particularly marks the assimilation into general knowledge of particular interpretations of the results of ecological studies of wetlands. One consequence of the increased awareness of wetland values has been a substantial upsurge in research activity. Nevertheless, I would argue that, while the case for wetland conservation is very strong, the development of wetland policies has often presumed that science provides a firmer foundation for decision making than either is, or can be, the case.

Wetlands are widely distributed throughout the world. My focus in this review is generally on Australia, but the issues are not unique to a single continent. Lewis (1995) discusses similar issues from a North American perspective.

What is a wetland?

While it is easy to talk in general terms about the importance of wetlands and to support their protection, do the various parties to such discussions share the same image of wetlands?

'Wetland' is not a term of great antiquity. The *Oxford English Dictionary* defines it as 'an area of land that is usually saturated with water, often a marsh or swamp', and cites it as being first used in the eighteenth century by M. Catesby in his *Natural History of Carolina, Florida and the Bahamas*. However, it is only in the last 30 years that the term has enjoyed wide usage.

Various types of wetland have long been recognized and given vernacular names, sometimes reflecting an extremely sophisticated classification of habitats (Boulé 1994; Mitsch, Mitsch and Turner 1994; Gopal and Sah 1995; Pakarinen 1995; Scott and Jones 1995). However, if it is decided to recognize a broader inclusive category of 'wetland', how many of these distinct habitat types should be included and should 'wetland' be defined in greater detail than is provided for by the *Oxford English Dictionary*?

The most widely quoted definition of wetland is that adopted by the Convention on Wetlands of International Importance (the Ramsar Convention – UNESCO 1971): 'areas of marsh, fen, peatland or water, whether natural or artificial, permanent or temporary, with water that is static or flowing, the depth of which at low tide does not exceed six metres' (Ramsar Convention Article 1 para 1). This definition is much broader than that of the *Oxford English Dictionary*.

Scott and Jones (1995) suggested that, as well as being widely quoted, the Ramsar definition is the most accepted in that many countries are signatories to the Convention, and in becoming parties to the Convention they have adopted this definition of wetland. In the case of Australia it was the Ramsar definition which was adopted in the Wetlands Policy of the Commonwealth Government of Australia (Environment Australia 1997), but with certain significant exceptions. 'Rocky marine shores, including rocky offshore islands and sea cliffs and "the main in-channel elements of permanent rivers and streams, including waterfalls" are not considered wetlands; and "human-made" or purpose-built wetlands should *not* be considered as replacement, or compensation, for natural wetlands proposed for destruction without expert supporting advice' (Environment Australia 1997: 29). While the Ramsar definition is repeated in official documents referring to the Convention and its implementation, working definitions and wetland regulations in many countries clearly apply only to a subset of the broad, inclusive Ramsar definition.

Probably the second most-quoted definition of wetland is that of the US Fish and Wildlife Service (USFWS) (Cowardin *et al.* 1979: 3): 'Wetlands are lands transitional between terrestrial and aquatic systems here the water table is usually at or near the surface or the land is covered by shallow water. For the purpose of this classification wetlands must have one or more of the following attributes: (1) at least periodically, the land supports predominantly hydrophytes; (2) the substrate is predominantly undrained hydric soil; and (3) the substrate is nonsoil and is saturated with water or covered by shallow water at some time during the growing season of each year.'

This definition, although broad, is narrower than that of the Ramsar Convention as sites that 'are permanently covered with tidal water are considered deepwater habitats, regardless of water depth' (Cowardin and Golet 1995: 141).

Within Australia different agencies, even within a single state, employ different definitions of wetland to meet particular purposes and the requirements of individual pieces of legislation. Inconsistencies in definitions may give rise to conflict between agencies and are a source of confusion to the public (Pressey and Adam 1995). Many Australian wetland scientists have advocated adoption of the Ramsar definition because of its wide use in international fora, and to promote consistency (Pressey and Adam 1995). Nevertheless, acceptance of the appropriateness of the Ramsar definition is far from complete. Concerns arise from the inclusion within the Ramsar definition of artificial wetlands (most Australian wetland surveys have specifically excluded artificial systems) and from the implicit inclusion of at least some coral reefs (coral reefs, although widely acknowledged as important ecosystems of high conservation significance, have seldom been regarded as wetlands). The arbitrary six-metre depth limit also is a source of difficulty as it creates an artificial boundary within ecosystems; both corals and seagrass beds may extend into deeper waters. Very extensive areas of what are now regarded as wetlands in Australia occur in the inland arid zone and are only intermittently wet. The Ramsar definition makes reference to 'marsh, fen, peatland' – terms not regularly used in Australia and difficult to apply to arid zone intermittent and episodic wetlands (it is also difficult to apply the USFWS definition to these wetlands). In practice, the Ramsar Convention has been applied to intermittent wetlands. The Ramsar definition clearly applies to all freshwaters, regardless of depth, and so includes lakes and rivers as wetlands. Whether or not this is viewed as desirable may depend on both prejudice and background. Wetland ecologists with a limnological background may concentrate on the 'wet' of wetland; terrestrial ecologists, while recognizing the links between open water and surrounding areas, may stress the 'land' and recognize wetlands as being transitional between terrestrial and aquatic systems (as in the USFWS definition).

Despite concerns over the implications of adopting the Ramsar definition, alternatives proposed specifically for Australia have also failed to gain total acceptance. Paijmans *et al.* (1985: 2) observed that 'Wetlands are quirks and local aberrations of the hydrological cycle which differ from their surroundings by the persistent presence of free water.' Their definition of wetlands is broad, including land permanently or temporarily under water or waterlogged, with temporary wetlands having surface water or waterlogging of sufficient frequency and/or duration to affect the biota. In the case of arid zone basins, Paijmans *et al.* (1985) argued that, even if they fill only once every several decades, they should be regarded as wetlands because they support a distinctly aquatic biota when filled.

The New South Wales Government Wetlands Management Policy (DLWC 1996: 12) defines wetland as land that is:

- inundated with water on a temporary or permanent basis
- inundated with water that is usually slow moving or stationary
- inundated with water that may be fresh, brackish or saline

The inundation determines the type and productivity of the soils and the plant and animal communities.
The Policy covers all natural wetlands.

This definition differs from the Ramsar definition in excluding artificial wetlands and at least some rivers.

'Shallow' is not defined, but would exclude some lakes, although it is unclear whether seagrass is included. The policy acknowledges difficulties in definition: 'there may be some problem in defining wetlands. Many experts have attempted precise definition, but no definition has been universally accepted. There is agreement about most wetlands, and the disputes are confined to marginal cases. A commonsense approach will be adopted when assessing whether or not an area should be classified as a wetland . . .' Unfortunately, when push comes to shove common sense is likely to be the first casualty.

Ecologists would recognize the essentially Procrustean nature of any classification of ecosystems (such as wetlands or rainforest) and would not be surprised that there was lack of consensus as to definition. However, such uncertainty creates difficulty for implementing any policy or legislation which presumes that there is an agreed definition, while any definition developed for a particular purpose is likely to attract criticism.

I would argue that attempts to invest wetland definitions with apparent precision are counterproductive and in the long term may be used to discredit science in the eyes of the public. For particular purposes a precise definition may be both desirable and possible, but a precise general-purpose definition is an unattainable dream. It would be better to explain to the public the spatial and temporal complexity of the natural world than to engage in pseudoscientific exercises of imposing artificial limits on natural continua, exercises which serve more to confuse than to enlighten.

Regardless of how the term 'wetland' is defined, the very breadth of the concept limits its usefulness. The various types of wetland have little in common, either in terms of biota or ecological processes. A mangrove is very different from an alpine sphagnum swamp, even though there would be agreement that both are unequivocally wetlands. The generalizations which could be applicable to all wetlands are few in number; a wetlands policy which attempts to cover all wetlands risks being little more than a set of motherhood statements, appealing but not operational.

Is it possible to devise a generally acceptable classification of wetland types within the overall concept of wetland? An agreed classification would aid communication and permit comparisons, and could allow refinement of policy by particular provisions for individual wetland types. Vernacular classifications may recognize fine distinctions between wetland types but in any given language do not encompass the global range of wetlands. Additionally, with an increasingly urbanized population, the usage of many terms applied to wetlands has become imprecise and the nuances of distinctions have been lost. Nevertheless, terms which are unique to particular countries become part of the national identity, unlikely to be surrendered lightly to the cause of international uniformity. For example, 'billabong' is a source of mystification to international audiences but will continue to be used by Australians (e.g. Fairweather and Roberts 1996). A feature of most vernacular classifications is that they are not consistently hierarchical. Hierarchical classifications have been developed for the USFWS (Cowardin *et al.* 1979) and the Ramsar Convention Bureau (Scott and Jones 1995), but an informal survey of Australian wetlands scientists (Pressey and Adam 1995) showed that fewer than half of those scientists approached supported the adoption of a global classification scheme.

Wetland boundaries

Regulatory regimes which treat wetlands differently from other components of the landscape not only require that wetlands be defined, but that their boundaries can be precisely determined. The determination of the position of a wetland can become crucially significant in legal proceedings. Two main approaches to establishing wetland boundaries have been employed: prescriptive mapping or prescriptive definition. For the purposes of State Environmental Planning Policy 14 (Coastal Wetlands) in New South Wales, wetlands were mapped on the basis of air photo interpretation (Adam *et al.* 1985), with wetland being defined by vegetation type. The policy applies only to the areas mapped, so that the application of ·the policy requires the translation of the boundary as mapped to the field. The second approach does not rely on prior inventory and mapping but provides definitions and criteria for boundary limitation to be used as required. The first approach, although arbitrary, limits the scope for argument but is impractical for any policy covering large numbers of wetlands without very expensive mapping and surveys. The second approach does not require regulatory agencies to invest in expensive inventory and mapping but does not guarantee certainty as both the criteria for determining boundaries and their implementation attract controversy.

The USFWS classification (Cowardin *et al.* 1979) stated that wetlands can be identified by: vegetation dominated by hydrophytes; occurrence of hydric soils; or hydrology. Subsequently, considerable effort has been devoted to developing criteria for assessing these features (Cowardin and Golet 1995).

Carter, Gammon and Garrett (1994) determined the boundary of a large wetland using hydrological, soil and vegetation data. The boundary position differed according to what parameter was used, though, given the size of the wetland, Carter *et al.* (1994) suggested that the variation (38m horizontally and 1.1m vertically) was relatively small. However, a land developer might think differently. In addition to variation in boundary position according to which particular parameter is used for definition at one point in time, boundaries will also fluctuate over time (although the extent of fluctuation may depend on the parameter used).

The question to be asked is whether those requiring the determination of wetland boundaries can seek comfort from science. As Carter *et al.* (1994: 199) observed:

> To ecologists interested in ecotones and continuums, placement of exact boundaries may seem presumptuous. The designation of discrete wetland boundaries is more a regulatory problem than an ecological decision. The definition of wetland is sufficiently vague and circular to preclude easy determination of such boundaries. Attempts to refine the definition and impose numerical criteria have so far failed, partly because of the great variability of wetlands nationwide, the difficulty of measuring wetland

hydrology, which is spatially and temporally variable and is controlled to a great extent by weather, and the broad water-tolerance of many species. This failure is probably due also to a dearth of good scientific information and the reluctance of scientists and managers to compromise on the boundaries of what they consider to be a valuable national resource. Very few existing wetlands have not been affected by disturbance, especially near their edges, and alteration of their species composition, soils, or hydrology makes placement of boundaries more difficult.

As the wetland boundary zone is frequently one of low topographic variation, a small vertical change in the position of the wetland boundary may be expressed as a large horizontal shift (Adam 1992). With changing climatic and hydrological conditions, many wetland boundaries will be temporally unstable. This is likely to be particularly the case in much of inland Australia.

Cowardin and Golet (1995: 148) argued that the USFWS definition of wetland 'was proposed as an ecological, not a regulatory, definition', and suggest that 'the definition of wetland is a scientific issue; the scope and intensity of wetland protection are policy issues. Although policy is of critical importance, we do not believe that it should overrule science in scientific matters'.

The problem with this approach is that by focusing scientific attention on the precise delineation of wetland boundaries, policy makers and administrators have, perhaps unintentionally, not been made aware of the inherent variability of wetland margins. While policy issues should not overrule science, greater informed dialogue between scientists and policy makers might assist in focusing policy.

If a fixed wetland boundary is, from an ecological perspective, a legal fiction, one solution to developing regulations to protect wetlands would be to define buffer zones. However, the basis for defining effective buffers is likely to vary between wetlands; additionally, the width of buffers required to protect wetlands from particular actions in the catchment might vary according to the activity concerned (see Mulamoottil *et al.* 1996).

The value of wetlands

Until recently wetlands were regarded, at least in western countries, as wastelands. The conversion of wetlands to dry land was seen as a public good, creating an economically valuable resource at the same time as removing a threat to human health and welfare. The process of conversion was even given the name 'reclamation', with connotations of restoring the land to some former state of grace from which it had regrettably lapsed. (With the shift in appreciation of wetland values reclamation has begun to be disfavoured – the current terminology in Britain is 'land claim', which, in Australia, would carry different meanings, so another term will be required.) While local communities may have valued and benefited from the bounty of wetlands, large landholders and central governments saw only economic gain from wetland destruction. Examples of wetland reclamation are discussed by, among others: van Veen (1955), Wagret (1969), Williams (1970; 1990 a, b), Richards (1990), Gosselink and Maltby (1990) and Whitney (1994).

From the start of the modern conservation movement in the late nineteenth century, certain wetlands were recognized as 'special' places worthy of protection, but concern for wetlands in general can be dated to the early 1960s. Since then various accounts have argued that wetlands have numerous values of direct and indirect benefit to humankind. (See for example, Adam *et al.* 1985; Maltby 1985; Mitsch and Gosselink 1986; Williams 1990 a, c; Dugan 1993.) Although there are variations in emphasis between these and other accounts the most frequently recognized values include:

- Functional values – Wetland ecosystem functions are of value to humanity and to other components of the biosphere. Particularly important functions are
    i)   Productivity
    ii) Chemical cycling and retention (including both nutrients and pollutants)
    iii) Hydrological and geomorphological functions.
- Provision of habitat (which contributes to the importance of wetlands for the conservation of biodiversity)
- Recreational value
- Educational value
- Scientific value
- Aesthetic value

The documentation of these values has provided the impetus for the development of wetland protection policies in a number of countries. However, the diversity of wetlands means that not all values apply equally to all types of wetland, and even within a particular wetland type the extent to which findings from particular sites can be extrapolated to the whole class is still uncertain. The number of wetland types for which reliable quantitative data on functional and habitat values have been collected is small.

While some of the non-science-based values have been significant factors in the move towards wetland protection (for example the value of some wetlands for duck hunting and fishing creates a considerable constituency in favour of protection and generates major economic activity), it is supposedly science values which have, in general, been the impetus for policy development. These values are created by the functional attributes of wetlands and through the provision of habitat. Of the functional attributes the first to achieve prominence was productivity.

The development of the systems approach to ecology stimulated an interest in the study of productivity. In the case of saltmarshes the pioneering study of Teal (1962) led to the 'outwelling hypothesis' of Odum (1968; 1980) and initiated a continuing controversy. Teal (1962) studied productivity in an Atlantic coast *Spartina alterniflora* saltmarsh and concluded (p. 624) that 'the tides remove 45% of the production before the marsh consumers have a chance to use it and in so doing permit the estuaries to support an abundance of animals'. This conclusion received support from studies by Odum and de la Cruz (1967). The outwelling hypothesis developed from these studies posits that saltmarshes 'produce more material than can be degraded or stored within the system, and that the excess is exported to coastal waters where it supports ocean productivity' (Lefeuvre and Dame 1994: 172). This intuitively appealing hypothesis rapidly turned into dogma and was the major argument used to justify measures to protect saltmarshes, initially in the United States (Nixon 1980) and subsequently globally.

Nixon (1980) critically reviewed the evidence for the outwelling hypothesis and suggested that it was based on a very uncertain foundation, supported by few quantitative data, which had been overinterpreted. What reliable data existed did not permit a consistent interpretation and Nixon (1980) concluded that 'chaos reigns and that we dare not make any statement at all about what marshes are importing or exporting' (p. 447).

Subsequently, there have been a great many studies of aspects of the outwelling hypothesis (see Adam 1990; Lefeuvre and Dame 1994; Dame and Lefeuvre 1994). The majority of studies have been from American estuaries, with relatively few studies from Europe and even fewer from elsewhere. These studies have confirmed the variability between estuaries indicated by Nixon (1980), but patterns of behaviour are beginning to emerge which suggests that net fluxes between marsh–estuarine systems and the sea are a function of the stage of geomorphological evolution of estuaries (Dame 1994; Dame and Lefeuvre 1994).

The appeal of the outwelling hypothesis in the 1960s was such that it was often taught as 'gospel' in basic ecology courses, and it formed a cornerstone of many arguments in favour of saltmarsh conservation (Nixon 1980). The hypothesis provided the logical underpinning for legislation protecting saltmarshes and in the state of Rhode Island a form of the hypothesis was formally written in the legislation (Nixon 1980).

Nixon (1980) suggested that the story of the outwelling hypothesis had important lessons for scientists involved on the interface between science and policy:

> A relatively few researchers, however well-meaning and able, failed to make and maintain a firm distinction between what they thought was happening, or what they thought 'ought' to be happening, and what they had good data to show was happening. Because they told their story so often (and at times so eloquently), because at least some of them were very well-known and respected, because the credibility of the printed scientific literature was so strong, perhaps because it sounded like such a 'good' story, too many of us failed for too long to examine the concept critically. Instead, we passed it on eagerly as one of the accomplishments of marine ecological research. And we passed it on very effectively, to students, to managers, to legislatures, to funding agencies, and to each other. It occurs to me that if the early papers had clearly presented 'outwelling' as an exciting hypothesis, it might not have taken so long for the data necessary to evaluate it to become available. It is reassuring that the scientific process has prevailed, and that we have begun to obtain some of the measurements necessary to evaluate the hypothesis, but it is humbling to realize how quickly and completely an idea became implanted in the literature, and in our minds, with so few data to support it. (p. 442)
>
> But before I finish, I want to go back to the theme of the introduction to this review. I do so because I think there is a larger lesson running through the past twenty years of work than is apparent in a summary of our present knowledge. During the last few months I have had a number of vigorous discussions with various groups over my opinions about the history of ecological research on the question of marsh–estuarine inter-actions. In a number of these discussions, a common sentiment was expressed that it was the responsibility of the ecological community to help in the 'battle' to gain time while environmental awareness developed among the public and the regulatory agencies. The momentum of the developers was so great that an atmosphere of certainty and consensus was necessary for the voice of the ecologists to be heard. The essence of the argument is that, 'Yes, perhaps we overstated the case a bit, but it was important to help save the marshes. Now that is done, or at least well along, and we can go back and work on getting our science right.'

I do not agree. It is a bad bargain to trade our credibility for political advantage. Science is a social enterprise, we communicate through the scientific literature, and must do nothing to undermine the integrity of that communication. Both in sending and in receiving information, we must remain skeptical. Reading the literature on marsh–estuarine interactions convinces me that we have been too willing to trust our own preconceptions, and too eager to believe what other people are saying about their data when they agree with those preconceptions. Ecology is a young science, and we are still about the business of learning some of the basics. (p. 510)

The outwelling hypothesis rapidly became part of accepted wisdom, still appears in elementary textbooks and is one of the unspoken arguments justifying wetland protection policies. The more complex and still emerging models of the interactions between saltmarshes and coastal waters (Dame 1994; Dame and Lefeuvre 1994) remain in the province of science and have not deposed the outwelling hypothesis (which is accepted as fact and not hypothesis) in the broader public mind.

In Australasia the outwelling hypothesis has been implicitly accepted in a number of accounts addressed to the general public (see, for example Lear and Turner 1977; Milledge, Parker and Campbell 1988; Queensland Department of Primary Industries 1989; Crisp, Daniel and Tortell 1990). There have been few studies of ecosystem functioning in Australian saltmarshes, although more attention has been given to mangroves. Quantitative studies of fluxes in mangroves suggest that, currently, generalizations would be unwise (Robertson and Alongi 1995) but that the models emerging from studies of saltmarshes (Dame 1994) may be relevant to mangroves also. The data on which the outwelling hypothesis was based were from *Spartina alterniflora* marshes, a type of saltmarsh very different from most others (Adam 1990); these differences have not prevented the extrapolation of productivity estimates from American marshes to Australian marshes (Larkum 1981). These estimates, although not supported by data, have nevertheless become widely accepted as fact. The outwelling hypothesis was explicitly applicable to coastal saltmarsh and not to all wetlands. Nonetheless, in popular belief wetlands in general are frequently claimed to be highly productive (comparable with *Spartina alterniflora* marshes), and much of this productivity is exported.

In presenting a critique of the outwelling hypothesis I am not arguing that intertidal wetlands are unimportant. The modern synthesis clearly indicates that fluxes between intertidal wetlands and coastal waters are biologically significant, and there is increasing evidence for the importance of intertidal wetlands as nursery habitat for fish, including many of commercial significance.

Given that there is still ample justification for intertidal wetland protection does it matter that 'we have done the right thing for the wrong reason'? I would argue that it is necessary for scientists to seek to inform the public, policy makers, bureaucrats and politicians of the current understanding of wetland functions; the persistence of belief in the outwelling hypothesis indicates that this will not be an easy task. If policies are based on incorrect assumptions the actions sanctioned by the policies may have undesirable outcomes, and, importantly, if scientists do not correct past inaccurate interpretations the credibility of both policies and science may be undermined. (With the tendency to increased regulation of land use the basis for regulation is likely to come under sustained rigorous examination, both through the political process and via litigation.)

The other aspect of wetland function which has entered into public consciousness is the role of wetlands in nutrient and pollution assimilation, giving rise to the metaphor of wetlands as the landscape's kidneys and to the development of artificial wetlands for wastewater management (see Reddy and D'Angelo 1994; Brix 1994; Mitsch 1994). Again it needs to be stressed that different wetlands are likely to have different assimilatory capacity and that broad generalizations are dangerous.

Assessing wetland functions

If protection of functional attributes of wetlands is recognized as important justification for wetland protection then implementation of wetland policies may require that functional value be assessed. For example, one of the requirements of the Canadian Federal Policy on Wetland Conservation is that there be no net loss of wetland function (Rubec 1994). The success of this policy can only be determined if wetland function can be measured.

Considerable effort has been devoted to the assessment of wetland function (Brinson et al. 1994; Larson and Mazzarese 1994; Maltby et al. 1994). However, there are currently no methodologies which could be used to assess all functions in all types of wetlands; at present we have indicators of functions, but quantification is likely to be difficult. Nevertheless, it would appear that some of those responsible for the administration of wetlands believe that assessment of functional values is both possible and practicable.

In 1995–96 a Commission of Inquiry, constituted under New South Wales planning law, was held into a proposal to develop a marina at Shellharbour on the NSW South Coast. The development would involve the loss of an area of saltmarsh, included within State Environmental Planning Policy 14 – Coastal Wetlands. One of the issues before the inquiry was whether the saltmarsh was of value. The Department of Urban Affairs and Planning in its submission to the inquiry argued that:

> the value of a wetland is the sum of its functional characteristics as a habitat for fauna and flora, its bioproductivity, its contribution to biodiversity, its flood retention role, and its water polishing, scientific and educational usefulness. Inclusion in the Policy on the other hand is based on a number of mapping criteria which are not in the strictest sense measurement of values. They are useful in defining a boundary which encompasses a minimal disturbed wetland comprising indigenous wetland vegetation communities. The true value (of) a wetland can only be demonstrated through detailed studies such as an EIS [Environmental Impact Statement]. (DUAP 1995: 17)

The inclusion of wetlands in SEPP14 (New South Wales State Environmental Planning Policy 14) was based on mapping criteria which did not, in any sense, let alone the strictest, constitute a measure of values. The need for separation of the processes of wetland identification and delineation from analysis of wetland function was one of the major recommendations in Lewis (1995). SEPP14 was predicated on a general recognition of wetland values (Adam et al. 1985) and my interpretation of the policy does not lead me to anticipate a rigorous quantitative evaluation of every single wetland. The suggestion that 'The true value (of) a wetland can only be demonstrated through detailed studies such as an EIS' (environmental impact statement) is particularly surprising. The number of EISs which have involved rigorous collection of site-specific data addressing the supposed components of wetland value is vanishingly small, while I am unaware of any EIS (for any habitat) which has involved specific studies of 'bioproductivity'.

If the Department's submission is a true indication of what is expected of an EIS then the assessment process will become both lengthy and extremely expensive. The Department has been required to study more EISs than other agencies or individuals. It is therefore puzzling that such a misleading claim as to the role and content of an EIS, in practice as distinct from theory, could be made.

The Commissioners (Simpson and Train 1996) recognized that there were many attributes of wetlands which could not be quantified:

> The Commission considers that many wetland values can only, at this stage, be considered subjectively. Some may consider that freshwater wetlands, because they provide habitat for amphibians, because of the

bird species they attract, because of their aesthetic values and other factors are more valuable than saltwater wetlands. Conversely others consider that saltmarsh wetlands because of the nature of the plants they support as a result of their rarity, because of the habitat they provide for migratory birds of the foreshore, because of their drought resistance and other factors are more valuable. The Commission is not in a position to value the merits of each wetland type. On the evidence few, if any authorities, could make such valuation. What is obvious is that wetlands are important, are different, and at this stage in scientific understanding the Commission cannot value these attributes quantitatively, only acknowledge their existence.

As a result of the above, the Commission considers that the only criteria which it can adopt in assessing relative values of the two wetlands are related to potential for water quality improvement and areas of wetlands. (p. 84)

The suggestion that the only criteria on which assessment could be based are those which can be quantified has implications for any planning decisions involving wetlands. (Of the two criteria the Commissioners considered one – area – is trivial, while the other – potential for water quality improvement – perhaps involved a greater degree of speculation than was acknowledged by the Commissioners.) Historically many (most) planning decisions have ultimately required subjective judgement. While the range of attributes to be considered may have increased as a result of scientific investigations, the absence of site-specific data is not necessarily an impediment to considering the whole range of attributes in the evaluation process. To use the absence of such data as an excuse for failing to exercise professional judgement will not improve the quality of decisions.

The positions adopted by the Department of Urban Affairs and Planning and the Commissioners is perhaps indicative of a breakdown in communication and understanding between wetland scientists and both the bureaucracy and the wider community. The expectations of the quantitative advice that scientists can provide are both unrealistic and undesirable. In response to the report of the Commissioners the Minister for Urban Affairs and Planning granted approval for the development.

An approach to assessing wetlands which has attracted much attention is the attempt to assign economic values to some or all of wetland attributes. A much-quoted early attempt to ascribe a monetary value to Atlantic coast *Spartina* marsh estimates they were worth (in 1974) US$2500 per acre per year for their nutrient uptake function (Gosselink, Odum and Pope 1974). This estimate provided much ammunition to those fighting to protect wetlands.

Nixon (1980) pointed out that 'not one real marsh–estuarine nutrient uptake study' (p. 443) was cited in defence of the assigned value. He also suggested that there was an element of 'double think' in promoting simultaneously the outwelling hypothesis and a nutrient sink role for saltmarshes. Nixon (1980) nevertheless considered that estimating the economic worth of wetlands was an interesting and challenging scientific problem.

Most estimates of wetland monetary value have been impressively large and have thus supported the case for wetland protection. Estimates have been provided for a range of different wetland types, but all must be regarded as rubbery. Costanza *et al.* (1997) have produced estimates of the global value of ecosystem services provided by major biomes; these figures suggest that the total value of ecosystem functions is about twice current global gross national product. Broken down by biome, the estimates suggest that wetlands are major providers of ecosystem services. Leaving aside problems with the methodology by which the figures were derived, Common (1997) suggested that they were 'meaningless and useless' and that the idea that assigning monetary value to ecosystem functions was the only way of persuading politicians and the public to give serious consideration to protecting ecosystem functions was incorrect. Given the difficulty in identifying and quantifying wetland attributes, let alone in assigning dollar values to attributes, the refinement of economic evaluations is unlikely to occur quickly, although research currently in progress on inland wetlands in New South Wales is showing promise in assessing the relative value of utilization of water resources for irrigation and wetland conservation (Morrison and Kingsford 1997). Additionally, the variability of wetlands in space and time is likely to render extrapolation from single studies difficult. In many cases wetland values benefit society at large rather than individual landholders. We could, for example, come up with an estimate of the value of estuarine wetland-dependent fish in a whole fishery (although dividing that estimate by the area of wetland to produce a value per hectare would be of little utility in the context of site-specific management decisions in view of the variability between sites). This value is of interest to fishers and consumers, but may be of little direct relevance to landholders who gain no return for providing the basis of the livelihood of fishers.

Although absolute economic values cannot be placed on either wetlands in general or on specific sites, it is justified to assert that wetlands are of considerable economic benefit (Mitsch and Gosselink 1986) and this should be taken into account in decision-making processes. The production of artificially precise figures to support decisions is unwise – but the pressure to do so is great when the benefits of activities which destroy or degrade wetlands can be quoted to the nearest dollar and politicians are exhorted to display economic rationality in their decision making.

The urgency of action

Accepting that wetlands collectively provide ecological services of great value to human society (even if these services cannot as yet be quantified with any great precision), how urgent is the need to develop and implement wetland protection policies? The development of wetland policies has in large part been predicated on a belief that there has been a historic and continuing loss of wetland.

That there have been extensive losses of wetlands is undeniable (Tiner 1984; Mitsch and Gosselink 1986; Dugan 1990; Gosselink and Maltby 1990). However, in many parts of the world estimates of the extent of wetland resources are still only preliminary and documentation of destruction incomplete. Even for one of the most distinctive wetland types, mangroves, Field (1995) regarded the available data on current extent and recent losses as extremely rough.

There has been a tendency in the public and political debate over wetland protection to hang arguments on particular figures of wetland loss rather than to use more general phases such as 'many wetlands have been lost'. This can result in original data being taken out of context and misinterpreted. A case in point is provided by Goodrick's (1970) study of coastal wetlands in New South Wales. This was one of the first studies to draw attention to the loss of wetland in the state and has been extensively used to promote wetland protection. Goodrick (1970) estimated that by 1969 some 60 per cent of wetland habitat of high value to waterfowl had been destroyed or had its value as habitat much reduced. The total figure was not broken down into areas actually destroyed and areas degraded. Goodrick's figure also applied only to habitats of high value to waterfowl, which is only a subset of total coastal wetland. Nevertheless, Goodrick's estimate has entered common knowledge as an estimate of wetland destruction and is frequently extrapolated to all wetland types in the state. The interpretation of loss was repeated in the draft Commonwealth wetland policy (ANCA 1996). In Western Australia the basis of much-quoted estimates of the loss of wetlands on the Swan Coastal Plain has been shown to be fallacious by Giblett (1996). It is undeniable that there have been substantial losses, but the exact area remains unknown. Giblett (1996: 136) suggested that 'This is a salutary instance of the abuse of statistics and how errors can easily be perpetu-ated behind a facade of scientific reliability.'

In Australia wetland losses in urban or near-urban areas have been high (as elsewhere in the world), but the national loss of wetlands is probably lower (as a proportion of the pre-European area) than in many other developed countries. The low population density and the use of much of the land for pastoralism rather than crop production has limited total losses, although some particular wetland types may have suffered severe losses. However, very many wetlands have experienced degradation, e.g. from the spread of weeds, increased nutrients, extra sedimentation, modifications to hydrological regimes and altered fire regimes. Degraded wetlands are still wetlands and may retain most wetland functions, although the relationship between degree of degrada-tion and maintenance of functions has rarely been investigated. Active management may provide rehabilitation for many degraded wetlands, although reversing the effects of eutrophication and sedimentation may not be possible.

While it is clearly important to control wetland destruction, the con-centration of attention on loss has limited awareness of the extent of degradation.

Conflicting agendas

The development of policies for wetland protection was driven by recog-nition of the importance of the ecological functions of wetlands. The Canadian Federal Wetlands Policy aims to ensure that there is no net loss of wetland function (Rubec 1994); in New South Wales State Envir-onmental Planning Policy 14 – Coastal Wetlands was introduced so that 'the coastal wetlands are preserved and protected in the environmental and economic interests of the State' (the economic interests arise from functions such as provision of fish habitat, nutrient assimilation and flood mitigation).

The relationship between wetland function and area is poorly known – production will increase with increasing area, although probably not linearly, while habitat value may vary considerably between sites. Re-gardless of the details of the relationship, maximizing the protection of wetland function requires measures to conserve the maximum area of wetland. In Australia, as elsewhere, very many wetlands are not managed by public authorities. Wetland conservation must therefore necessarily be achieved in different ways from conservation of other high-profile ecosystems, such as forests, where much of the land is publicly owned and is managed by, or for, public authorities.

It is my impression, however, that amongst both environmental groups and the wider public there is a view that conservation is about the preservation of the 'best' in some form of reserve. Wetlands are certainly underrepresented in formal reserve networks and there are good reasons for advocating greater reservation. Nevertheless, the wetland conservation paradigm which arises from a consideration of the importance of ecosystem functions must go far beyond the declaration of a few conservation reserves.

The tension between a classical nature conservation approach and a broader environmental perspective is also seen in the working of the Ramsar Convention. The convention seeks to enhance the protection and management of outstanding wetlands through establishing a register of wetlands of international significance. At the same time signatories to the convention are required to promote wise use of all wetlands. The listing of wetland under the convention has a fairly high public and media profile, the requirement to protect all wetlands is less often recognized.

Arising out of the Ramsar Convention have been attempts to prepare regional inventories of wetlands (Scott 1989). In Australia there has been the production of a *Directory of Important Wetlands* (Blackley, Usbeck and Langford 1996), with various criteria established to the inclusion of particular sites. The directory reflects a traditional nature conservation approach, more concerned with 'state' than with 'process'. Wetlands included within formal conservation reserves, or recognized as 'special' through listing on registers such as those of the Ramsar Convention, the *Directory of Important Wetlands*, the National Estate or by National Trusts, clearly require protection and management. However, I would contend that the compilation of such registers and the maintenance of wetland function are two different processes, although they have become confused in the public mind. There is a great danger that wetlands not recognized as being 'important' will be regarded as being of no consequence and hence not protected. The attempt at the Shellharbour Inquiry (Simpson and Train 1996) to rank the wetland under contention against others (so as to demonstrate that it was of lesser value and thus not deserving of retention) illustrates this danger, and I would argue, was contrary to the intent (Adam *et al.* 1985) of SEPP14.

To aim only for the protection of a few outstanding wetlands, difficult as even that may be to achieve, is a recipe for loss in the longer term of most of the wetland resource, and ignores the wider goal of retaining wetlands as functional components of the landscape.

Mitigation

A feature of a number of regulatory regimes, but most notably that of the United States, is provision for approval of activities which damage or destroy existing wetlands when the proponent intends some form of mitigation. Mitigation may take the form of restoration or rehabilitation of some existing degraded wetland or the creation of a new artificial wetland. In the USA permits for wetland destruction are issued under Section 404 of the Clean Water Act 1977, and the permits lay down requirements for mitigation. The guidelines for mitigation indicate that damage to ecosystem functions should be mitigated. 'This means that the compensatory wetland must equal or exceed the performance of the damage site' (Zedler 1996: 33). Mitigation must either be contemporaneous with the destruction or through the establishment of a mitigation bank. 'Mitigation banking can be achieved through the creation, restoration, enhancement or preservation of other wetland areas of equivalent value generally located outside the immediate area of wetlands loss or alteration.'

> Wetland mitigation banks are normally relatively large blocks of wetlands whose estimated tangible and intangible values, termed credits, are similar to cash deposits in a regular checking account. As anticipated development takes place, credits equivalent to the estimated unavoidable wetland losses are withdrawn or debited from the bank to compensate for the losses incurred. (Reppert 1992: 1)

This concept, although attractive in principle, presents difficulties in establishing equivalency of value, given our imperfect knowledge both of wetland functions and of means of assessing them. In practice this difficulty has generally been avoided – 'However, the methodology which is most commonly used for valuation and accounting purposes is a non-analytical (and non-functional) one which merely tabulates credits and debits according to a range of various wetland types' (Reppert 1992: 1).

Advocates of mitigation argue that compensatory wetlands are successful. However, there are few data which support this view, not necessarily because of failure, but because there has been little long-term monitoring. Even when monitoring has occurred, failure to establish a priori explicit criteria for success renders assessment of results difficult. Mere establishment of a wetland is no guarantee that desired functions are adequate. Zedler (1991) discusses the case of a Spartina foliosa saltmarsh established as a habitat for the endangered light-footed clapper rail. Although a *S. foliosa* marsh was established, the project could not be regarded as successful as it 'lacks that species [i.e. light-footed clapper rail] and is deficient in food chain support, plant productivity, and nutrient supply functions' (Zedler 1991: 35), leading Zedler to argue that 'mitigation policies must be put on hold until constructed wetlands are proved capable of attracting and sustaining the full complement of native species'.

Proponents of mitigation seek support from the scientific literature for predictions of the likely success of particular projects. Support is, however, sought selectively. 'Outdated ideas about community development appear to have found new advocates among authors of mitigation plans. The presumption that a site can be modified to replace a specific ecosystem, or to perform in predictable ways, is at odds with current ecological understanding of ecosystem complexity (cf. Pickett and White 1985; Botkin 1990). The old ideas of balance, equilibrium, homogeneity, and determinism die hard (Levin 1989)' (Zedler 1996: 33). Chapman and Underwood (1997) are similarly critical of most wetland rehabilitation and restoration projects, pointing to the lack of clearly articulated and achievable goals in many projects and the general failure to treat projects as experiments which should be designed, conducted and studied as such.

The way ahead

It has not been my intention to argue against the protection and conservation of wetlands. Rather it is to suggest that the case is sufficiently strong not to require pseudoscientific precision in the estimation and assessment of wetland values.

The increased awareness of wetlands, and the focus of attention on ecosystem functions, has led to an increased research effort. This research clearly demonstrates the importance of particular functions in particular wetlands, but also highlights the variability within and between wetlands. In the absence of site-specific data we should not be tempted into unwise extrapolation and attribution of spuriously precise estimates to functional values. I would propose that we should argue from the particular to the general, and use those few rigorous studies that have been reported to suggest that wetlands in general have a range of values, and invoke the precautionary principle to promote the protection of all values.

The evidence of substantial loss and degradation of wetlands is overwhelming, as is the evidence that wetlands provide numerous ecosystem services. There is an urgency to the need to develop and implement effective mechanisms for wetland protection and management. If the arguments used to promote wetland conservation are poorly founded, or are over-interpretations of the evidence they may provide ammunition to those who would seek to discredit the case for conservation (what Ehrlich and Ehrlich (1997) have termed the 'brownlash').

We need many more studies of the whole range of wetlands to fully determine the values of wetlands, but this will be a long-term task. Whether or not this endeavour will yield simple, predictive assessment of ecological value remains to be seen. Given the variation between sites, and variation in methodologies and objectives of individual studies, it will be difficult to search for valid generalizations about wetland functions through a synthesis of existing data. Carefully designed studies addressing specific hypotheses will be required. Economic assessment techniques may also be refined with further research.

The greater role for science in future development and implementation of policy probably lies in the provision of a firmer basis for management. The majority of wetlands are, to varying degrees, degraded and require active management for their restoration and rehabilitation. If mitigation remains a policy option then we need a much stronger scientific basis for creation of new wetlands, while the development of wetlands for wastewater management will also require continuing scientific input. It is important, however, to acknowledge that, while management prescriptions should be made on the basis of the best available knowledge, our current understanding of wetland functioning is still very incomplete. Management must be treated as an experiment, and the success, or failure, of particular exercises should be assessed through properly designed studies planned in advance of any action (Chapman and Underwood 1997). Only if management is approached in this way will adaptive management (Dovers and Mobbs 1997) be possible.

A number of wetland management manuals have been produced (for example, Payne 1992; Brouwer 1995; Hull and Beovich 1996) but these have an empirical rather than theoretical basis. A major task for the future is to synthesize the rapidly increasing experience of wetland managers so as to develop a firmer base for future management.

Wetlands have been singled out for special attention by both regulators and the environmental movement. While recognizing the value of wetland, is this appropriate? In Australia the tendency has been for public and political attention to be focused on particular ecosystems – rainforest, old growth forests, wetlands. Does this mean that other ecosystems are of lesser importance? The answer is clearly no, and the failure to recognize this is a reflection of our still imperfect perception of the environment. As Pressey and Adam (1995: 84) argued:

> If management and conservation were applied equally to all components of the landscape, wetlands would not have to be separated from the other components with which they interact physically, chemically and biologically. Moreover, the many habitats often grouped together as wetlands are not a natural, homogenous group. Many have more in common with non-wetland habitats than with each other.

The continuing and unresolvable debates regarding wetland definition and the delineation of boundaries would become less important if a landscape approach to management were adopted. The trend towards catchment management is a welcome move, under which the importance of wetlands to the catchment can be recognized, while the links between wetlands and other landscape elements can be protected.

Despite the general acceptance of the importance of protecting wetlands (both Presidents Bush and Clinton strongly supported the US no net loss policy – Mossop 1992; Zedler 1996), not all are convinced, either in specific instances or more generally, and wetland protection policies, particularly where they impact on private landholders, remain controversial. In New South Wales the NSW Farmers' Association is still critical of SEPP14 more than 10 years after its introduction (Motion 20. NSW Farmers' Association General Council April 23–24, 1996), while in the USA some landholders argue that the imposition of regulations over use of wetlands is a form of taking of property without compensation and is hence unconstitutional (McDonough 1994).

Although absolute unanimity is unattainable, the divide between the regulators and the regulated must be breached so that the community at large has ownership both of the problems and the solutions. The solutions must address political, social and economic issues as well as providing technical fixes, and, even when underpinned by community support, will require adequate funding. Given that many wetlands will remain in private ownership, policies must provide mechanisms which adequately recompense landholders (financially or in kind) for their stewardship of wetland resources.

If current political rhetoric about the need to protect and manage wetlands is to be converted into reality then not only may we require more and better instruments to achieve that goal, but science has to provide more and better management tools and a firmer underpinning of the basis for wetland protection policy.

References

Adam, P. (1990) *Saltmarsh Ecology*. Cambridge: Cambridge University Press.

Adam, P. (1992) Wetlands and wetland boundaries: problems, expectations, perceptions and reality. *Wetlands (Australia)* 11: 60–7.

Adam, P., Urwin, N., Weiner, P. and Sim, I. (1985) *Coastal Wetlands of New South Wales. A Survey and Report Prepared for the Coastal Council of New South Wales*. Sydney: Department of Environment and Planning.

Australian Nature Conservation Agency (ANCA) (1996) *Draft Wetlands Policy of the Commonwealth Government of Australia*. Canberra: ANCA.

Blackley, R., Usbeck, S. and Langford, K. (1996) *A Directory of Important Wetlands in Australia*. 2nd edn. Canberra: ANCA.

Botkin, D.B. (1990) *Discordant Harmonies: a New Ecology for the Twenty-first Century*. New York: Oxford University Press.

Boulé, M.E. (1994) An early history of wetland ecology. In: Mitsch, W.J. (ed.) *Global Wetlands: Old World and New*. Amsterdam: Elsevier, 57–74.

Brinson, M.M., Krurzynski, W., Lee, L.C., Nutter, W.L., Smith, R.D. and Whigham, D.F. (1994) Developing an approach for assessing the function of wetlands. In: Mitsch, W.J. (ed.) *Global Wetlands: Old World and New*. Amsterdam: Elsevier, 615–24.

Brix, H. (1994) Constructed wetlands for municipal wastewater treatment in Europe. In: Mitsch, W.J. (ed.) *Global Wetlands: Old World and New*. Amsterdam: Elsevier, 325–34.

Brouwer, D. (1995) *Managing Wetlands on Farms*. Paterson: CB Alexander Agricultural College 'Tocal'.

Carter, V., Gammon, P.T. and Garrett, M.K. (1994) Ecotone dynamics and boundary determination in the Great Dismal Swamp. *Ecological Applications* 4: 189–203.

Chapman, M.G. and Underwood, A.J. (1997) Concepts and issues in restoration of mangrove forests in urban environments. In: Klomp, N. and Lunt, I. (eds) *Frontiers in Ecology: Building the Links*. Oxford: Elsevier Science, 103–14.

Common, M.S. (1997) Roles for ecology in ecological economics and sustainable development. In: Klomp, N. and Lunt, I. (eds) *Frontiers in Ecology: Building the Links*. Oxford: Elsevier Science, 323–34.

Costanza, R., d'Arge, R., de Groot, R., Farber, S., Grasso, M., Hannon, B., Limburg, K., Naeem, S., O'Neill, R., Paruelo, J., Raskin, R., Sutton, P. and van der Belt, M. (1997) The value of the world's ecosystem services and natural capital. *Nature* 387: 253–60.

Cowardin, L.M. and Golet, F.C. (1995) US Fish and Wildlife Service 1979 wetland classification: a review. *Vegetatio* 118: 139–52.

Cowardin, L.M., Carter, V., Golet, F.C. and LaRoe, E.T. (1979) Classification of wetlands and deepwater habitats of the United States. Washington: US Fish and Wildlife Service. FWS/OBS 76/09.

Crisp, P., Daniel, L. and Tortell, P. (1990) *Mangroves in New Zealand: Trees in the Tide*. Wellington: GP Books.

Dame, R.F. (1994) The net flux of materials between marsh-estuarine systems and the sea: the Atlantic coast of the United States. In: Mitsch, W.J. (ed.) *Global Wetlands: Old World and New*. Amsterdam: Elsevier, 295–302.

Dame, R.F. and Lefeuvre, J.C. (1994) Tidal exchange: import–export of nutrients and organic matter in new and old world salt marshes. Conclusions. In: W.J. Mitsch (ed.) *Global Wetlands: Old World and New*. Amsterdam: Elsevier, 303–5.

Davidson, I. and Gauthier, G. (1993) *Wetland Conservation in Central America*. Ottawa: North American Wetlands Conservation Council (Canada).

Department of Land and Water Conservation (DLWC) (1996) *The NSW Wetlands Management Policy*. Sydney: New South Wales Government.

Department of Urban Affairs and Planning (DUAP) (1995) *Submission – Commission of Inquiry into the Shellcove Boatharbour Marina*. Sydney: DUAP.

Dovers, S.R. and Mobbs, C.D. (1997) An alluring prospect? Ecology and the requirements of adaptive management. In Klomp, N. and Lunt, I. (eds) *Frontiers in Ecology: Building the Links*. Oxford: Elsevier Science, 39–52.

Dugan, P.J. (1990) *Wetland Conservation: a Review of Current Issues and Required Action*. Gland: IUCN.

Dugan, P. (1993) *Wetlands in Danger*. London: Mitchell Beazley.

Ehrlich, P.R. and Ehrlich, A.H. (1997) *Betrayal of Science and Reason*. Washington DC: Island Press.

Environment Australia (1997) *Wetlands Policy of the Commonwealth Government of Australia*. Canberra: Environment Australia.

Fairweather, P.G. and Roberts, J. (1996) Multiscale patterns of vegetation occurrence in billabongs along the Murrumbidgee River. In: *Ecological Society of Australia Inc. 1996 Open Forum Abstracts*. Townsville: James Cook University of North Queensland, 42.

Field, C.D. (1995) *Journey amongst Mangroves*. Okinawa: ITTO/ISME.

Giblett, R. (1996) A city and its swamp sett(l)ing. In: Giblett, R. and Webb, H., *Western Australian Wetlands*. Perth: Black Swan Press, 127–46.

Goodrick, G.M. (1970) *A Survey of Wetlands of Coastal New South Wales*. CSIRO Division of Wildlife Research Technical Memorandum. Canberra: CSIRO Division of Wildlife Research.

Gopal, B. and Sah, M. (1995) Inventory and classification of wetlands in India. *Vegetatio* 118: 39–48.

Gosselink, J.G. and Maltby, E. (1990) Wetland losses and gains. In: Williams, M. (ed.) *Wetlands: a Threatened Landscape*. Oxford: Blackwell, 296–322.

Gosselink, J.G., Odum, E.P. and Pope, R.M. (1974) *The Value of the Tidal Marsh*. Baton Rouge: Louisiana State University.

Government of Canada (1991) *The Federal Policy on Wetland Conservation*. Ottawa: Environment Canada.

Hull, G. and Beovich, E. (1996) (eds) *Manual of Wetlands Management*. Melbourne: Department of Conservation and Natural Resources.

Larkum, A.W.D. (1981) Marine primary productivity. In: Clayton, M.N. and King, R.J. (eds) *Marine Botany: an Australasian Perspective*. Melbourne: Longman Cheshire, 369–85.

Larson, J.S. and Mazzarese, D.B. (1994) Rapid assessment of wetlands: history and application to management. In: Mitsch, W.J. (ed.) *Global Wetlands: Old World and New*. Amsterdam: Elsevier, 625–36.

Lear, R. and Turner, T. (1977) *Mangroves of Australia*. St Lucia: University of Queensland Press.

Lefeuvre, J.G. and Dame, R.F. (1994) Comparative studies of salt marsh processes in the new and old worlds: an introduction. In Mitsch, W.J. (ed.) *Global Wetlands: Old World and New*. Amsterdam: Elsevier, 169–79.

Levin, S.A. (1989) Ecology in theory and application. In: Levin, S.A., Hallam, T.G. and Gross, L.J. (eds) *Applied Mathematical Ecology*. New York: Springer Verlag, 3–8.

Lewis, W.M. (chair) (1995) *Wetlands. Characteristics and Boundaries*. Washington, DC: National Academy Press.

Lynch-Stewart, P., Rubec, C.D.D., Cox, K.W. and Patterson, J.H. (1993) *A Coming of Age: Policy for Wetland Conservation in Canada*. Ottawa: North American Wetlands Conservation Council (Canada).

Maltby, E. (1985) *Waterlogged Wealth. Why Waste the World's Wet Places?* London: International Institute for Environment and Development.

Maltby, E., Hogan, D.V., Immirzi, C.P., Tellam, J.H. and van der Peijl, M.J. (1994) Building a new approach to the investigation and assessment of wetland ecosystem functioning. In: Mitsch, W.J. (ed.) *Global Wetlands: Old World and New*. Amsterdam: Elsevier, 637–58.

McDonough, F.M. (1994) Wetlands: New Jersey's legal swamp. In: Mitsch, W.J. (ed.) *Global Wetlands: Old World and New*. Amsterdam: Elsevier, 879–85.

Milledge, D., Parker, P. and Campbell, F. (1988) *The Border Zone. Wildlife of the Far North Coast*. Byron Bay: Byron Environmental and Conservation Organisation.

Mitsch, W.J. (1994) The nonpoint source pollution control function of natural and constructed riparian wetlands. In: Mitsch, W.J. (ed.) *Global Wetlands: Old World and New*. Amsterdam: Elsevier, 351–62.

Mitsch, W.J. and Gosselink, J.G. (1986) *Wetlands*. New York: Van Nostrand Reinhold.

Mitsch, W.J., Mitsch, R.H. and Turner, R.E. (1994) Wetlands of the Old and New Worlds: ecology and management. In: Mitsch, W.J. (ed.) *Global Wetlands: Old World and New*. Amsterdam: Elsevier, 3–56.

Morrison, M.D. and Kingsford, R.T. (1997) The management of inland wetlands and river flows and the importance of economic valuation in New South Wales. *Wetlands (Australia)* 16: 83–98.

Mossop, D. (1992) Coastal wetland protection law in New South Wales. *Environmental and Planning Law Journal* 9: 331–59.

Mulamoottil, G., Warner, B.G. and McBean, E.A. (1996) (eds) *Wetlands, Environmental Gradients, Boundaries and Buffers*. Boca Raton: CRC Press.

Nakashima, S. and Khan, M.H. (1994) *A Basic Guide to Understanding the Environmental Impact of Rural Roads on the Wetlands of Bangladesh*. Dhaka: CARE International.

Nixon, S.W. (1980) Between coastal marshes and coastal waters. Twenty years research in salt marshes. In Hamilton, P. and Macdonald, K.B. (eds) *Estuarine and Wetland Processes with Emphasis on Modeling*. New York: Plenum, 437–525.

Oates, N. (1994) *Managing your Wetland. A Practical Guide for Landholders*. Melbourne: Victorian Wetlands Trust Inc. and Department of Conservation and Natural Resources.

Odum, E.P. (1968) Energy flow in ecosystems: a historical review. *American Zoologist* 8: 11–18.

Odum, E.P. (1980) The status of three ecosystem-level hypotheses regarding saltmarsh estuaries: tidal subsidy, outwelling and detritus-based food chains. In: Kennedy, V. (ed.) *Estuarine Perspectives*. New York: Academic Press, 485–95.

Odum, E.P. and de la Cruz, A.A. (1967) Particulate organic detritus in a Georgia salt marsh-estuarine ecosystem. In: Lauff, G.H. (ed.) *Estuaries*. Washington DC: AAAS, 383–8.

Paijmans, K., Galloway, R.W., Faith, D.P., Fleming, P.M., Haantjens, H.A., Heyligers, P.C., Kalma, J.D. and Loffler, E. (1985) *Aspects of Australian Wetlands*. Canberra: CSIRO Division of Water and Land Resources. Technical Paper 44.

Pakarinen, P. (1995) Classification of boreal mires in Finland and Scandinavia: a review. *Vegetatio* 118: 29–38.

Payne, N.F. (1992) *Techniques for Wildlife Habitat Management of Wetlands*. New York: McGraw-Hill.

Pickett, S.T.A. and White, P.S. (1985) *The Ecology of Natural Disturbance and Patch Dynamics*. New York: Academic Press.

Pinder, D.A. and Witherick, M.E. (1990) Port industrialization, urbanization and wetland loss. In: Williams, M. (ed.) *Wetlands: a Threatened Landscape*. Oxford: Blackwell, 234–66.

Pressey, R.L. and Adam, P. (1995) A review of wetland inventory and classification in Australia. *Vegetatio* 118: 81–101.

Queensland Department of Primary Industries (1989) *Our Mangroves*. Brisbane: QDPI.

Reddy, K.R. and D'Angelo, E.M. (1994) Soil processes regulating water quality in wetlands. In: Mitsch, W.J. (ed.) *Global Wetlands: Old World and New*. Amsterdam: Elsevier, 309–24.

Reppert, R. (1992) *Wetlands Mitigation Banking Concepts*. IWR Report 92-WMB-1. Fort Belvoir: US Army Corps of Engineers.

Richards, J.F. (1990) Agricultural impacts on tropical wetlands: rice paddies for mangroves in south and southeast Asia. In: Williams, M. (ed.) *Wetlands: a Threatened Landscape*. Oxford: Blackwell, 217–33.

Robertson, A.E. and Alongi, D.M. (1995) Mangrove ecosystems in Australia: structure, function and status. In: Zann, L.P. and Kailola, P. (eds) *State of the Marine Environment Report.* Technical annex I. The marine environment. Canberra: DEST, 119–33.

Rubec, C.D.A. (1994) Canada's federal policy on wetland conservation: a global model. In: Mitsch, W.J. (ed.) *Global Wetlands: Old World and New.* Amsterdam: Elsevier, 909–18.

Scott, D.A. (1989) *A Directory of Asian Wetlands.* Gland: IUCN.

Scott, D.A. and Jones, T.A. (1995) Classification and inventory of wetlands: a global review. *Vegetatio* 118: 3–16.

Simpson, W. and Train, W. (1996) *Report to the Honourable Craig Knowles. Minister for Urban Affairs and Planning and Minister for Housing on Shellharbour Boatharbour and Marina Development Proposed by Shellharbour City Council.* Sydney: Commission of Inquiry for Environment and Planning.

Teal, J.M. (1962) Energy flow in the salt marsh ecosystem of Georgia. *Ecology* 43: 614–24.

Tiner, R.W. (1984) *Wetlands of the United States: Current Status and Recent Trends.* Washington: US Fish and Wildlife Service.

Van Veen, J. (1955) *Dredge, Drain, Reclaim! The Art of a Nation.* The Hague: Martinus Nijhoff.

Wagret, P. (1969) *Polderlands.* London: Methuen.

Whitney, G.C. (1994) *From Coastal Wilderness to Fruited Plain. A History of Environmental Change in Temperate North America from 1500 to the Present.* Cambridge: Cambridge University Press.

Williams, M. (1970) *The Draining of the Somerset Levels.* Cambridge: Cambridge University Press.

Williams, M. (1990a) Understanding wetlands. In: Williams, M. (ed.) *Wetlands: a Threatened Landscape.* Oxford: Blackwell, 1–41.

Williams, M. (1990b) Agricultural impacts in temperate wetlands. In: Williams, M. (ed.) *Wetlands: a Threatened Landscape.* Oxford: Blackwell, 181–216.

Williams, M. (1990c) Protection and retrospection. In: Williams, M. (ed.) *Wetlands: a Threatened Landscape.* Oxford: Blackwell, 323–53.

Zedler, J.B. (1991) The challenge of protecting endangered species habitat along the Southern California coast. *Coastal Management.* 19: 35–53.

Zedler, J.B. (1996) Ecological issues in wetland mitigation: an introduction to the forum. *Ecological Applications* 6: 33–7.

11 | Acid rain and critical loads: science policy processes in the United Kingdom

Andrew Tickle

Introduction

The acid rain debate was one of the dominant environmental themes of the 1980s in the United Kingdom (UK) and, as such, gained widespread currency in most strata of society. Despite a perception among 'expert' environmental observers that acid rain was a problem of the 1980s, in the mid-1990s research showed that 46 per cent of a sample group of the public still thought acid rain to be a 'very serious' problem (Boehmer-Christiansen 1995: 92). The factors promoting the rise of acid rain as a social problem were multifaceted but centred on its novelty as a European transboundary environmental issue with identifiable polluters (normally defined in national terms) on the one hand and victims (countries and specific ecosystems, e.g. forest and freshwaters) on the other. The latter aspects (whether 'scientifically proven' or not) also gave an oppositional saliency to the issue which proved most attractive to media interests. At the same time, environmental concern was rising rapidly in the UK, as evidenced by the burgeoning growth in environmental groups (McCormick 1991: 156). By the end of the decade this 'green wave' had also affected governmental politics, with the then UK Prime Minister, Margaret Thatcher, giving her celebrated 'green speech' to the Royal Society of London. This, among other things, catalysed the comprehensive environmental White Paper *This Common Inheritance* (Secretary of State for the Environment *et al.* 1990) which established the environment as a key policy theme in UK political rhetoric.

Although there have been few detailed studies of science policy interactions in the acid rain debate, most authors concur that one of the primary defining characteristics of the issue has been its scientific-technological aspect (Irwin 1995; Boehmer-Christiansen and Skea 1991; Boehmer-

Christiansen 1995; Hajer 1995). However, the controversy over the science of acid rain and technologically related aspects was not related solely to the UK debate of the 1980s and early 1990s; such arguments were a common feature of the European debate from its inception in the early 1970s. The role of scientific investigation in the political resolution of environmental conflicts is often necessary and crucial although this fact may be obscured by contingent interpretations (of either a political or scientific nature, or both) engendered by trying to characterize complex natural systems where uncertainty (in its various guises) remains a key feature. In this sense the meaning of such ecological knowledges to the different social groups within the policy process is as crucial as the worth of the scientific data itself (Eden 1998). The case study presented in this chapter attempts to follow the political resolution of uncertainty within the science of acid rain (or its scientific analogue 'acid deposition') in relation to science policy processes in the UK. This is explored in relation to two contemporary concepts drawn from the domain of environmental policy analysis: epistemic communities and ecological modernization.

Epistemic communities feature in institutional accounts of international environmental regimes (see for example Haas, Keohane and Levy 1993) and have been defined as 'a specific community of experts sharing a belief in a common set of cause and effect relationships as well as common values to which policies governing these relationships will be applied' (Haas 1990). Thus, in addition to their institutional-political context within regimes such as the United Nations (UN), it is important to note the socio-cultural (discursive) basis of their constitution. They therefore have an important function in framing key knowledges and assumptions within ecological debates for decision makers. This point is emphasized by others who suggest that epistemic communities 'serve as informal linkages of knowledge that collectively influence policy actors, but in a micro-political manner' (O'Riordan and Jordan 1996: 87). As such, their political power should usually be understood to be limited. However, as Haas' study of the UN Med Plan suggests, an epistemic community may also constitute an influential policy community in its own right, being able to exert external power over individual nation states (Haas 1990).

Ecological modernization refers to the emergence, in the last decade or so, of a new style of environmental politics and policy where resource conflicts can be accommodated within current socio-political frameworks, possibly even creating new drivers for economic development (see, for example, Weale 1992). A key feature of ecological modernization is the recasting of the role of science, in terms both of its nature (more integrative) and its enhanced position in the policy process (Hajer 1995: 27).

Epistemic communities can be seen as playing an important role in both these respects.

The central proposition of this chapter is that the emergence, at a European level, of an epistemic community of scientists and policy makers has been a necessary condition for enhanced (more 'ecologically modern') UK acid rain policies in the 1990s. This advance appears to have been brought about mainly through the application of the UN-derived 'critical load' concept from the late 1980s onwards – an approach adopted by the UK Department of the Environment, and supported by scientists in government and academia and a number of environmental groups in opposition to other sectors of government and industry. This is an arguable proposition: more realist interpretations suggest that UK policy in this field was principally driven by domestic political processes (especially energy policy) and would be unlikely to be subject to such external influence (Boehmer-Christiansen 1995). However, even if it is judged incorrect or incompletely argued, the case study may provide a useful 'insider' insight from an 'environmentalist' perspective into the opportunities and limits of environmental science in UK political decision making.

The evolution of UK acid rain science

The term acid rain was neologized in the mid-nineteenth century through the pioneering scientific work of Britain's first pollution inspector, Robert Angus Smith (Smith 1852, 1872), but detailed investigations of the phenomenon had to wait a further century. Despite early work based in the English Lake District (Gorham 1955, 1958) and the United States (Likens, Bormann and Johnson 1972), Scandinavian scientists soon became the research leaders in acid deposition, following the identification of freshwater acidification problems in the 1960s (Odén 1968, 1976). Initial studies rapidly developed into nationally funded multidisciplinary programmes which, in various guises, ran for the best part of two decades in Sweden, Norway, Finland, France, Germany, the Netherlands, the United States, Canada and the UK (e.g. Overrein, Seip and Tollan 1980; NAPAP 1990). In addition, inter-country programmes funded through such supranational agencies as the Organisation for Economic Co-operation and Development (OECD) and the United Nations' Economic Commission for Europe (UN ECE) played a major part in the development of the new science and the policy consequent upon it (OECD 1979; United Nations 1984; Posch et al. 1999). It is contended that these initiatives were instrumental in creating an epistemic science policy community which has been influential in taking forward the political negotiations on acid rain in Europe. The same process can also be seen on a national scale and can be illustrated by following the phased evolution of UK acid rain policy from 1970 to the present day.

Phase 1: opposition and denial (1970–1982)

From the outset, the modal political response of the British government was defensive in response to suggestions (notably from Sweden and Norway) that air pollution emitted in the UK was causing damage to freshwaters and soils in Scandinavia. These claims surfaced first in an international political context at the United Nations Conference on the Human Environment in Stockholm in 1972 where the Swedish government demanded an immediate 50 per cent cut in regional emissions and a longer-term target of 90 per cent 'towards the end of the century' (Anon. 1972). By this time Swedish scientists had garnered enough evidence to show that the problem was significant in extent though detailed causes and processes were yet to be elucidated. This then became the task of the Norwegian SNSF project (running from 1972 to 1980) which held a conference of preliminary results in 1976. Recommendations from this 'scientific' conference which urged political action 'by all governments' showed the extent to which research scientists were already enmeshed in a highly political arena (Anon. 1976).

Research into acid rain per se was initiated in Britain in 1973 but was, in the main, rather limited and disparate. Most work was being carried out in governmental laboratories (funded mainly through the Department of the Environment) or by research teams within the dominant state electricity generating company, the Central Electricity Generating Board (CEGB). By 1976 a government review concluded that similar levels of acid deposition occurred in the UK as in Norway but that evidence of effects in the UK was lacking (Department of the Environment 1976). By the late 1970s Scottish research had already shown deterioration in water quality, but such problems (acidic waters with sparse or absent fish populations) were thought to be local problems, possibly connected with land use changes, including conifer afforestation. However, a survey carried out in 1979 by visiting SNSF scientists (with the active collaboration of scientists at the Freshwater Fisheries Laboratory at Pitlochry) showed a much broader pattern of acidity in both south western and central Scotland – similar to that seen in Norway – and which was unrelated to forestry cover (Wright and Henriksen 1980).

Over a similar period (1976–1980) CEGB scientists had been trying to collaborate with the SNSF programme but with less success. Several interesting points emerge from this aspect of the CEGB research laboratories' work. Firstly, despite the fact that much research work on UK freshwaters was ongoing, the CEGB's primary concern during this period appeared to be disproving that a problem existed in Scandinavia or, if it did, proving that the cause was not acid air pollution sourced from the UK. Thus, active collaboration with the SNSF project (in particular, joint studies and publications with the Norwegian Water Institute, NIVA) was consistently refused, although scientific co-operation was usually extended on a personal basis (Webb 1982) and a number of SNSF personnel came to scientific meetings in the UK at the invitation of the CEGB (Howells 1982). Throughout this period the interpretation given to CEGB research in external fora continued to play down the role of acid deposition, in the face of mounting Scandinavian data to the contrary (e.g. Chester 1982: 333).

But although the CEGB's stance essentially occupied a hegemonic position which represented official UK views, it was far from unchallenged nationally. Even at the 1982 Stockholm conference, other British scientists (e.g. Professor Fred Last, a senior scientist at the Institute of Terrestrial Ecology) were impressed enough by the state of the scientific knowledge to regard immediate remedial action as being justified. This view was supplemented by a BBC documentary, *A Killing Rain*, which underscored Scandinavian concerns over the CEGB's use of political, as opposed to scientific interpretations of data (the CEGB's scientific director, Dr Peter Chester, appeared to suggest in the programme that acid rain was good for Norwegian water quality). In tense correspondence that followed (Howells 1982; Rosseland 1983), one of the Norwegian scientists interviewed in the programme highlighted a clear difference between the UK and Scandinavia in the attitude to scientific uncertainty around which the policy issue revolved:

> As a scientist, I do of course work for the opportunity to do more research, but we must never use that as an overall excuse to take decisions based on a 'lower platform of knowledge'. I wonder if the need for the ultimate scientifically (*sic*) proofs are just that high for other decisions taken by your Parliament in cases like this. If they are, your politicians have my greatest respect. (Rosseland 1983: 3)

Much of the CEGB's attitude to the acid rain problem was a simple reflection of the energy policy values of its political master: the UK government, as represented by the Department of Energy. Indeed, the then Chairman of the CEGB, Walter Marshall, was a former chief scientist at the Department of Energy, demonstrating the closeness of the two institutional cultures. Science and, by implication, environmental science, also had an important role in governmental policy, but the framing of such knowledges for policy use was usually derived from the norms and practices of narrower disciplines such as physics (Wynne and Simmons 1996). In the case of acid rain, this meant a commitment to make decisions based on an objective knowledge base ('sound science') and a well-established burden of proof (of harm) – two norms that are almost impossible to satisfy given the scientifically complex nature of the acid deposition problem.

As early as 1972 in Stockholm the UK indicated its willingness to act on the acid rain issue *if* harm were proven, thus setting the stage for the political controversy that ensued – which was to be decided by science-based arguments. As such, UK participation in international treaty-making on transboundary air pollution was strongly tempered by extant scientific evidence. Thus, the 1977 OECD study proving the transboundary nature of European air pollution paved the way for the UK to sign the 1979 United Nations Economic Commission for Europe (UN ECE) Convention on Long-Range Transboundary Air Pollution (CLRTAP) – a framework treaty setting out generalized principles and responsibilities but no legally binding targets for reduction.

Further movement occurred internationally in 1982 at the Stockholm acidification meeting which, additionally to the expert meetings, encompassed a ministerial conference to which the scientists would forward their best advice. Their advice was threefold: that sufficient proof did exist to justify action; that reductions in emissions would reduce damage and that extant control technologies would be cost-effective. On the advice of the CEGB, the UK broadly resisted these conclusions and, as a result, was badly isolated at the conference (by this time Germany, which had previously opposed Scandinavian claims, had undergone a rapid about-face due to the perceived threat to their forests from acid rain). However, the experience was salutary for UK Department of the Environment (DoE) officials, who were already beginning to question the CEGB's reassurances on the topic and determined to enhance their own research programme, mostly carried out through the Natural Environment Research Council (NERC), its Institute of Terrestrial Ecology (ITE) and various university research groups. The Stockholm meeting was also influential in reinstating a £120 000 grant to the ITE and others to work on acid rain.

Phase 2: Research as policy action (1983–1986)

The Stockholm experience also persuaded the then UK environment minister, Giles Shaw, and his officials of the need to rescue the UK's poor
international reputation on pollution matters. However, ensuing strategies to improve international relations were not without other problems;
for example, to avoid more damaging relations with Norway, the UK
dropped its usual anti-whaling stance at the 1983 International Whaling
Conference, rousing fury elsewhere (Caufield 1983). Nationally, Stockholm also catalysed the DoE to review its policy stance, leading to a
proposal to Cabinet to fit a number of large coal-fired power stations
with abatement technology. Unsurprisingly, this was opposed successfully within government, notably by the Department of Energy (on behalf of the CEGB) and the Treasury on the grounds of cost allied with a
lack of scientific justification. The sole political outcome was a further
allocation by the DoE of £500 000 for acid rain research (Park 1987:
224), clearly a preferable sum to the estimated £1000 million needed for
the original DoE clean-up proposal (Caufield *ibid.*).

The failed DoE initiative of 1983 marked an important initial break in
intra-governmental cohesion on environmental issues, heralding the much
more overt policy conflicts which became typical of interdepartmental
battles in the late 1980s as the DoE sought to establish itself as an
environmental 'leader' per se rather than act as a transparent agent of
governmentally agreed policy (Wynne and Simmons 1996). In relation
to the acid rain issue and concomitant with a move towards this position
of environmental leadership, the DoE made clear moves to overtake the
CEGB in acid rain research, thus 'ensuring that it received the science it
needed for policy advocacy inside government' (Boehmer-Christiansen
1995: 27).

To this end the DoE also appointed a series of scientific advisory
groups (e.g. the UK Review Group on Acid Rain, UK RGAR) to provide
'up-to-date perspectives on the extent and effects of air pollution in the
United Kingdom' (UK RGAR 1987). The composition of these review
groups was determined primarily by scientific expertise but tended to
correspond to individuals whose research groups were funded in some
measure by DoE grants (usually provision of advice was a contractual
requirement for those in receipt of DoE funds). Excepting RGAR, representatives from the power industry (the CEGB and its privatized suc-
cessor companies, National Power and PowerGen) tended not to be well
represented. The DoE provided the secretariat functions, thus being able
to exercise some influence on the political tenor of the final reports and,

more importantly, their time of publication. These functions became important as the review groups quickly became an authoritative source for policy-oriented information, often implicitly critical of current governmental positions. The publication of one review group report in 1989 even occasioned the dissociation of a former CEGB scientist's name from its findings due to her non-concurrence with some of the overall conclusions. (Based on computerized catchment models, the group advised 'that in order to restore most surface waters to something approaching their pristine condition, acid deposition would have to be reduced by about 90 per cent' (UK AWRG 1989: 44) – a position clearly at odds with CEGB interpretations.)

Within the period in question (1983–1986) two such review groups on acid rain reported (UK RGAR 1983; UK AWRG 1986). The earlier RGAR report was far more significant, being the first evidence that the UK also suffered rainfall acidity comparable to affected areas in Scandinavia. The latter interim Acid Waters Review Group (AWRG) report on acid waters suffered from a lack of suitable time trend data and was thus fairly conservative in its conclusions, noting only that 'local evidence of acidification was identified in certain distinct geological regions' (UK AWRG 1986: 37). The fact that such findings mirrored the earlier report of the SNSF research in Scotland (Wright and Henriksen 1980) gives a clear indication that for any report to have political impact in the UK science policy arena, the 'Made in the UK' research tag was to prove almost obligatory. This necessity for UK 'branded' findings was obviously a major factor in the DoE setting up its *pro-active* research programme and its ancillary network of review groups. This essentially state-led approach contrasts interestingly with the more autonomous and entrepreneurial roles accorded by Zehr (1994) to US scientists involved in the National Acid Precipitation Assessment Programme (NAPAP).

However, the major research initiative of this second phase in UK acid rain science was the setting up of the Surface Waters Acidification Programme (SWAP). This five-year £5 million project represented the major *reactive* response of industry, namely the CEGB and their main fuel supplier the National Coal Board (NCB). From the start it was perceived principally by its critics as a political attempt to 'park the issue'. In setting up the study, Walter Marshall had recognized that CEGB research, predominant in the 1970s, was becoming devalued politically and astutely used his political and scientific credentials to persuade the Royal Society (of which he was a Fellow) and the corresponding bodies in Sweden and Norway to run the affair independently of the sponsors, with no interference in the dissemination of the findings.

Many scientists in Scandinavia were ambivalent as to the value of the SWAP project and were minded not to co-operate; the Norwegian government's response stated they would be 'unhappy if the study were to delay measures which we now know are needed to reduce emissions' (Anon. 1983). The antipathy of Swedish and Norwegian officials and scientists was not helped by the overall tone of the CEGB and the NCB announcements of the project. The NCB's comments, for example, were particularly undiplomatic:

> There is a great deal of public interest in acid rain, considerable speculation about its effects *but few scientific facts to rely on.* This impartial investigation by The Royal Society with the Norwegian and Swedish academies into the causes of the acidification of surface waters will establish scientific facts to assist the three participating countries. *It will also benefit other countries by enabling them to have a clearer understanding of the complex issues involved.* (NCB 1983, emphasis added)

A later statement issued by the NCB Chairman, Ian McGregor, interpreting an initial Royal Society study group report, also gave a good indication of the NCB's attitude to the state of scientific proof and attempted to shift responsibility away from industry to natural causes, such as volcanic eruptions:

> There is too much at stake for decisions to be related to inadequate information about the causes and effects of industrial air pollution. *The truth is there is no scientific consensus.* . . . If all industrial activity ceased tomorrow there would still be acid rain for natural reasons and there always will be. (NCB 1984, emphasis added)

Although the SWAP project was nominally an industry initiative to be carried out by independent scientists, links with the government through its informal policy culture were close and had a major bearing on the interpretation of the programme's results. Although Marshall's role was originally as facilitator of the study, Margaret Thatcher's close reliance on Marshall's scientific advice meant that SWAP findings (and other research results, e.g. from the DoE's own programme) tended ultimately to be viewed through the lens of Marshall's narrow physics-based scientific paradigms, in addition to his structural opposition based on his role as Chairman of the CEGB. Similarly, the Director of SWAP, Sir John Mason FRS (also a physicist), was another establishment scientist and acid rain sceptic recently retired from heading the UK's Meteorological Office.

It is therefore not surprising that the government, using SWAP, was able to maintain its political stance on acid rain despite burgeoning criticism from such diverse and important sources as parliamentary select committees (European Communities Committee 1984; Environment Committee 1984), expert bodies (Royal Commission on Environmental Pollution 1984), scientists and environmental groups. This has led Hajer to suggest that 'the inauguration of SWAP was the single most important discursive interpellation in the acid-rain controversy ... [i]t deferred decision-making, closed the discursive standards, depoliticized the issue and precluded any open debate both among scientists or even in larger fora' (Hajer 1995: 174). Whilst the latter elements of this analysis may be questioned, relative control of the domestic debate allowed the UK to resist external pressures from other European countries and supranational political institutions such as the UN ECE and the European Community (EC). In both these fora the UK had successfully delayed internationally binding legislation (the proposed sulphur protocol to the 1979 UN ECE LRTAP Convention and the draft EC Directive on Large Combustion Plants, respectively) on the basis of both unproven science and the failure of other countries to recognize UK pollution reduction achievements made before the baseline year of 1980. However, despite the UK's refusal to sign the 1985 UN ECE sulphur protocol, the DoE became increasingly involved in more international co-operation, both scientific and political, thus heralding the final and current phase of acid rain science.

Phase 3: internationalization, critical loads and epistemic communities (1986–present)

The critical load concept had its origins in Canadian acid rain policies of the early 1980s, where 20 kilogrammes of wet sulphate deposition per annum (20 kg SO_4 per ha per year) was set as an overall 'target load' to protect the most sensitive aquatic ecosystems (Brydges and Wilson 1991). By the time of the 1982 Stockholm acidification meeting, the allied term 'acceptable acid loading' had begun to creep into the European literature, with figures of 0.3–0.5 grammes of sulphur per square metre (g.S square metre) being quoted against a typical deposition figure of 2–3 g.S per square metre in southern Scandinavia (Aniansson 1982: 225; Dickson 1983: 270). Although the full political ramifications of the nascent 'critical load' concept were still to be realized, such biologically driven figures were the first indication of the magnitude of the emission reduction task to be faced. This challenge was quickly taken up by an international coalition of NGOs (representing 293 organizations and several million members in the European Community, Scandinavia, the United States, Africa and elsewhere) urging 'all ECE countries to so reduce their total sulphur load that this level [0.5 g.S square metre] is not exceeded' (Anon. 1982: 3). Following Stockholm, the Nordic Council of Ministers initiated an expert working group on critical loads for sulphur and nitrogen which defined formally the term 'critical load' as

> (t)he highest load that will not cause chemical changes leading to long-term harmful effects on the most sensitive ecological systems (Nilsson 1986: 4)

and gave differentiated quantitative values for different systems (forest soils, groundwater, surface waters). Nevertheless, it was clear that more work needed to be carried out and a further workshop, under the joint auspices of the Nordic Council and the UN ECE took place in 1988 (Nilsson and Grennfelt 1988).

The involvement of the UN ECE was crucial for two main reasons. Firstly, it brought the critical loads work under the wing of the one of the two supranational political institutions active in the field (the other being the EC), thus assuring its continuing policy relevance in negotiations on emission reductions. In doing so, however, political power over critical loads moved from the Nordic countries to working groups within the UN ECE, which led to increased political accommodation. This is shown best by the change made by the UN ECE Working Group on Nitrogen Oxides in the definition of critical loads from that cited above to

> (a) quantitative estimate of an exposure to one or more pollutants below which significant harmful effects on specified sensitive elements of the environment do not occur according to present knowledge (Nilsson and Grennfelt 1988: 9)

which has been criticized as less stringent and more subjective, especially in relation to what may be interpreted as being 'significant' harmful effects (Tickle 1995). Secondly, the community of scientific and policy experts involved in critical loads was widened to all countries (parties) participating in the LRTAP Convention. In time, this was to prove a significant vehicle for the co-option of less environmentally progressive countries to the stringent reduction targets inherent in the critical load calculations. Part of such policy advances may be attributable to such countries' involvement in inter-country data quality assurance programmes and perceived pressures to play their role within what has been described as the 'tote-board' diplomacy of the CLRTAP (Levy in Haas *et al*. 1993).

Indeed, widespread national collaboration in the critical loads project was a necessity if the UN ECE was to realize its early aim '. . . to map the geographical areas experiencing higher than critical loads with respect to the sensitivity of different soil types' (UN ECE Working Group on Effects 1986 quoted in Nilsson and Grennfelt 1988: 8). This aim was formalized in 1988 when the critical loads approach was written into the preambular text of the Sofia (NO_x) protocol as '. . . an effect-oriented scientific basis to be taken into account when . . . deciding on further internationally agreed measures to limit and reduce emissions . . .' with the specific instruction in Article 6 to 'determine the geographical distribution of sensitive areas' (UN ECE 1988: 13–20). Mapping was also important in a wider political and economic sense as it might allow for more cost-effective control strategies whereby the most threatened areas identified could be targeted for the earliest remedial action. However, reaching the stage of considering abatement scenarios based upon critical load data was to require a further five years of national and international research work, much of it of a strongly interdisciplinary and collaborative nature. This was primarily achieved by an intensive sequence of collaborative projects, scientific review workshops and science policy meetings organized under the auspices of the UN ECE and resourced 'in kind' by the participating countries. It was through this network that the epistemic acid rain science policy community was established as a meaningful force in the negotiations that followed.

To some extent, the UK was rather slow in becoming a part of the critical loads community. Participation in the peer review of the initial critical load programme (the Oslo workshop in 1986) was minimal (one UK scientist) and Waterton (1993) has characterized the UK critical loads debate at the time as immature, with the only policy discourses utilizing the term 'critical loads' being those of the NGOs (e.g. Rose 1986; Elsworth and Ågren 1987; Markham, quoted in Ashmore, Bell and Garretty 1988: 119). By the time of the Skokloster critical loads workshop in 1988, five scientists representing a number of differing acid rain subdisciplines were in attendance, together with a scientific civil servant from the UK Department of the Environment (DoE), underscoring the increasing political importance of the process. Shortly after, the House of Commons Environment Committee noted that evidence submitted – in particular from the Nature Conservancy Council (NCC), the Department of the Environment, and the UK NGOs Friends of the Earth (FoE) and the World Wildlife Fund (WWF) – showed

> ... a general consensus that critical load theory was important both as a means of identifying the significance of different factors in ecological damage, and in setting limits for particular pollutants. (House of Commons Environment Committee, 1988: para.64, xxiii)

Evidence presented by both the NCC and WWF also gave preliminary indications of likely critical load exceedances in key areas of Britain (Gee and Stoner 1988; Tickle and Ashmore, 1988), particularly in relation to potential impacts upon terrestrial ecosystems of nature conservation value. This was reflected in one of the Environment Committee's key recommendations that critical loads be a crucial part of the consideration of nature conservation requirements in emission reduction strategies (House of Commons Environment Committee, 1988: para.11(5), xi).

By 1989, with the Skokloster critical load figures relatively well established, the next phase of work was for each country participating in the UN ECE Convention to map critical loads. At the same time, the UK DoE now began explicitly to align its policy stance to critical loads, stating, '(t)he Government believes that the development of an approach based on ... critical loads is central to further progress' with the objective 'to reduce acid emissions to a rate that is sustainable by the natural environment' (Secretary of State for the Environment *et al.* 1990: 143–144). To this end the DoE research programme was reorganized around a national critical loads programme, including the provision of new advisory groups to give multidisciplinary inputs and integration of data through mapping and modelling work (Critical Loads Advisory Group 1994: 1).

The establishment of this programme was important in two major respects. Firstly, it marked a decisive break with the dominant influence of CEGB research in policy formulation. This was a reflection of two processes: the continued build-up of the DoE's 'independent' research programme with the research councils and the universities, and the break-up of the CEGB's research laboratory consequent upon electricity privatization. Secondly, a policy stance explicitly related to critical loads implied a diminution in the ability of the government to set arbitrary reduction targets for acid emissions – now, at least rhetorically, policy should be guided by science. This thus increased the political importance of the scientific deliberations, both in the UK and abroad. In addition, as the last point implies, the adoption by the UK of a policy discourse based upon critical loads drew it much closer to the overall political thrust of the more environmentally progressive countries within the UN ECE. This again signalled a decisive break with the past where the UK had been politically isolated since 1985 by its refusal to sign the Helsinki protocol. Thus the UK acid rain science and policy community became 'internationalized', opening itself to greater external influence (i.e. beyond the traditional nexus of opposition from the CEGB and allied government departments) which would bear more explicitly on the national policy process. It is clear that the DoE – or at least those within the DoE closely involved in acid rain policy – were quite well aware of this point.

Although the change in stance may not have been recognized by other countries involved in the UN ECE work, the point was not lost on the UK NGOs, which now saw clear opportunities in the new policy framework for the type of more ecologically driven targets they had been calling for since the early 1980s. Together with the increasing scientific capacity of the NGOs (notably within Greenpeace, Friends of the Earth and WWF), this allowed for greater dialogue with the DoE, which was keen to garner political support from all quarters. However, support for UK critical load policies by the NGOs (albeit essentially strategic – see Tickle 1995), was by no means uncritical. For example, fierce debate surrounded UK mapping methods, which the NGOs believed underestimated the proportion of land at risk, which in turn would reduce the amount of pollution reduction required (Tickle and Sweet 1993; Pearce 1993). The argument also related to the degree of autonomy each country should have in determining criteria for mapping and the methods involved. Most European countries used a system based on cumulative percentile curves which were usually predicated on protecting 95 per cent of ecosystems; the UK by comparison aimed only at protecting the modal sensitivity class – an approach that the NGOs saw as illustrative of the UK's political expediency in implementing the critical loads concept. Part of the problem related to cultural perceptions of maps per se which, in relation to many physical phenomena, are seen as being absolute. Critical load maps, in contrast, were highly contingent on the statistical abstractions of the underpinning data set and the NGO critique was successful in underscoring this point.

The timescale for the UK critical loads programme was principally driven by the UN ECE and, in particular, the data requirements of the Working Group on Effects (WGE) and the Task Force on Integrated Assessment Modelling (TFIAM) which had to be fed into the renegotiation of the 1985 Helsinki (Sulphur) Protocol due by the end of 1993. This was to prove an extremely tight schedule (in fact the new protocol was not signed until June 1994) and was to put severe pressures on the British scientists involved, who – on occasion – were dissatisfied with the quality of data that could be provided in such short timescales. Some former CEGB scientists, who were now working mainly for the largest privatized electricity utility (National Power), were unhappy at their perceived exclusion from the critical loads work. They also sought to criticize its practices and results, focusing principally on the assertion that results neither appeared in the 'open literature' (i.e., peer-reviewed academic journals) nor were being 'presented fairly and accurately to the public and policymakers' (Skeffington 1993, 1995). Despite such claims, it was clear that most of the scientific community involved regarded critical loads as the best tool available for the job, with imperfections in methodologies and data uncertainties likely to be rectified by ongoing research. In addition, it garnered valuable research funds at a time of increasing funding difficulties for UK scientific institutions.

In terms of specific ecosystems, the principal focus of the UK critical loads programme was the effect on soils and waters which had been supposed to be the most sensitive 'receptors' (Department of the Environment 1991). In this sense, critical loads were set in a precautionary manner: if soils or waters were being acidified beyond their innate buffering capacity, there existed the potential for adverse effects on processes and organisms within those ecosystems. However, the consequences for UK species and habitats, and particularly those of conservation value, were far from clear. To this end, the country conservation agencies (English Nature, the Countryside Council for Wales, and Scottish Natural Heritage) under the auspices of the national Joint Nature Conservation Committee (JNCC) set out to evaluate the likely impact on Sites of Special Scientific Interest (SSSIs) in England, Scotland and Wales. This was done by comparing geographical areas where critical loads were exceeded under future deposition scenarios (envisaging either a 60 per cent cut or 80 per cent cut in UK acid emissions from 1980 levels) in relation to SSSI locations (Farmer 1993). Results from this study were highly significant, showing that nearly a quarter of the land area of SSSIs would remain 'at risk' from acid deposition even when current UK policy targets were met (a 60 per cent cut in sulphur emissions by 2005). This was seen as unacceptable and the study concluded testily that '(p)rotection of Britain's natural environment would . . . require a very significant reduction in acid emissions beyond that currently agreed to' (*ibid.*: 2).

Looking further afield to the UN ECE, national country data were assimilated into a common European map to be used as the basis for negotiating further Europe-wide sulphur reductions. This data could then be used in integrated assessment models such as RAINS (the Regional Acidification INformation and Simulation model – Alcamo *et al.* 1990) to calculate pollution reduction scenarios which would optimize (at a given level of reduction) between ecological impact (critical load exceedance) and economic costs. Initially, targets were defined in terms of widely differing national 'target loads' which were then aggregated to form provisional scenarios which indicated the need in industrialized countries for cuts of 75 per cent and upwards (TFIAM 1992: 8). This 'target load' approach was soon seen as unequal and consequently infeasible (equity in multilateral agreements clearly being an important initial objective – Sand 1990: 6–9) and parties to the Convention instead sought relative parity in terms of equitable reductions in the gap between current deposition figures and critical load values, the so-called 'gap-closure' scenarios. In this sense, the ecological benefit in each country would be equal (the same reduction in impact) though the effort (and cost) to be undertaken by each country would be different, based upon their relative import and export of sulphur and the varied costs of clean-up in each country (determined by country specific cost-curves). The final scenario settled upon after a year of negotiations offered a 60 per cent gap closure between 1990 sulphur deposition levels and extant critical load values.

For the UK, this translated into a required reduction of 79 per cent of 1980 levels of sulphur emissions by 2000. This ran considerably beyond current national policy which, based upon commitments made in the 1988 European Commission Large Combustion Plant Directive, was likely to deliver a 50 per cent reduction by the same date. As the final negotiations approached in the summer of 1993, the UK's best offer was a 70 per cent cut by 2005, a compromise position hammered out by a Cabinet Committee in July. Contrary to policy rhetoric, however, this position was not informed by critical loads information as no 70 per cent scenarios had been calculated in terms of UK ecological impacts – the figure was a simple political fudge between 60 per cent (clearly too low) and 80 per cent, favoured by the DoE on the basis of the SSSI evidence but opposed by the Department of Trade and Industry (Dti) on behalf of the electricity generators and British Coal as too costly. Research comparing the effects of a 70 per cent versus 80 per cent cut was swiftly carried out by the conservation agencies. It showed that a further 10 per cent reduction was particularly important to protect large upland areas of England and Wales from further risk (Farmer and Bareham 1993). This maintained the DoE's case for an 80 per cent reduction but the lack of a dynamic element in the modelling work meant that no scientific case for a particular timescale could be made, except on precautionary grounds. A classic political compromise was then reached with an 80 per cent cut – but by 2010, a date that was unlikely to hamper the newly privatized electricity industry's cost-effectiveness and hence profits. In this sense, although the science of critical loads informed the decision, it did not win the overall argument.

Issues for uncertainty, sustainability and policy

Clearly, the long controversy over acid rain science and policy in the UK was dominated by the issue of uncertainty, not only in relation to the science but also political uncertainty. A taxonomy of uncertainty that has been offered in relation to acid rain seeks to distinguish between uncertainty that may be pragmatic (to be overcome by further work and resources), theoretical (needing new paradigms) or related to ecological complexity (Cramer, quoted in Irwin 1990: 555). All three forms can be seen in the current discussion, although the first two forms tended to be reduced over time. The most visible form of uncertainty was that emphasized in the early and middle phases of the debate by industry and government, relating principally to the unproven relationship between emissions and their effects, both abroad and in the UK. Here, both pragmatic and theoretical forms of uncertainty were dwelt upon as a key policy discourse (especially by elite political actors), shielding the electricity industry from the necessity for action and avoiding incurring substantial public expenditure to be redeemed either through taxation or electricity bills. As others have noted, costly policy action on acid rain was in direct contravention to prevailing political ideology, which sought to drive down public spending and hence tax burdens (Weale 1992: 87). In this phase, scientific uncertainty (and hence science itself) was being used as a 'fig leaf' for policy inaction, a strategy that could be questionable politically in terms of future reliance on new research and results (Boehmer-Christiansen 1988: 147). This science-based policy approach has been shown to be deeply ingrained within pragmatist policy discourses in the UK and was originally intended to reduce unwanted influence of both interest groups and corporatist actors in the policy process (Hajer 1995).

The point has already been made that complex ecosystem uncertainty also served as an obstacle to policy resolution, given the inappropriate use of narrow physical sciences-based paradigms to judge the efficacy of the scientific (environmental) case being presented. This judgement varied, however: at an early stage, a number of influential ecomodernist policy actors (*sensu* Hajer 1995) showed that the wider scientific community largely accepted the validity of evidence as sufficient for action (e.g. the 1982 Stockholm meeting, the 1984 House of Commons acid rain enquiry), whereas this was resisted institutionally (within industry and government) until a much later date. It is nevertheless clear that the policy debate, both in the UK and Europe, generated an impressive research effort that did much to reduce scientific uncertainty in a relatively short period of time. In relation to the UK, much of the theoretical uncertainty was broken down by the new paradigm of critical loads, which served as a multi-disciplinary focus for the science policy debate from the late 1980s onwards. Under some definitions critical loads had already crossed the natural science/social science boundary by defining 'thresholds of acceptable impact or damage' that might more properly be seen as a matter for 'societal judgement' and thus beyond the normal or pure remit of science (Boehmer-Christiansen 1992: 141). But how were these decisions being made and how did the concept relate to other environmental management principles?

Critical load values were usually derived in two major ways, either empirically from the interpretation of field or laboratory evidence or through the use of mass balance relationships, usually calculated using computer models. Clearly, these methods contained uncertainties and this was highlighted often in both scientific and policy/political fora. For example, the determination of the so-called 'empirical' critical loads for nitrogen effects on vegetation was a highly subjective process with proposed alterations to set values being contingent on the data available, the interpretation of this in background papers intended to inform scientific decision making and, most contentiously, the cultural (national) attitudes and discourses of the individual scientists, policy makers or NGO observers attending any particular workshop. Although such difficulties could later be resolved by the activities of the policy secretariat of the UN ECE (usually delegated to Working Group or Task Force chairmen), such negotiations revealed an ad hoc element of a process that was usually promoted as quantitative and objective. Modelled data tended to be less contentious, although the use of competing integrated assessment models in the final negotiations provided a basis for certain countries to introduce uncertainty into previously allocated reduction targets (e.g. the UK in championing its own Abatement Strategy Assessment Model (ASAM) as opposed to the RAINS model adopted by the UN ECE's Task Force on Integrated Assessment Modelling).

In terms of the relationship of critical loads to other environmental management principles, the concept is clearly derived from that of assimilative capacity except in the important distinction that critical loads are generally used in a normative, rather than permissive manner (Chadwick and Nilsson 1993). In this generalized sense, critical loads are not precautionary. However, they can be argued to have highly precautionary aspects, a fact that may have diminished their forcefulness in some UK policy circles (notably the DTI). For example, many of the most stringent critical loads defined have their basis in mass balance chemical calculations where the defined adverse effect may simply denote a change in input–output relationships with no proven ecological ramification, save the change itself. Thus, assuming the parameter chosen is the most sensitive in a particular system (a judgement that cannot be proven), the threshold criteria selected (if met) should offer an extremely low risk of damage. The stringency of the concept has also been demonstrated in a more empirical sense by modelling studies which have shown that meeting critical loads in Europe would equate with a zero emissions approach – also the presumed endpoint of a precautionary approach to acid air pollution.

Critical loads have also been argued by the UK DoE to 'have particular significance for sustainable development, since a deposition above critical load is not . . . sustainable in the long term' (CLAG 1994: 1). To a great extent, this mirrors the suggestion that DoE policy discourses took on ecomodernist (as opposed to pragmatic, traditional) values throughout the 1980s in response to documents such as the World Conservation Strategy and the Brundtland report (Hajer 1995). In this study of UK acid rain policy up to 1988, Hajer shows that ecomodernist rhetoric (including precautionary and sustainability discourses) failed to influence policy outcomes (partly due to strategy paradoxes within ecomodernist actor coalitions) which remained strongly pragmatist. Hajer's interpretations also prompt key questions such as the extent to which the concept of critical loads is an ecomodern one, and whether its deployment in policy discourses and policy making has caused any movement away from previous institutional practices.

In relation to environmental precaution, it can certainly be postulated that critical loads are more closely related to current tenets of ecological modernization (Hajer 1995: 27) than those of previous practices, despite the close epistemological links to assimilatory capacity. However, the adoption of critical loads within the UK 'science-based policy approach' has meant that the policy outcome of critical loads has been deeply enmeshed within extant governmental practices, dominated by economic, political and industrial influences. This led essentially to the stringent reductions in sulphur emissions demanded by critical load criteria being delayed until they fell sharply owing to economic (market) factors engendered in the main by electricity privatization. In this sense, previous policy practice prevailed at the seeming expense of ecological modernization. This, however, disregards small – yet key – changes that the adoption of critical loads as a policy discourse and practice was able to engender.

Firstly, it formed a substantive element of the DoE's ability to reposition itself as an ecomodernist actor in national policy, underpinned by the broad support of the UK scientific community, the wider international epistemic science policy community and a sometime tacit alliance with environmental NGOs through a shared commitment to ecologically driven targets. Indeed, the participation of new actors such as NGOs in the epistemic community can be seen as a modernizing feature within environmental politics, principally stemming from the recognition of their role within supranational regimes such as the UN (see Princen and Finger 1994; Wapner 1995; Arts 1998). Pressure from NGOs in the mid-1980s had contributed to the perception of the UK as an environmental 'foot dragger' (the 'dirty man of Europe') in international circles, particularly in relation to the failure to sign the 1985 Helsinki protocol. This, among other things, reinforced the determination of the UK to participate fully in the critical loads process and instilled a fierce will not to be left out of the second sulphur protocol. The accession of the UK to this international treaty has thus strengthened the place of the critical loads approach in UK pollution control policy, also evidenced in national practice by recent impact-related regulatory regimes for individual power stations based on critical load criteria.

In terms of the model of epistemic community action suggested by Adler and Haas (1992), this suggests that the acid rain science policy community has been reasonably successful in piloting the critical loads concept through three key stages of policy evolution – innovation, diffusion and persistence – to the point where it has become established as new policy orthodoxy within the institutional sphere of the UN ECE. In terms of policy diffusion and persistence, the ongoing challenge for the critical loads concept and UK acid rain policy is within the European Community (EC), which has also adopted the policy discourse of critical loads within its sustainability strategies with the aim, as stated in the Fifth Environmental Action Programme, of 'no exceeding ever of critical loads and levels'(Commission of the European Communities 1992: 42). As the EC has been traditionally seen as having more influence over UK acid rain policy than the UN ECE (Boehmer-Christiansen 1995), this has now become the new testing ground of the policy durability of the critical loads concept and its associated epistemic community.

The debate surrounding a critical loads-based approach to EC acid rain policy has been ongoing since December 1995 when the Environment Council requested the Commission to develop an acidification strategy by early 1997. The resulting communication (Commission of the European Communities 1997) was based strongly on critical loads and adopted the same RAINS model that had prevailed in the negotiation of the UN ECE second sulphur protocol. Although it was recognized in the strategy that the full goal of nil exceedance was not possible in the immediate future, an interim goal of a 50 per cent reduction in ecosystem impacts by 2010 (using 1990 as a baseline) was proposed. The strategy gained broad support from the European Parliament and the Environment Council in 1998 and its key proposal of national emissions ceilings reached fruition in June 1999 when the Commission agreed a proposed directive on national ceilings for emissions of acidifying and ozone-forming air pollutants, which was then the subject of debate within the European Parliament and the Council of Ministers. Adopted formally in late 2000, the cuts become legally binding within two years.

The diffusion of the critical loads concept to the sphere of the EC has meant a parallel policy process developing between it and the UN ECE. In the latter forum, attention shifted – after the signing of the second sulphur protocol – to the renegotiation of the 1988 NO_x protocol. Again, based on the critical loads approach, this has now led to a novel 'multi-effects' protocol (to the Convention on Long-Range Transboundary Air Pollution) which will address all major acidifying and eutrophicating air pollutants (sulphur dioxide, nitrogen oxides, volatile organic compounds and ammonia) simultaneously. In a 'reverse' diffusion of the EC policy process, the new UN ECE protocol has aims that strongly echo those within the newly agreed EC legislation, including binding national emission ceilings and the same target year of 2010. However, the UN ECE protocol is more innovative in attempting to deal with eutrophication – an approach omitted in the new EC directive.

These policy directions within the two major environmental decision-making regimes in Europe suggest that critical loads have now become embedded as a central paradigm in strategies to combat acid rain and other major forms of air pollution. It is suggested in this chapter that the diffusion and persistence of the critical loads concept has been supported strongly by an epistemic community of scientists and policy makers whose advocacy at key stages (e.g. the adoption of the EC acidification strategy) has led to this new orthodoxy. However, whilst the evidence for this assertion is relatively clear, it is not sufficient to counter more realist claims that larger structural factors (economic and political) are more important drivers for the final emission reductions adopted, as has been illustrated both in this research and elsewhere (Boehmer-Christiansen 1995) in the case of the UK.

Acknowledgements

I am grateful to Christer Agren, Rick Battarbee, Sonja Boehmer-Christiansen, Richard Clarke, Andrew Farmer, John Murlis, Richard Skeffington, Ian Welsh, Bob Wilson and Emma Wilson for their useful comments on the original manuscript. Needless to say, they bear no responsibility for the shape of the final analysis. An earlier version of this chapter was published in *Energy and Environment* 10(6) (1999) and appears here by kind permission of the editor.

References

Adler, E. and Haas, P.M. (1992) Epistemic communities, world order, and the creation of a reflective research program. *International Organisation* 46(1): 367–90.

Alcamo, J., Shaw, R. and Hordijk, L. (1990) (eds) *The RAINS Model of Acidification: Science and Strategies in Europe.* Dordrecht: Kluwer.

Aniansson, B. (1982) (ed.) *Acidification Today and Tomorrow.* Stockholm: Swedish Ministry of Agriculture Environment '82 Committee.

Anon. (1972) *Air Pollution across National boundaries: the Impact on the Environment of Sulfur in Air and Precipitation.* Sweden's case study for the United Nations Conference on the Human Environment, 1972. Stockholm.

Anon. (1976) Report from the international conference on the effects of acid precipitation in Telemark, Norway, June 14–19, 1976. *Ambio* 5: 201–2.

Anon. (1982) *Stop Acid Rain! An International NGO Statement to the 1982 Stockholm Conference on Acidification of the Environment.* Stockholm: The Swedish NGO Secretariat on Acid Rain.

Anon. (1983) Britain funds independent research on acid rain. *New Scientist* 8 September: 671.

Arts, B. (1998) *The Political Influence of Global NGOs. Case Studies on the Climate and Biodiversity Conventions.* Utrecht: International Books.

Ashmore, M., Bell, J.N.B. and Garretty, C. (1988) (eds) *Acid Rain and Britain's Natural Ecosystems.* London: Imperial College Centre for Environmental Technology.

Boehmer-Christiansen, S. (1988) Black mist and acid rain – science as a fig leaf of policy. *Political Quarterly* 59(2): 145–60.

Boehmer-Christiansen, S. (1992) How much science does environmental performance really need? In: Lykke, E. (ed.) *Achieving Environmental Goals.* London: Belhaven Press.

Boehmer-Christiansen, S. (1995) *The Domestic Basis of International Agreements: Modelling National/International Linkages. The British Case Study:*

Policy Formulation and Implementation. SPRU contribution to EC Contract EVSV-CT 930.85. Brighton: Science Policy Research Unit, University of Sussex.

Boehmer-Christiansen, S. and Skea, J. (1991) *Acid Politics: Environmental and Energy Policies in Britain and Germany.* London: Belhaven Press.

Brydges, T.G. and Wilson, R.B. (1991) Acid rain since 1985 – times are changing. *Proceedings of the Royal Society of Edinburgh,* 97B: 1–16.

Caufield, C. (1983) Treasury vetoes action on acid rain. *New Scientist* 15 September: 747.

Chadwick, M.J. and Nilsson, J. (1993) Environmental quality objectives – assimilative capacity and critical load concepts in environmental management. In: Jackson, T. (ed.) *Clean Production Strategies: Developing Preventive Environmental Management in the Industrial Economy.* Boca Raton: Lewis Publishers.

Chester, P.F. (1982) Acid rain, catchment characteristics and fishery status. In: *Ecological Effects of Acid Deposition.* Report and background papers 1982 Stockholm Conference on the Acidification of the Environment: Expert Meeting I. Report snv pm 1636. Solna: National Swedish Environment Protection Board.

Commission of the European Communities (1992) *Towards Sustainability: a European Community Programme of Policy and Action in Relation to the Environment and Sustainable Development.* Vol.II Com (92) Final. Brussels: CEC.

Commission of the European Communities (1997) *Communication to the Council and European Parliament on a Community strategy to combat acidification.* COM(97) 88 Final. Brussels: CEC.

Critical Loads Advisory Group (CLAG) (1994) *Critical Loads of Acidity in the United Kingdom.* Penicuik: Institute of Terrestrial Ecology.

Department of the Environment (1976) *Effects of Airborne Sulphur Compounds on Forests and Freshwaters.* Pollution Paper No. 7. London: Her Majesty's Stationery Office.

Department of the Environment (1991) *Acid Rain: Critical and Target Load Maps for the United Kingdom.* London: Department of the Environment.

Dickson, W. (1983) Water acidification – effects and countermeasures. Summary document. In: *Ecological Effects of Acid Deposition.* Report and background papers 1982 Stockholm Conference on the Acidification of the Environment: Expert Meeting I. Report snv pm 1636. Solna: National Swedish Environment Protection Board.

Eden, S. (1998) Environmental issues: knowledge, uncertainty and the environment. *Progress in Human Geography* 22(3): 425–32.

Elsworth, S. and Ågren, C. (1987) (eds) *The Limits to Nature's Tolerance.* Göteborg: Swedish NGO Secretariat on Acid Rain.

Environment Committee (1984) *Acid Rain. Volume 1.* Session 1983–84, fourth report, House of Commons. London: HMSO.

European Communities Committee (1984) *Air Pollution.* Session 1983–84, 22nd report, House of Lords paper 265. London: HMSO.

Farmer, A.M. (1993) *SSSIs at Risk from Soil Acidification,* JNCC Report No. 156. Peterborough: JNCC.

Farmer, A.M. and Bareham, S.A. (1993) *The Environmental Implications of UK Sulphur Emission Policy Options for England and Wales.* JNCC Report No. 176. Peterborough: JNCC.

Gee, A. and Stoner, J.H. (1988) Effects of acidification on upland freshwaters. In: Usher, M.B. and Thompson, D.B.A. (eds) *Ecological Change in the Uplands.* Oxford: Blackwell.

Gorham, E. (1955) On the acidity and salinity of rain. *Geochimica et Cosmochimica Acta* 7: 231–9.

Gorham, E. (1958) The influence and importance of daily weather conditions in the supply of chloride, sulphate and other ions to fresh water from atmospheric precipitation. *Philosophical Transactions of the Royal Society of London. Series B* 241: 147–78.

Haas, P.M. (1990) *Saving the Mediterranean.* New York: Columbia University Press.

Haas, P.M., Keohane, R.O. and Levy, M.A. (1993) *Institutions for the Earth.* Cambridge, Mass: MIT Press.

Hajer, M.A. (1995) *The Politics of Environmental Discourse: Ecological Modernization and the Policy Process.* Oxford: Clarendon Press.

House of Commons Environment Committee (1988) *First Report: Air Pollution.* Volume 1. London: Her Majesty's Stationery Office.

Howells, G.D. (1982) Letter to B. Rosseland, Directorate for Wildlife and Freshwater Fish, Norway, 28 November, mimeo.

Irwin, A. (1990) Acid pollution and public policy: the changing climate of environmental decision-making. In: Radojevic, M. and Harrison, R.M. (eds) *Atmospheric Acidity: Sources, Consequences and Abatement.* Amsterdam: Elsevier.

Irwin, A. (1995) *Citizen Science: a Study of People, Expertise and Sustainable Development.* London: Routledge.

Likens, G.E., Bormann, F.H. and Johnson, N.M. (1972) Acid rain. *Environment* 14: 33–40.

McCormick, J. (1991) *British Politics and the Environment.* London: Earthscan.

National Acid Precipitation Assessment Programme (NAPAP) (1990) *1990 Integrated Assessment Report.* Washington, DC: NAPAP.

National Coal Board (NCB) (1983) £5 million 'acid rain' research. NCB welcome 'major joint programme'. Press release, 5 September. London: National Coal Board.

National Coal Board (NCB) (1984) Acid rain – no scientific consensus. Coal Board chairman says UK not a major source. Press release, 27 January. London: National Coal Board.

Nilsson, J. (1986) (ed.) *Critical Loads for Nitrogen and Sulphur.* Miljørapport 1986: 11. Copenhagen: Nordic Council of Ministers.

Nilsson, J. and Grennfelt, P. (1988) (eds) *Critical Loads for Sulphur and Nitrogen.* Miljørapport 1988: 15. Copenhagen: Nordic Council of Ministers.

Odén, S. (1968) *The Acidification of Air Precipitation and its Consequences in the Natural Environment.* Energy Committee Bulletin 1. Stockholm: Swedish Natural Sciences Research Council.

Odén, S. (1976) The acidity problem – an outline of concepts. *Water, Air and Soil Pollution* 6: 137–66.

Organisation for Economic Co-operation and Development (OECD) (1979) *The OECD Programme on Long-range Transport of Air Pollutants: Measurements and Findings.* 2nd edn. Paris: OECD.

O'Riordan, T. and Jordan, A. (1996) Social institutions and climate change. In: O'Riordan, T. and Jager, J. (eds) *Politics of Climate Change: a European Perspective.* London: Routledge.

Overrein, L.N., Seip, H.M. and Tollan, A. (1980) (eds) *Acid Precipitation – Effects on Forest and Fish. Final Report of the SNSF Project 1972–1980.* Research report 19/80. Oslo-Ås: SNSF.

Park, C.C. (1987) *Acid Rain: Rhetoric and Reality.* London: Methuen.

Pearce, F. (1993) How Britain hides its acid soils. *New Scientist* 27 February: 29–33.

Posch, M., de Smet, P.A.M., Hettelingh, J.-P. and Downing, R.J. (1999) (eds) *Calculation and Mapping of Critical Thresholds in Europe: Status Report 1999.* Bilthoven: RIVM/UN ECE Cordination Center for Effects.

Princen, T. and Finger, M. (1994) *Environmental NGOs in World Politics: Linking the Local and the Global.* London and New York: Routledge.

Rose, C. (1986) *Acid Rain: the Ecological Imperative for Pollution Controls.* Gland, CH: WWF International.

Rosseland, B.O. (1983) Letter to Dr G.D. Howells, Central Electricity Research Laboratory, 28 January, mimeo.

Royal Commission on Environmental Pollution (1984) *Tenth Report.* London: HMSO.

Sand, P.H. (1990) *Lessons Learned in Global Environmental Governance.* Washington DC: World Resources Institute.

Secretary of State for the Environment *et al.* (1990) *This Common Inheritance: Britain's Environmental Strategy.* Cm 1200. London: HMSO.

Skeffington, R.A. (1993) Problems with the critical loads approach: a view from industry. In: Hornung, M. and Skeffington, R.A. (eds) *Critical Loads: Concepts and Applications,* ITE Symposium no. 28. London: HMSO.

Skeffington, R.A. (1995) Critical loads and energy policy. In: Battarbee, R.W. (ed.) *Acid Rain and its Impact: the Critical Loads Debate.* London: Ensis Publishing.

Smith, R.A. (1852) On the air and rain of Manchester. *Memoirs and Proceedings of the Manchester Literary and Philosophical Society* 2: 207–17.

Smith, R.A. (1872) *Air and Rain: the Beginnings of a Chemical Climatology.* London: Longman Green.

Task Force on Integrated Assessment Modelling (TFIAM) (1992) *Integrated Assessment Modelling: Progress Report by the Chairman of the Task Force.* EB.AIR/WG.5/R.27. Geneva: United Nations Economic Commission for Europe.

Tickle, A. (1995) Critical loads and NGO policy. In: Battarbee, R.W. (ed.) *Acid Rain and its Impact: the Critical Loads Debate.* London: Ensis Publishing.

Tickle, A. and Ashmore, M. (1988) *Critical Loads in the United Kingdom*. London: Imperial College Centre for Environmental Technology.

Tickle, A. and Sweet, J. (1993) *Critical Loads and UK Air Pollution Policy*. London: Friends of the Earth.

UK Acid Waters Review Group (AWRG) (1986) *Acidity in United Kingdom Fresh Waters. Interim Report*. London: Department of the Environment.

UK Acid Waters Review Group (AWRG) (1989) *Acidity in United Kingdom Fresh Waters. Second Report of the UK Acid Waters Review Group*. London: HMSO.

UK Review Group on Acid Rain (RGAR) (1983) *Acid Deposition in the United Kingdom*. Stevenage: Warren Spring Laboratory.

UK Review Group on Acid Rain (RGAR) (1987) *Acid Deposition in the United Kingdom 1981–1985. A Second Report of the United Kingdom Review Group on Acid Rain*. Stevenage: Warren Spring Laboratory.

United Nations (1984) *Airborne Sulphur Pollution: Effects and Control*. Air Pollution Studies 1. New York: United Nations.

United Nations Economic Commission for Europe (UN ECE) (1988) *Executive Body for the Convention on Long-range Transboundary Air Pollution. Report of the Sixth Session of the Executive Body*. Geneva: United Nations Economic and Social Council.

Wapner, P. (1995) Politics beyond the state: environmental activism and world civic politics. *World Politics* 47: 311–40.

Waterton, C. (1993) *The UK Case Study for Acidification and Transboundary Air Pollution – a Preliminary Survey*. Contribution no. H.9 to the project on Social Learning in the Management of Global Environmental Risks. Lancaster: Centre for the Study of Environmental Change, Lancaster University.

Weale, A. (1992) *The New Politics of Pollution*. Manchester: Manchester University Press.

Webb, A.H. (1982) Collaborative research with Norwegian SNSF project. Internal CERL Memorandum to Dr P.F. Chester, 26 August 1982, mimeo.

Wright, R.F and Henriksen, A. (1980) *Regional Survey of Lakes and Streams in Southwestern Scotland, April 1979*. IR 72/80. Oslo-Ås: SNSF.

Wynne, B. and Simmons, P. (1996) Institutional cultures and the management of global environmental risks in the U.K. Final U.K. contribution to the project on Social Learning in the Management of Global Environmental Risks. Lancaster: Centre for the Study of Environmental Change, Lancaster University.

Zehr, S.C. (1994) The centrality of scientists and the translation of interests in the US acid rain controversy. *Canadian Review of Sociology and Anthropology*, 31(3): 325–53.

12 | Uncertainty, epistemic communities and public policy

K.J. Walker

Introduction

Policy making almost always takes place in an uncertain environment. The future cannot be predicted accurately; policies may have unexpected outcomes; and often policies which appear successful in the short term generate deleterious effects over a longer timescale.

Environmental problems are particularly susceptible to uncertainty. In the first place, those issues defined as 'environmental' typically derive from human-induced disruptions to ecological processes. Often their impacts are initially invisible, but manifest themselves in the form of threats to livelihood, safety, health, or even cherished aesthetic values (McMichael 1993; Boyden 1987; Dubos 1965, 1970). Tracing the sources of disruption, and determining how they might be dealt with, creates still further uncertainty. The long chain of connections from biophysical processes to public policy introduces uncertainties at every juncture. Scientific uncertainty means that even attested scientific data does not yield concrete knowledge. Selective allocation of resources and selective attention to findings both add further uncertainty. Once 'out of the lab', scientific findings must run a maze of political, bureaucratic and social obstacles before they can be interpreted into policy. And interpretation itself may introduce still further distortions. It is with these political issues that this chapter is primarily concerned.

Political processes, whether in democratic systems or more repressive ones, typically deal most decisively with issues over which there is little argument, or where there is general agreement as to the best course of action. This state of affairs is often referred to as 'closure': a situation in which debate has been concluded and action can be undertaken.

Unfortunately, the very nature of ecological processes, and even more, human knowledge about them, can conspire to defeat closure. Worse, because decisions are often taken with incomplete information, they may turn out at later dates to have been mistaken. Furthermore, the dominant problem-definition within an existing political order – the 'policy regime' – may itself rest on a misdiagnosis.

Virtually all nation states are committed to 'growth' and 'development', by which is generally meant more humans engaged in more intensive exploitation of their environment, diverting a greater share of its productivity to themselves. The pressures of statecraft tend to push nations of any kind in this direction, in part for sheer survival. But the logic of this process becomes increasingly perilous once it starts to clash with considerations of ecological stability (Walker 1989). States attempting to resolve such conflicts often veer unpredictably between particular policy directions, or indulge in displacement and denial behaviour.

The resulting dissonance between statecraft and biology has not escaped notice. The Ehrlichs, in a recent chapter, comment that:

> Decision makers, too, have a tendency to focus mostly on the more obvious and immediate environmental problems – usually described as 'pollution' – rather than on the deterioration of natural ecosystems upon whose continued functioning global civilization depends. Indeed, most people still don't realize that humanity has become a truly global force, interfering in a very real and direct way in many of the planet's natural cycles.
>
> For example, human activity puts ten times as much oil into the oceans as comes from natural seeps, has multiplied the natural flow of cadmium into the atmosphere eightfold, has doubled the rate of nitrogen fixation, and is responsible for about half the concentration of methane (a potent greenhouse gas) and nearly a third of the carbon dioxide (also a greenhouse gas) in the atmosphere today – all added since the Industrial Revolution, most notably in the past half-century. Human beings now use or co-opt some 40 percent of the food available to all land animals and about 45 percent of the available freshwater flows (Ehrlich and Ehrlich 1996).

But the uncertainty of science is commonly exploited for purposes extraneous to the problems at issue, or to cast doubt on its findings. And since science often cannot offer guidance on the socio-political implications of specific issues, it frequently fails to find an exact mirror in policy. In part, this is because what matters in science may not be important to policy.

Issues exist on which there is intense scientific debate, but which may nonetheless point to a single public policy measure. Other issues about which scientists are in substantial agreement may precipitate acrimonious disagreement in the process of public debate. For example, the probability and extent of an extended greenhouse effect has stimulated considerable scientific controversy; but in public policy terms the sorts of measures which are required by any significant greenhouse effect are similar. All revolve about 'cautionary' strategies. They include curbs on the emission of 'greenhouse' chemicals – especially carbon dioxide, methane, and CFCs – and a search for less damaging alternatives. That, in turn, implies the exploration or development of alternative, sustainable energy capture and delivery systems, and more conservative use of all resources. While this search continues, policy makers may need to consider constraints on the utilization of low-lying lands, and arrangements for the resiting, not only of human settlements, but of floral assemblages and faunal populations (Lowe 1989; Falk and Brownlow 1989; Hekstra 1969). On the other hand, there is little scientific disagreement about the nature of AIDS, its mode of dissemination, and the available treatments; yet there is considerable public dissension about predominantly non-scientific questions such as whether to make expensive treatments available through the public health service, measures to protect health workers from the risk of acquiring the disease, and which groups in society to target for the most effective delivery of necessary services. In many societies, measures considered desirable and even essential for the prevention or treatment of AIDS run head-on into custom, prejudice, and social power relationships. Such problems involve economic and ethical issues rather than scientific ones, but may well involve scientists by virtue of their positions and knowledge.

A further, and at first blush surprising, source of uncertainty is the *persistence* of government policy. Its effect is to distort the acceptability of scientific findings; it is one of the mechanisms which are seen as making policy responses excessively conservative.

In the present state of knowledge, it is possible neither to predict nor resolve such conflicts reliably. The study of policy processes by political scientists and scholars in related fields is a relatively new phenomenon. In consequence, there is still considerable controversy about the methodology to be followed, and even about the philosophical underpinnings of the enterprise; and there is as yet a relatively small body of data. Where broad general conclusions have been reached by particular scholars, they are often hotly contested by others.

This chapter seeks to expose some of the techniques currently favoured for the analysis of policy, especially where concerned with the environment. It pinpoints some of the characteristics that help to make environmental questions particularly intractable. It attempts to show what useful research has been done in the policy studies area. It draws particularly on

Australian experience, highlighting the critical weaknesses of Australian policy-making institutions. It considers the role scientists might play in the policy process.

Certain initial assumptions are taken more or less for granted. Most informed people nowadays agree that there is a global 'environmental crisis'. While nowhere near as dramatic as predicted by some writers of the late 1960s and early 1970s, it is nonetheless widespread (Commoner 1971; Simmons 1989; Smith *et al*. 1999; Dryzek 1987). Furthermore, whereas in the 1960s environmental problems were constructed very largely in terms of resource shortages and population growth, recent studies have added epidemiological, social, and politico-economic dimensions, the pervasiveness of which is as yet underappreciated in policy-making circles.

Globally, and especially in the temperate colonies, the crisis takes a particularly interesting form, since it calls sharply into question the patterns of 'development' – or, more correctly, resource exploitation – which have characterized the two centuries since the Industrial Revolution (Boyden *et al*. 1990; Smith 1990; Bolton 1981, 1992; Lines 1991). Its impact has been two-pronged: overexploitation has created local crises, which are deepened by global trends. The existence of such stresses has direct implications for some widely and deeply held convictions and shibboleths.

Understanding the relationship between science and policy is further complicated by the fact that those who practise science often see science in a very different light from those who explore its social impact. This in itself reflects a substantial, multifaceted and ongoing debate about the political control and social definition of science, which cannot be explored in depth here. This is not because it lacks importance, but because of the tremendous scope of the topic. However, numerous issues touching on the relation of science to policy must be addressed. They will be dealt with further below, but it is important at this stage to note the tension between a positivist view, in which science is perceived as socio-politically neutral, 'hard' and predictive, and the socially constructionist view, in which science is a social activity, and that fact determines not merely how science is done, but also what its findings are (Chalmers 1982). At its most extreme, the latter tips over into the mindless relativism of postmodernism, in which nothing has any consistency or meaning. For our purposes, it is important to appreciate that, on the one hand, the social determination of science does not automatically invalidate its findings, but, on the other, the allocation of resources and the consequent 'training' of scientific research in particular directions is important in policy terms, as are the orthodoxies within the scientific community, and selective recognition and rejection of important findings, whatever the apparent motivation (Martin 1981, 1986).

Finally, this chapter's broad general approach is *critical*. The critical exploration of political and social processes is the foundation of social science just as surely as critical scientific method gives value to science. Critical thought is one of the most powerful tools, not merely for the acquisition of knowledge, but for the continual reassessment which is essential to it. This stance may be shocking to those who imagine that all the realities are comfortably settled, and 'someone up there' – in politics or the sky – is guarding our best interests; but reality is more typically confusion and indirection. Secondly, in the more casual sense of the word, there is much to criticize in the record of governments everywhere, despite frequent claims to leadership in environmental awareness and change (Walker 1995; Paehlke and Torgerson 1990; Paehlke 1989; Doyle and Kellow 1995).

Policy and policy processes

Many, if not all, environmental policy problems arise from the interaction between inherent characteristics of ecological systems and biological phenomena. Virtually none is unique to environmental issues, being encountered in various forms in other policy fields. But they cluster more markedly at the human–environment interface, thus increasing synergies and stresses. Scientists, as a group, tend to be familiar with the technical aspects of these problems, often without fully appreciating their disturbing implications for policy.

Environmental policy arises from issues which frequently lie outside the normal concerns of day-to-day politics, and even of statecraft. Coping may require significant institutional change and modification of policy processes. It thus places unusually heavy stresses on policy makers and calls into question conventional approaches to the study of policy, since the fundamental nature of environmental questions means that their preconceptions are likely to need re-examination. The record suggests that in neither field has the challenge been handled very successfully.

Environmental policy

The fundamental nature of ecology is itself policy-relevant. A large animal living on a small planet and appropriating increasing amounts of the output of its natural productive processes will need to manage its impact with progressively greater effectiveness as that impact increases. As many scientists have pointed out, humanity is now high on the exponential growth curve, which means that many impacts are growing in criticality. In consequence, sustainability looms ever larger as an overriding goal.

Dovers comments that:

> sustainable development has been only a little too grandly described as the universally agreed goal of human progress. Seeking to integrate concerns of human development, ecological integrity, and security over the long term, sustainability presents a suite of interrelated policy problems of a complexity and magnitude sufficient to profoundly challenge existing modes of policy analysis and policy formulation. Larger policy problems in sustainability – such as greenhouse, land degradation, biodiversity, or population-environment linkages – display a number of attributes found more often, and more often in combination, than problems in more traditional policy fields. (Dovers 1999)

He identifies ten critical issues:

1. problematic spatial and temporal scales;
2. possibly absolute ecological limits;
3. irreversibility and urgency;
4. connectivity and complexity;
5. pervasive risk, uncertainty and ignorance;
6. cumulative effects;
7. new moral dimensions (future generations, other species);
8. 'systemic' problem causes (deeply embedded in patterns of production and consumption and governance);
9. requirements (substantive and political) for community participation; and
10. sheer novelty.

The need for thoroughgoing policy reform is not merely implied; in particular, the necessity of systematic attention to considerations of ecological balance and sustainability at all levels of policy making is spelt out. Yet, especially in the English-speaking world, the political trend of the last two decades of the twentieth century has been away from ecology and towards abstract economic theories with only tenuous connections to reality. This trend is by no means so radical as it might appear, and in the broader ecological context, really represents a reaffirmation of indiscriminate developmentalism, often with the open sponsorship of the nation state. Any transfer of power away from the state has tended to be toward multinational corporations, leading to greater fragmentation of activity, and more massive failures of necessary co-ordination (Walker 1999; Self 1993). Economic rationalism, internationalization and the emphasis on spurious 'competition' have in fact introduced new sources of anti-ecological rigidity (Smith *et al.* 1999; Baran and Sweezy 1966).

On the international stage, it has been observed that the grander that stage, the more general and sweeping the commitments, and the less attention given to their actualization. Predictably, the string of pledges, from Toronto to Rio to Kyoto, has produced a series of commitments which are neither adequate in sum to abate the problems they address, nor equitable in the distribution of the burden. Even in the key case of greenhouse gas emissions, 'business as usual' ensures that what is done is less than adequate (Taplin 1994). The resultant failures often lead to major political controversies. The controversies over genetically modified seed, patenting of seed varieties, and the impact of modern agribusiness on more traditional forms of agriculture illuminate several such issues. They include inadequate regulation, the pitfalls of leaving major decisions to commercial interests, the separation of political elites from the interests of those they rule, and such interesting but as yet not extensively explored questions as the ecological impacts of global monoculture and the risks associated with 'terminator' genes.

The study of policy processes has also, in general, preferred to ignore the fundamental nature of biological and ecological issues. At their worst, it has treated ecological questions as single issues similar to abortion, gay rights, or legalizing marijuana (Warhurst 1983). The tendency to treat environment as 'just another' issue has led to neglect and misunderstanding of its unique features, and in Australia especially, to a construction of Green politics which substantially distorts its motivations and appeal, and may conceal a specific political agenda (Walker 1995 and forthcoming).

Torgerson has argued that the 'administrative state' tends to justify its existence in terms of a world-view of rationally directed progress. The administrator gains legitimacy from the claim to be the impartial co-ordinator, to whose judgement others should submit. Such attitudes, he claims, cope poorly if progress cannot be assumed; they are ill-equipped to deal with insubordinate humans or recalcitrant nature (Torgerson 1990). Worse, they have difficulty accepting that their own rationality is limited and incomplete.

These are among the reasons that environmental policy issues tend to multiply, and often dramatize, the inherent uncertainties of the policy process. Often urgency, irreversibility, and indivisibility combine with ignorance to exacerbate the problem. The interests of non-human species are often involved; compromise may be impossible; and few environmental issues are anything but long-term. Scientific data is frequently uncertain or incomplete.

The sociology of public policy

Many professionals, even when quite sophisticated, tend to equate policy with legislation and government regulation. This is partly because many – lawyers, planners, managers, bankers, safety officers – work within that framework. It also happens because legislation and regulations are the visible *outputs* of government. But they are far from being the sum total of the policy process. In the first place, the processes leading up to legislation and regulation – the inputs to government – have an important effect on their content; in the second, implementation of laws and regulations inevitably requires interpretation, adjudication, and even modification. Legislation does not just happen: it is itself the outcome of highly complex interactions involving powerful political pressures. Most of these are informal – that is, they are not embodied in prescribed, legally or constitutionally defined processes – and, to a large degree, invisible. But they influence the policy outcomes far more critically than does rational argument. The effectiveness of trade union work-to-rule action demonstrates the second point tidily: if all the rules and regulations are obeyed to the letter, the system grinds to a halt. They have to be interpreted flexibly and intelligently. Long before issues of interpretation reach the courts, those who have to apply legislation are informally making sense of it. In the process, they may substantially distort and even frustrate its purpose.

Thus the formal model of political process taught in civics classes in school, of elections, cabinet deliberations, legislation, implementation through government departments, etc, is only the visible part of an immense iceberg. Politics has to be understood as the outcome of a myriad highly informal but often quite decisive processes, most invisible to the public, which shape visible outcomes such as legislation, budgets, and administrative behaviour.

Weak government

Anglo-Saxon political systems in particular have often displayed a pattern of 'weak' government, generally lacking strong programmatic policy orientations. This policy vacuum has meant that pressure groups, or vested interests, have a significant influence on policy. Additionally, bureaucrats often become dominant. They frequently play an important role not merely in determining actual policies, but even in defining policy directions and the underlying philosophies.

Weak government means 'garbage-can' policy making, in which policies are taken on board from various sources urging them, rather than stemming from a strong programmatic orientation or specific capabilities. Under these conditions, strong vested interests gain a major say. Atkinson and Coleman, in an important study, classified interactions between government and interest groups into six categories. They found that in all categories involving 'weak' government, pressure groups had significant policy impact (Atkinson and Coleman 1989a, 1989b).

Where multiple governments exist, as in federal political systems, their interaction with the myriad interests in the community serves to deepen the complexities. The more parties are involved, the more likely it is that policies will be compromised, and the process of policy making will be riddled with short cuts, ad hoc approaches, and exclusion of parties with legitimate interests.

Bureaucracy

Bureaucracies play a crucial role in policy making in all democratic societies. This is partly because of their direct professional involvement with the issues covered by their departmental responsibilities, and partly because of their close, preferential access to ministers and Cabinet. Politicians' perceptions are influenced, if not determined, through the bureaucracy. In Australia, with its 'characteristic talent for bureaucracy', this power is subtly enhanced (Davies 1964: 4).

Bureaucracies filter reality, advise, interpret, and implement. All too often, they are 'captured' in whole or part by their client groups – it is widely agreed, for example, that this is a major problem with forestry – and they may also be dominated by a particular view of policy which is derived from quite specific attitudes and world-views. All these processes provide opportunities for the exercise of power, through the control of information and, above all, interpretation. This may not always result in outcomes which are environmentally undesirable. But the priorities of statecraft are not favourable.

Bureaucracies also suffer from fragmentation, interdepartmental conflict, and pecking orders reflecting significant differences in power. These mean that the outcomes of policy processes are often not rational even in political or economic terms; conflict, often played out by proxy through Cabinet, may result in wins for some departments and losses for others. Sometimes, these can drastically affect the ability of particular departments to do their jobs; in the English-speaking world, departments concerned with environment or conservation have been markedly lacking in ability to hold their own against encroachment from 'developmentalist' departments or the Treasury (Russ and Tanner 1978; Christoff 1998). Such departments have typically been low in status. This has severely constrained their ability to participate in the overall policy process, or to influence its general direction.

This long-term persistence of power relationships is matched by a corresponding persistence of government policy. For Australia and India at least, this persistence is well documented; there is every evidence that it is a common phenomenon elsewhere. The most remarkable case in Australia is the dogged adherence of successive Tasmanian governments to a brilliant economic development strategy which, due to conservative opposition when first proposed, was essentially stillborn (Tighe 1992; Davis 1986; Blakers 1994). A second example is the persistence of large dam proposals and their associated irrigation projects (Walker, 1999, 1992; Goldsmith and Hildyard, 1985). This has been a serious problem in India, too, where undesirable and even dangerous dam proposals frequently enjoy strong backing from government and from groups likely to benefit, often to the detriment of other less powerful communities. The Narmada scheme in India is one such example, and has been particularly controversial (Kalpavriksh 1985; Gadgil and Guha 1995; Goudie 1990). Particularly well documented globally is deforestation (Palmer 1992; Gadgil and Guha 1992). Bureaucracies often play a significant role in persistence of this kind, most importantly by institutionalizing existing practices, resisting change, and (paradoxically) by acting as channels for powerful forces with particular agendas. The outcome, frequently, is differential attention to scientific information according to its acceptability.

Policy distortions

Consequently, it is not surprising to find that policy makers frequently ignore and distort scientific data.

In Western Australia, the government entomologist discovered as long ago as 1945 that all known major pests of cotton were present at the site of the proposed Ord River irrigation scheme. The state government, throughout the 1950s and 1960s, steadfastly ignored this finding. Cotton was chosen as the preferred crop when the scheme opened in 1963; cultivation collapsed in 1974 (Walker 1992). Scientific evidence suggesting that unacceptable contamination of Bass Strait and its fisheries would occur if the Wesley Vale pulp mill were built was criticized by the proponents (Economou 1992; Chapman 1992). Kellow concluded that technical decisions in the electricity industry in Australia and New Zealand were rarely, if ever, based on technological factors; rather, arguments from technological necessity were used to justify decisions reached on other grounds (Kellow 1986).

Governments frequently attempt to control or interpret scientific data. In 1989, fish caught in the vicinity of Sydney's notorious sewer outfalls were found to have very high levels of pesticide and heavy metal contamination, the pesticides being on average 122 times the National Health and Medical Research Council (NH&MRC) limits. The state government through its Minister for Agriculture, the State Pollution Control Commission, and the Sydney Water Board, attempted to prevent the publication, not of raw scientific data, but of discussion and analysis which would make it intelligible to lay persons, on the ground that they might unduly alarm the public! (Beder 1990, 1989). Beder has also cast serious doubt on existing sewage disposal practices.

Examples of suppression and ignoring of scientific data in other contexts are rife. Scientific findings are eagerly exploited when they support a desired course of action; they are as often ignored when they do not. The public reverence and respect accorded to science may be severely damaged. Jasanoff (1997: 353) comments tellingly on 'the unforgiving power of memory and the fragility of trust', going on to say that 'the perceived unreliability of ruling institutions undermined the foundations for collective confidence in science' in some controversies in the United States.

Policy communities and networks

Despite this apparent anarchy, some tools for the systematic analysis of policy processes, and for understanding the patterns of interaction among them, do exist.

The notion of the *policy regime* helps to clarify what might be called the *terms of reference* upon which policy makers rely. It defines a state of broad agreement about a set of basic policy concerns, which may include development, economic growth, social welfare, national self-definition, and many others. The scope of a regime may be local, regional, national, or international. The specifics are often defined by the most powerful and effective players in the policy process; and mapped to the area of agreement – often called a 'contract zone' – between these players. Policy regimes are the orthodoxy of the policy process. Changing the regime, and its fundamental assumptions, is normally very difficult. This has been a major concern in environmental policy terms, because in most countries the dominant policy regimes have long been focused on economic growth and development (Walker 1989).

Most modern political systems are unavoidably heterogeneous, and democratic ones are frequently characterized by 'weak' systems of government. Their lack of autonomous power means that their policy systems are fragmented, and their policy space constrained by their limited ability to direct and shape political, social, and economic change (Atkinson and Coleman 1989a). Globalization further undermines national policy autonomy. Under such conditions, the various circles which influence public policy are of great importance. Not only do various sectors of the economy and the larger society tend to develop their own reference points; their collective actions and ideas often form important themes in policy for the community at large. Immigrant groups, for example, tend to favour further immigration; gun owners frequently mouth rhetoric about weapon-based 'freedom'; and such groups have means of communication, such as regular journals – and nowadays the Internet – which keep them in touch. Some groups are supranational in scope; some, such as the large multinational corporations, are extremely powerful.

The classification of such groups has long attracted attention, but their categorization in policy-oriented terms has been more recent. Specifically, analysts now distinguish policy communities and policy networks. An interesting, influential, and correlated phenomenon is the epistemic community, of which the scientific community is often an exceptionally important element (Wright 1988; Haas 1989).

Communities and networks gain importance because not all issues involve the same people: particular issues have their own 'public'. National policy debates commanding a widely attentive audience eagerly following press and media are at best rare. Individuals as well as groups follow their own preferred issues and problems with markedly more attention than others: attention to policy is itself fragmented.

Policy communities

Government policy and administration unavoidably falls into sectors, often corresponding to economic, social or geographical divisions. Each sector generates one or more policy communities – farmers, graziers, industrialists, supermarket owners, teachers, scientists, migrants, indigenous peoples – with interests which may be shared or conflict. Sector boundaries are fuzzy. Membership of policy communities embraces government ministers, public servants, and others directly concerned with the making and implementing of policy; members of firms with a direct interest in the policy issues in question, for example, road hauliers, where road transport policy is concerned; journalists involved both in the specialist press and in national reporting; and, possibly, members of the public who may belong to special interest societies or may act as individuals. Some have obvious pecuniary interests; others may have no economic interests, but are engaged in scholarship, have an interest for reasons of curiosity or public concern, or are outsiders likely to be affected by the activity. Thus the policy community for town planning activities often includes affected residents and businesses as well as academics, bureaucrats, politicians and those with a general interest.

The crucial characteristic of policy communities is communication among the members, serving both to identify the central issues of relevance, and to explore proposed policies and courses of action. Where memberships are dispersed, as with farmers and other rural groups, there may be highly formal, established channels. Organizations such as the National Farmers' Federation, the Cattlemen's Union, the various trades unions, and many others publish journals expressing their communities' views, keeping the membership in touch and presenting their claims in politics. There is commonly an established orthodoxy or viewpoint which dominates such discourse. This finds expression in professional journals or news media, sometimes restricted in circulation to putative members of the community.

In addition, of course, there is informal communication, which may include matters left unsaid or concealed from the wider community. Sometimes, these views emerge when they are perceived to be acceptable. In early 1996, when the then-new Australian Liberal (actually conservative) government moved to restrict native title and review the operations of the Aboriginal and Torres Strait Islander Council (ATSIC, a formal representative body) various racists apparently judged it acceptable to air their views. Similar phenomena are associated with most policy changes, and are not new; Victorian England's plague of self-declared Utilitarians bears an uncanny resemblance to the antics of modern economic rationalists of the political Right.

Policy communities are frequently linked to existing government departments or sections within them and often the institution provides a focal point or a point of reference for the community concerned. Activity is devoted to changing policy, gaining concessions or hand-outs, maintaining and improving political and administrative contacts, and so on. Though policy communities necessarily focus on government, many organizations, religious and secular, participate in the identification and definition of policy concerns. In recent years, international non-governmental organizations (NGOs) such as Greenpeace have become increasingly important.

Policy communities often debate policy options in an ongoing and generally amicable fashion. This leads to definitions of their central concerns, and frequently a consensus about desired policies. These may be embodied in a series of informal 'rules of the game'. But these rules often define whose ideas will be considered sympathetically; members in good standing are more likely to be attended to, and their policies adopted. In short, the 'ins' dominate *even if* one or more 'outs' exist who offer better analyses and better policies.

Often policy communities are especially intolerant of attempts to redefine their central problems. A doggedly defended orthodoxy may be used as a criterion for determining who is 'in'. For example, a road lobby comprising hauliers, truck manufacturers, oil companies, and a froth of sycophantic journalists may well ridicule and exclude 'crackpots' arguing for major changes in road policy to save fuel, cut congestion, or reduce accident rates. Consequently, individuals or organizations seeking to raise questions which are either in the public interest, or which clearly ought to be considered, may simply be ignored. Pluralistic political systems tend to legitimize vested interests by assuming that they have a claim on government's time and attention directly proportional to the financial gains or losses they expect. Policy biases inevitably result. Government reliance on vested interests for information, feedback on policy options, and numerous other functions confers privilege and influence on favoured members of the policy community.

Environmental groups, both local and international, frequently find that their status as members of the policy community is precarious. Though they gain some attention through the media, their full message is rarely heard or comprehended. Even practicable and worthwhile proposals are likely to be ignored. Consultation on major issues is fitful and incomplete, largely because the environmentalist construction of environmental questions does not fit easily into the prevailing growthist, developmentalist mentality. Furthermore, some heavily promoted decision strategies, especially those associated with the absurdly named 'wise use' doctrine, are explicitly designed to exclude environmental and other broadly based community interest groups (Helvarg 1994; Ehrlich and Ehrlich 1996;

Doyle 1999). But exclusion of the conservation lobby and the environment movement in general from the policy community sharply reduces opportunities for sympathetic, constructive debate. The consequent unnecessarily poor problem analysis yields constrained policy options.

Policy communities, in addition, can often have limited lives, or vary appreciably in size with the salience of an issue. By contrast, 'permanent' interests are more enduring; the Institute for Public Affairs, an Australian think-tank and support group associated more or less intimately with the Liberal Party and its predecessors, has existed since the 1920s.

Policy networks

Not all policy communities, and not all their members, have power. To distinguish between groups which directly affect decisions and those which do not, the concept of a policy *network* is used. This consists only of those people who actually make or directly influence decisions: those with their hands on the levers of power. This includes, most obviously, politicians, business people, trade union functionaries and bureaucrats. This does not mean that they are necessarily members of the government, or even of extended state instrumentalities such as statutory corporations or advisory boards. In road haulage, major trucking firms or unions may have a very direct input into policy, especially if government consults them directly, or if 'garbage can' policy making is in train. Advisers to political parties, in or out of government, may be network members; well-connected journalists and privileged private individuals may also influence policy. Network members are, by definition, members of policy communities; but the reverse is not necessarily true.

Policy networks, by their nature, are highly fluid, even evanescent. The network which is relevant to one decision may be significantly different for another, even on a related issue, if it is removed even slightly in place, time or focus. Membership changes, depending on the issues involved. The forest industry, for example, may deeply influence forest management, but it is the tourist industry that is dominant in coastal development.

A policy network may embrace one or more policy communities; it may exclude some members of a particular community, but include others. It may focus on a specific policy problem, or a cluster of related problems; or it may form about a particular policy process 'such as budgeting, auditing or planning' (Wright 1988: 606). A policy network is best defined by describing the interactions among its members.

It matters who is 'in' and who is 'out'. The policy networks which determine environmental policy, for example, typically exclude major environmental pressure groups. They tend to be treated by government as single-issue groups to be given some sops, but not integral participation in the 'important' business of economic growth. Yet environmental groups frequently have substantially larger memberships than major political parties, let alone lobby groups.

Exclusion can be very dramatic when an issue is highly charged politically. Caufield shows this exceptionally clearly in her account of the nuclear safety issue (Caufield 1990). The nuclear establishment in the USA resisted very strongly criticisms from scientists both within and outside its own bureaucracy whenever these tended to impinge on the perceived viability of the weapons programme or on nuclear power. Yet, as medical evidence emerged, the positions of the parties tended to converge quite markedly, with the difference that the establishment remained less critical and less risk-averse in setting standards than the critics. The latter, however, often found that their position cost them dearly in employment opportunities and standing within the 'orthodox' scientific community. Interestingly, later validation of their views, partially or completely, did not recoup their positions.

In the nuclear case, the exclusion was multiple: the critics were isolated from the policy community, the network, and, most significantly, the epistemic community.

Epistemic communities

An epistemic community is a network of professionals with recognized expertise and competence in a particular domain and an authoritative claim to policy-relevant knowledge within that domain or issue-area. Although an epistemic community may consist of professionals from a variety of disciplines and backgrounds, they have (1) a shared set of normative and principled beliefs, which provide a value-based rationale for the social action of community members; (2) shared causal beliefs, which are derived from their analysis of practices leading or contributing to a central set of problems in their domain and which then serve as the basis for elucidating the multiple linkages between possible policy actions and desired outcomes; (3) shared notions of validity – that is, intersubjective, internally defined criteria for weighing and validating knowledge in the domain of their expertise; and (4) a common policy enterprise – that is, a set of common practices associated with a set of problems to which their professional competence is directed, presumably out of the conviction that human welfare will be enhanced as a consequence (Haas 1992a: 3).

They 'need not be made up of natural scientists; they can consist of social scientists or individuals from any discipline or profession who have a sufficiently strong claim to a body of knowledge that is valued by society' (Haas 1992a: 16).

Particularly in the international field, Haas argues, policy makers, while not anxious to relinquish control of the policy process, have tended to minimize uncertainty by turning to epistemic communities for advice 'especially in the wake of a shock or crisis'. He identifies four ways in which epistemic communities can help. Firstly, they can 'elucidate the cause-and-effect relationships and provide advice about the likely results of various courses of action'. Secondly, they can 'shed light on the complex interlinkages between issues'. Thirdly, they can 'help define the self-interests of a state or factions within it'. Finally, they can help in the formulation of policies. Haas notes that epistemic communities, by interpolating their own problem definitions and perceptions, may move 'towards goals other than those initially envisioned by the decision makers'. Therein lies both their ability to achieve international co-ordination (a point much stressed) and their ability to gain a stranglehold in national policy.

As they are more frequently transnational than are policy communities or networks, the extent of epistemic communities, their internal lines of communication, and their norms, values, or power bases often distance them from immediate policy processes. This can confer independence of control by any one government, and lend authority to the community's views.

An 'international ecological epistemic community' consisting of scientists in cognate disciplines was detected by Haas in his pioneer study of pollution control in the Mediterranean. He showed that willingness to act on pollution control was greatest in countries where a scientifically based ecological epistemic community existed, and had access to government. The worst respondents were countries such as Albania and Algeria, where scientific communities were weak or had minimal access to government. The epistemic communities were able to emphasize the need for international regional co-operation both to investigate Mediterranean pollution problems and to put into place a regime of co-operation and collaboration which would deal with them. The success of this regime was marked: even those countries which had initially rejected co-operation became involved by degrees, and often quite rapidly. A side-effect was to aid in the establishment of an epistemic community in each, and to make the governments markedly more receptive to their advice (Haas 1989, 1990). Haas has since turned his attention to damage to the ozone layer

(Haas 1992b). The global scale, predictably, means slower progress, due to the much larger numbers of participants in the process.

The influence of economists illustrates the potential for transformation of policy makers' initial goals. Economists' penetration of national politics in the English-speaking nations has been spectacularly successful. Pusey's finding that half Australia's senior federal public servants in the most powerful departments were university-educated economists owing allegiance to neoclassical economic orthodoxy helps to explain their dominance in an uncontested policy regime of deregulation, free market, and anti-welfare state ideas: so-called 'economic rationalism'. While concentrated in the Treasury, their influence spreads throughout the most powerful departments, giving them control over virtually all aspects of government policy. Most crucially, these bureaucrats – themselves part of a powerful policy network – derive their support from a community of international scope, embracing not only a large sector of the international academic economics community, but also their supporters and sycophants in the press and media. Pusey's brilliant analysis shows how this intellectual orthodoxy is both irrelevant and damaging to the interests of ordinary Australians, and how it also works against the national interest in overseas trade (Pusey 1991).

Where there is substantial agreement on the nature of a specific problem, a scientific epistemic community's authority is substantial; but dissension permits the casting of doubt and introduction of uncertainty. This is readily exploited by vested interests. Uncertainty over the effects of climatic modification in the international epistemic community is often seized upon by sceptics. By casting doubt on matters which all informed parties agree on, they may go beyond the scope of legitimate scientific disagreement. The British government was able to find scientists to assure the public that there was no danger from mad cow disease; but it could not control scientific opinion offshore, nor, ultimately, prevent the publication and discussion of adverse findings by British scientists.

Science and public policy

Haas' position, using his findings from the MedPlan study, was that epistemic communities could play a key role, not merely in defining major problems, but also in resolving them through depoliticization. But his position has been challenged. In particular, Jasanoff suggests that reliance on epistemic communities may blind policy makers to important cultural differences in the way scientific knowledge is interpreted, as well as making them vulnerable to an overly technical approach to the problems.

The authority of science in public debate is constrained by such uncertainties. When science emerges from the laboratory and enters public life, it undergoes an important change, from research science to what Jasanoff has called 'regulatory science', but which might be better and more broadly characterized as 'science for policy' (Jasanoff 1990). This kind of science is socially constructed, the product of extensive negotiation, often on a case-by-case basis, between scientific advisers and policy makers. Social construction is possible for two reasons. Firstly, laboratory or pure science does not conform to a uniform model either of practice or procedure. This is important in two ways: firstly, science, even in the positivist model, is never final. New findings, reinterpretations of existing data, novel theoretical formulations, all may change what science 'tells' us, and what its implications are for public policy (Chalmers 1982). Secondly, if we accept that science itself is a mechanism by which humans construct models which make reality intelligible to *them*, then we can accept that reality may generate unwelcome surprises at points of poor fit with the humans' models. These fundamental uncertainties mean that the boundaries of science – the things on which it can pronounce authoritatively – are indeterminate. Not only are there perfectly legitimate disagreements between scientific peers about method and interpretation, but there must also be disagreements about the areas in which scientific findings are authoritative.

The implications for interaction between science and public policy are far-reaching. The cruder models are immediately ruled out: it cannot be the case that authoritative science decides the 'hard' questions while social scientists and policy makers tailor social systems to fit the findings, for what scientists do, as well as how they do it, can itself be determined by social factors. The interface of science with society in itself creates uncertainty, mainly about the boundaries between 'hard' knowledge and educated guesses. In recent years, a whole new discipline, risk analysis and evaluation, has grown up around the problem. But risk analysis really only adds a new layer to the social construction of scientific questions, by offering a methodology for the systematization of one aspect.

Social construction of science becomes especially critical in policy making because the process typically works through committees and advisory bodies. Under these conditions, science *as policy* is no longer pure: it becomes contingent on its social construction. Jasanoff notes that scientists participating in such bodies are frequently aware of this (Jasanoff 1990: 229).

This fact has repercussions. In the laboratory, 'good' science conforms to *scientific* standards: reproducibility, corrigibility, accepted methods, and so on. While it is true that some science does not always conform to these standards and some is even faked, the standards still exist. It is also true, of course, that they do not address some of the major problems of the philosophy of science; in particular, no existing model of science can offer a universally valid model of scientific method. But these problems pale once science enters the policy arena and its definition becomes negotiable. Then the problem of defining 'good' science is greatly exacerbated.

Negotiation privileges policy networks, defining problems in ways acceptable to them. Dudley explicitly documents a split in the UK scientific community between personnel in the Central Electricity Generating Board (CEGB), the civil service, and the universities, over the effects of sulphur and nitrogen oxides in creating acid rain, both in Britain and abroad (see also Chapter 11 this volume). Further, UK scientists and policy makers tended to be at odds with continental opinion, especially in Scandinavia. Government policy was made on a 'least-effort' basis, using scientific dissension as the excuse (Dudley 1986). 'Least effort', as in the mad cow disease scandal in Britain, can lead to the flouting of accepted norms of laboratory science. So politicized, indeed, was the British establishment's reaction to the latter issue that one wonders if it should not be called 'Mad Tory Disease'. (Anon. 1996).

By isolating one favoured problem definition, and its attendant prescriptions, and excluding others, bureaucratic convenience is served. The problem is rendered manageable even if it is displaced rather than solved. Proliferation of specialized advisory committees, bureaux, and other similar bodies adds to the attractiveness of this technique.

The problem with such devices is the closed-door nature of the policy making involved. In Britain, a stubborn refusal to contemplate the health risks both of feeding livestock on animal tissues and of the possible connection between scrapie in sheep and Creuzfeld-Jakob disease (CJD) in humans was supported by rejection and suppression of contrary views. In retrospect, this has turned out to be doubly ironical, because investigation finally showed that these practices (however repulsive!) were not, in fact, the source of the problem, and attention turned to other pathways (Aldhous 1996; Mackenzie 1999). But failure to recognize and investigate the problem led to delays and avoidable deaths.

Closed-door approaches are not merely more error-prone; they may also ignore social, economic and other consequences which are not evident to scientists in prestigious consulting roles. Viable and even superior alternatives may never be considered, let alone offered to policy makers and the public. Democratic policy-making processes, however inefficient, incorporate various checks and balances, which make public scrutiny relatively easier. When these checks and balances are lacking, it may become impossible to cross-examine proponents and elicit reasons (let alone sound justifications) for their recommendations.

In Australia, and no doubt elsewhere, these problems are aggravated by several factors. Most state governments in Australia more nearly resemble plebiscitary dictatorships than democracies; their policy making is typically rigid and often hostile to criticism. At best, it excludes dissenters. Undesirable policy trends are consequently perpetuated, often for decades; for example, the continuing failure to limit clearance of native vegetation, or to appreciate the unviability of European-style broadacre agriculture in tropical and subtropical Australia. Gerontocracy, cronyism and bureaucratic empire-building act as multipliers.

Consequently, 'public' science cannot be evaluated by criteria which are purely scientific in research science terms. Instead, public scrutiny, not merely by scientific peers, but by affected members of the public or those with cognate interests, has to be admitted as legitimate. Under those conditions, science is 'good' when integrated into public debate and examination of alternatives, 'bad' when information is concealed and important questions, such as assessment of risks to the public, are determined by fiat.

Lack of consensus among scientists too may arise for non-scientific reasons. Active policy regimes, even institutionalization of scientific access to government, do not necessarily guard against it, nor guarantee the best advice. Informal old-boy networks often control succession in formal advisory institutions as well as informal advisory contacts, retiring members tending to nominate their own protégés. The new recruits, typically senior scientists retired from active research, may fossilize the past pattern of scientific research investment. It may also self-select for politically active scientists. To the tunnel vision of highly specialized scientists, with its accompanying inability to see the 'big picture' can be added the conservatism of the scientific establishment.

The place of science

By following the implications of the social construction of science ruthlessly through the policy process, Jasanoff has made a seminal contribution to the science policy debate. Her linked critique of the view that epistemic communities are capable of breaking policy bottlenecks by the application of authoritative scientific knowledge is very important.

Rejecting a positivist view of science, Jasanoff emphasizes the degree to which even 'laboratory' science is the outcome of a process of social construction. In her view, it is not merely the choice of subject matter and the priorities controlling the allocation of resources which are the outcome of social interactions; problem definition and assessment of data in the day-to-day practice of science are also the outcome of a social process. This is the more true, the 'softer' the data, and the larger the element of inference in its interpretation. Thus biology and ecology suffer more from social construction, though even physics is by no means immune. Much of Jasanoff's own work has been in relation to the US FDA and EPA, where 'soft' science is the norm (Jasanoff 1990; Wynne 1992).

The first consequence of this view is the acceptance that the area of uncertainty surrounding science is much greater than positivist and utilitarian views might allow. It is not simply a question of uncertainty appearing at the juncture between disciplines, organizations, or policy processes: the whole process, from the first scientific observation to the last detail of policy implementation, is permeated with uncertainty. The second is the enormously important suggestion that epistemic communities, '. . . far from fulfilling any independent agenda . . . are merely the instruments of a technological culture through which powerful industrial states impose a particular vision of natural and political order on the rest of the world' (Jasanoff 1996b: 193–194).

Scientists in public debate

Science, because it enjoys authority, is inextricably involved in the exercise of political power; that much is plain. But, because scientific authority is linked with power, it may, firstly, become tainted by the demands of the powerful, and second, lose authority to the extent that it is seen as partial.

Nation states currently face a major dilemma. The demands of statecraft continually push them towards a mutually competitive stance in which growth in population, industry and (especially) military capacity are perceived as imperatives. At the same time, humanity is increasingly coming up against barriers to unlimited and indiscriminate growth. States, paradoxically, are at once the engines of growth and the only institutions which are likely to have the capacity, in the foreseeable future, to regulate human environmental impact (Walker 1989).

Not only does this pose a policy dilemma; it also suggests that the future holds two mutually exclusive outcomes. One is of a world in which internecine squabbles and overexploitation lead to extensive misery of the kind that can already be seen in some Third World countries. The other is of a world in which limitation of human environmental impact, and the pursuit of ecological health as deliberate policy, have averted massive degradation and misery. The first outcome is very likely to emerge without prompting; the second would require behaviour from states which is more co-operative and which is built upon cautionary principles.

The second strategy in particular is likely to place very severe burdens on the capacity of nation states to make and implement decisions which have an environmental impact. It is for this reason that the scholarly controversy over the policy process is of far more than academic interest. It pivots, in particular, on perceptions of the relative difficulty of translating scientific knowledge into public policy, as well as the issue of what constitutes 'good' science in the policy context. In this context, reduction of uncertainty becomes important.

Haas' position, as sketched above, is broadly that the authority possessed by epistemic communities may permit them successfully to resolve policy deadlocks, *by the application of their specialized, authoritative knowledge*. In his studies of the Mediterranean and greenhouse gas cases he purports to demonstrate this thesis.

Jasanoff, on the other hand, questions whether such techniques work effectively unless there is already a substantial area of agreement on the authoritative status of the 'science' deployed, and other conditions for agreement are met. In particular, she suspects a conservative bias, making it more difficult for novel policy directions to gain acceptance, *even when* there is substantial agreement on the issue in the relevant epistemic community. For evidence, she cites the slow progress towards effective international agreement on a wide range of urgent, environmentally sensitive topics.

There are merits in both arguments. Haas' account of the process by which formerly recalcitrant regimes were entrained by MedPlan convincingly suggests that the depoliticization of the problem was clearly a critical factor in gaining acceptance, most especially in countries where a relevant epistemic community was weak or vestigial. This account tends to undermine the Jasanoff argument that depoliticization works only for conservative, bureaucratic forces.

However, the context assumes importance here. In the Mediterranean, and at the international level, there was no dominant bureaucracy in a position to commandeer and control the policy process, nor were there similarly capable political elites; hence the epistemic community gained greater prominence. In other, well-documented cases, dominant bureaucracies have been able to confine, channel, and even redefine such bodies of knowledge. Political elites have also been able to block or control policy processes so as to delegitimize or exclude particular issues. This has been particularly so for national politics. In 'weak' states, too, the deflection or redefinition of issues by powerful vested interests is not exactly unheard-of.

Thus, although their status as an epistemic community permits scientists to enjoy a degree of intellectual and policy authority, their capacity to participate decisively in political debate is contingent on elements of the policy environment which are outside their control. However, where an epistemic community is in substantial agreement over the nature of a problem it can generally offer a substantial reduction in uncertainty. This defines the 'decision space' more clearly, making a political resolution easier. This enhances, though it cannot guarantee, its chances of acceptance, and consequently its capacity to reduce uncertainty.

Conclusion: authoritative knowledge and public policy

Epistemic communities lend their members authority in the political arena. The Baconian conception of scientific knowledge as useful, contributing to the 'amelioration of the condition of mankind', is enshrined in the basic values of science, which emphasize the free communication of knowledge, objectivity, impartiality, corrigibility, and a critical intellectual stance. While science, and individual scientists, may at times fall short of all these goals, their existence, while not guaranteeing information, evaluation, and opinion emanating from it, greatly enhances its credibility. However, to the degree that scientific uncertainty permeates the whole enterprise, adherence to the standards cannot guarantee certainty.

But this is not surprising. In recent years, the 'Enlightenment enterprise' of solving all human problems by the application of pure reason has come increasingly under attack, most particularly from the postmodernist school, given to an anchorless relativism. This attack has been reinforced by harder-headed criticism, deriving from the persistent anomalies – 'impossibility results' – to be found in almost all formalized social choice processes (Smith *et al.* 1999; Arrow 1951, 1963; Sen 1970, 1982; Taylor 1976, 1982; Axelrod 1984; Ostrom 1990).

The attack on rationalism has tended to emphasize uncertainty and relativism. But despite her emphasis on the social construction of science, such an influential authority as Jasanoff herself rejects outright relativism. She suggests instead the utility of a healthy awareness of the socially determined nature of science in detecting attempts to disguise political programmes under the guise of scientific objectivity.

> Relativism no longer stands as a barrier against taking positions on matters of social moment or using one's capacities to advance one's arguments as forcefully as possible. There is, at bottom, a deeply normative project that runs through even the most playful narratives that we in science and technology studies construct about the common production of knowledge and social order. It is to render more visible the connections and the unseen patterns that modern societies have often taken pains to conceal, often by enlisting the unquestionable forces of the physical world as represented by the voices of scientist-seers or as hardened into obedient machines. (Jasanoff 1996a: 413)

Goodin, who advocates a broadly utilitarian approach to deciding major socio-political issues, also cautions against 'uncertainty as an excuse for myopia'. Narrowing the policy perspective can achieve closure by limiting the number of alternatives considered, and making the decision focus more instrumental. But it will often be achieved at the cost of ignoring worthwhile perspectives and possible opportunities (Goodin 1982).

Once it enters into public policy, science itself needs to be self-aware, both with regard to its limitations and to the uses for which it might be co-opted. This, in turn, requires a heightened ability to appreciate which uncertainties are being reduced, and for whose benefit.

References

Anon. (1996) The BSE scare: mad cows and Englishmen, *The Economist*, 30 March: 21–3.

Aldhous, P. (1996) Scrapie theory fed BSE complacency . . . , *New Scientist*, 13 April: 4.

Arrow, K.J. (1951, 1963) *Social Choice and Individual Values*. New York: Wiley.

Atkinson, M.M. and Coleman, W.D. (1989a) Strong states and weak states: sectoral policy networks in advanced capitalist economies, *British Journal of Political Science* 19(1) January: 47–67.

Atkinson, M.M. and Coleman, W.D. (1989b) *The State, Business, and Industrial Change in Canada*. Toronto: University of Toronto Press.

Axelrod, R.M. (1984) *The Evolution of Cooperation*. New York: Basic Books.

Baran, P.A. and Sweezy, P.M. (1966) *Monopoly Capital*. Boston: Monthly Review Press.

Beder, S. (1989) *Toxic Fish and Sewer Surfing*. Sydney: Allen & Unwin.

Beder, S. (1990) Science and the control of information: an Australian case study. *The Ecologist* 20(4) July/August: 136–140.

Blakers, A. (1994) Hydro-electricity in Tasmania revisited. *Australian Journal of Environmental Management* 1(2) September: 110–120.

Bolton, G. (1981) *Spoils and Spoilers: Australians Make their Environment, 1788–1980* (2nd edn 1992). Sydney: George Allen & Unwin.

Boyden, S. (1987) *Western Civilisation in Biological Perspective*. Oxford: Oxford University Press.

Boyden, S., Dovers, S. and Shirlow, M. (1990) *Our Biosphere Under Threat: Ecological Realities and Australia's Opportunities*. Melbourne: Oxford University Press.

Caufield, C. (1990) *Multiple Exposures: Chronicles of the Radiation Age*. London: Penguin.

Chalmers, A.F. (1982) *What Is this Thing Called Science?* 2nd edn. St Lucia, Queensland: University of Queensland Press.

Chapman, R.J.K. (1992) *Setting Agendas and Defining Problems: The Wesley Vale Pulp Mill Proposal*. Deakin Series in Public Policy and Administration No. 3. Geelong, Victoria.: Centre for Applied Social Research, Deakin University.

Christoff, P. (1998) Degreening government in the garden state: environment policy under the Kennett government. *Environmental Law and Planning Journal*, 15(1) February: 10–32.

Commoner, B. (1971) *The Closing Circle: Nature, Man, and Technology*. New York: Knopf, 1971.

Crosby, A.W. (1993) *Ecological Imperialism: the Biological Expansion of Europe, 900–1900*. Cambridge: Cambridge University Press.

Davies, A.F. (1964) *Australian Democracy*, 2nd edn. Melbourne: Longman.

Davis, B.W. (1986) Tasmania: the political economy of a peripheral state. In: Head, B. (ed.) *The Politics of Development in Australia*. Sydney: Allen & Unwin, 209–25.

Dovers, S. (1997) Sustainability: demands on policy. *Journal of Public Policy* 16: 303–18.

Dovers, S. (1999) Institutionalising ecologically sustainable development: promises, problems and prospects. Chapter 11 in Walker, K.J. and Crowley, K. (eds) *Australian Environmental Policy 2: Studies in Decline and Devolution*. Kensington, NSW: NSW University Press, 205–23.

Doyle, T. (1999) Roundtable decision-making in arid lands under conservative governments: the emergence of 'Wise Use'. Chapter 7 in Walker, K.J. and Crowley, K. (eds) *Australian Environmental Policy 2*. Kensington, NSW: NSW University Press, 122–41.

Doyle., T.J. and Kellow, A.J. (1995) *Environmental Politics and Policy Making in Australia*. Melbourne: Macmillan.

Dryzek, J. (1987) *Rational Ecology*. Oxford: Basil Blackwell.

Dubos, R. (1965) *Man Adapting*. New Haven and London: Yale University Press.

Dubos, R. (1970) *So Human an Animal*. London: Rupert Hart-Davis.

Dudley, N. (1986) Acid rain and pollution control policy in the UK. *The Ecologist* 16(1): 18–23.

Economou, N. (1992) Problems in environmental policy creation: Tasmania's Wesley Vale pulp mill dispute. Chapter 3 in Walker, K.J. (ed.) *Australian Environmental Policy*. Kensington, NSW: University of NSW Press, 41–57.

Ehrlich, P. and Ehrlich, R. (1996) 'Wise use' and environmental anti-science. In Ehrlich, P. and Ehrlich, R. *The Betrayal of Science and Reason*. Washington, DC: Island Press.

Falk, J. and Brownlow, A. (1989) *The Greenhouse Challenge: What's to be Done?* Ringwood, Victoria.: Penguin.

Gadgil, M. and Guha, R. (1992) *This Fissured Land: An Ecological History of India*. Delhi: Oxford University Press.

Gadgil, M. and Guha, R. (1995) *Ecology and Equity: the Use and Abuse of Nature in Contemporary India*. New Delhi: Penguin Books India, 68–76.

Goldsmith, E. and Hildyard, N. (1985) *The Social and Environmental Effects of Large Dams*. Camelford, Devon: Wadebridge Ecological Centre.

Goodin, R. (1982) *Political Theory and Public Policy*. Chicago and London: University of Chicago Press.

Goudie, A. (1990) *The Human Impact on the Natural Environment*. 3rd edn. Oxford: Basil Blackwell.

Griffiths, T. and Robin, L. (1997) *Ecology and Empire: Environmental History of Settler Societies*. Carlton, Victoria: Melbourne University Press.

Haas, P.M. (1989) Do regimes matter? Epistemic communities and Mediterranean pollution control. *International Organization* 43(3) Summer: 377–403.

Haas, P.M. (1990) *Saving The Mediterranean*. New York: Columbia University Press.

Haas, P.M. (1992a) Introduction: epistemic communities and international policy coordination. *International Organization* 46(1) Winter: 1–35.

Haas, P.M. (1992b) Banning chlorofluorocarbons: epistemic community efforts to protect stratospheric ozone. *International Organization* 46(1) Winter: 187–224.

Hekstra, G.P. (1969) Global warming and rising sea-levels: the policy implications. *The Ecologist* 19(1): 4–15.

Helvarg, D. (1994) *The War Against the Greens: The 'Wise Use' Movement, the New Right, and Anti-Environmental Violence*. San Francisco: Sierra Club Books.

Jasanoff, S. (1990) *The Fifth Branch: Science Advisers as Policymakers*. Cambridge, Mass: Harvard University Press.

Jasanoff, S. (1996a) Beyond epistemology: relativism and engagement in the politics of science. *Social Studies of Science* 26(2) May: 393–417.

Jasanoff, S. (1996b) Science and norms in global environmental regimes. Chapter 8 in Hampson, F.O. and Reppy, J. (eds) *Earthly Goods: Environmental Change and Social Justice*. Ithaca and London: Cornell University Press, 173–97.

Jasanoff, S. (1997) Public knowledge, private fears. Review of Irwin, A. and Wynne, B., *Misunderstanding Science: The Public Reconstruction of Science and Technology* (Cambridge University Press, 1996). *Social Studies of Science* 27: 350–5.

Kalpavriksh – the Environmental Action Group and the Hindu College Nature Club, Delhi University (1985) The Narmada Valley project – development or destruction? *The Ecologist* 15(5/6): 269–85.

Kellow, A. (1986) Electricity planning in Tasmania and New Zealand: political processes and the technological imperative. *Australian Journal of Public Administration* XLV(1): 2–17.

Lines, W.J. (1991) *Taming the Great South Land: a History of the Conquest of Nature in Australia*. Sydney: Allen & Unwin.

Lowe, I. (1989) *Living in the Greenhouse*. Newham, Victoria: Scribe.

Mackenzie, D. (1999) Vets may have spread mad cow disease. *New Scientist* 2199, 14 August: 24.

Martin, B. (1981) The scientific straitjacket: the power structure of science and the suppression of environmental scholarship. *The Ecologist* 11(1) January–February: 33–43.

Martin, B. (1986) *Intellectual Suppression*. North Ryde, NSW: Angus & Robertson.

McMichael, A.J. (1993) *Planetary Overload*. Cambridge: Cambridge University Press.

Ostrom, E. (1990) *Governing the Commons: The Evolution of Institutions for Collective Action*. Cambridge: Cambridge University Press.

Paehlke, R.C. (1989) *Environmentalism and the Future of Progressive Politics*. New Haven and London: Yale University Press.

Paehlke, R. and Torgerson, D. (1990) *Managing Leviathan: Environmental Politics and the Administrative State*. London: Belhaven Press.

Palmer, J.A. (1992) Destruction of the rain forests: principles or practice? Chapter 7 in Cooper, D.E. and Palmer, J.A. (eds) *The Environment in Question: Ethics and Global Issues*. London and New York: Routledge, 81–93.

Pusey, M. (1991) *Economic Rationalism in Canberra*. Cambridge: Cambridge University Press.

Russ, P. and Tanner, L. (1978) *The Politics of Pollution*. Camberwell, Victoria: Widescope.

Self, P. (1993) *Government by the Market? The Politics of Public Choice*. London: Macmillan.

Sen, A.K. (1970) *Collective Choice and Social Welfare*. San Francisco: Holden-Day.

Simmons, I.G. (1989) *Changing the Face of the Earth: Culture, Environment, History*. Oxford: Blackwell.

Smith, D. (1990) *Continent in Crisis: a Natural History of Australia*. Ringwood, Victoria: Penguin.

Smith, J.W., Lyons, G. and Sauer-Thompson, G. (1999) *The Bankruptcy of Economics: Ecology, Economics and the Sustainability of the Earth*. London: Macmillan; New York: St. Martin's Press.

Taplin, R. (1994) Greenhouse: an overview of Australian policy and practice. *Australian Journal of Environmental Management* 1(3) December: 142–55.

Taylor, M. (1976) *Anarchy and Cooperation*. London: Wiley.

Taylor, M. (1982) *Community, Anarchy and Liberty.* Cambridge: Cambridge University Press.

Taylor, M. (1987) *The Possibility of Cooperation.* Cambridge: Cambridge University Press.

Tighe, P.J. (1992) Hydroindustrialisation and conservation policy in Tasmania. Chapter 7 in Walker, K.J. (ed.) *Australian Environmental Policy.* Kensington, NSW: University of NSW Press, 161–95.

Tilly, C. (1992) *Coercion, Capital, and European States, AD 990–1992.* Cambridge, Mass, and Oxford: Blackwell.

Torgerson, D. (1990) Limits of the administrative mind: the problem of defining environmental problems. In: Paehlke, R. and Torgerson, D. *Managing Leviathan: Environmental Politics and the Administrative State.* London: Belhaven Press, 115–61.

Walker, K.J. (1989) The state in environmental management: the ecological dimension. *Political Studies* XXXVII, March: 26–39.

Walker, K.J. (1992) The neglect of ecology: the case of the Ord River Scheme. Chapter 9 in Walker, K.J. (ed.), *Australian Environmental Policy.* Kensington, NSW: NSW University Press, 183–202.

Walker, K.J. (1995) Environmental Policy: Adequacy of Government Response. Paper presented to the Annual Conference of the Australasian Political Studies Association, Melbourne, September 1995.

Walker, K.J. (1999) Statist developmentalism in Australia. Chapter 2 in Walker, K.J. and Crowley, K. (1999) *Australian Environmental Policy 2.* Kensington, NSW: NSW University Press, 22–44.

Walker, K.J. Marginalising environmentalism: pluralist orthodoxies in Australia (forthcoming).

Walker, K.J. and Crowley, K. (1999) *Australian Environmental Policy 2.* Kensington, NSW: NSW University Press.

Warhurst, J. (1983) Single-issue politics: the impact of conservation and anti-abortion groups. *Current Affairs Bulletin.* 60(2) July: 17–31.

Wright, M. (1988) Policy community, policy network and comparative industrial policies. *Political Studies* 36: 593–612.

Wynne, B. (1992) Carrying out science (and politics) in the regulatory jungle. Review essay, *Social Studies of Science* 22(4) November: 745–58.

13 | Managing ecosystems for sustainability:
challenges and opportunities

John Handmer, Stephen Dovers and
Tony Norton

Ecological science may be a threat to our lifestyles

The underlying assumption of this volume is that as a society, we seek, or should seek, to move towards a vision of sustainability (Brundtland 2000). In other words, we should move along a pathway of 'sustainable development' – recognizing that there will be many conflicting visions of a sustainable state which will shift dramatically with changing knowledge and attitudes – so that the most we can hope to achieve is to be moving in approximately the right direction (Dovers and Handmer 1992). To achieve this aim, we need policies for sustainable development which are 'ecologically sound, socially acceptable and politically supportable' (Harte and Gough, Chapter 8, this volume). We could add that they must be administratively feasible too. Unfortunately, it is all too often overlooked that well-articulated policies or regulations are not an outcome in themselves; they are a critical but early stage in achieving some desired outcome.

It may seem logical that policies should be based on state-of-the-art ecological knowledge, and that successful implementation would follow for such rationally derived policies. This is simply not the case. Apart from the reality that, at present, 'rationality' appears to be owned by a large faction of the economics discipline, there are some important reasons for implementation difficulties with most policies and in particular for policies purportedly based on ecological knowledge. First among these is that ecological science has not appeared to contribute directly and immediately to economic prosperity, security or identity: the dominant issues of most national politics. Even where ecological issues are high on the political agenda or where ecological science is required by law, ecological imperatives must compete with other, often apparently more immediate, social and economic priorities. Increasingly, the World Trade Organisation, for example, is demanding that policy and regulations designed for ecological protection must be justified by risk assessments that satisfy the Organisation. In practice, this is likely to make it more difficult to put these regulations in place and increase reliance on so-called self-regulation.

While the current policy and political discourse is dominated by a neo-liberal economic rationality, many other 'rationalities' exist and some are current and not unimportant in environmental debates (cf. Dryzek 1987). An 'ecological' rationality is one, although what is represented by it in policy debates may not strike ecologists as accurate or familiar. The meaning of 'ecology' and 'ecological' in policy debates is not only or even primarily defined by ecological scientists, but by lay activists, ethicists and others as well. Another, influential, rationality is a communicative or discursive one, stressing uncertainty and the contestability of knowledge and the requisite democratic discourse that must precede policy design and implementation. This has had great impacts on environmental policy, most apparently through the increase in community-based or participatory policy programmes. Such an approach wishes ecologists and other 'experts' to engage more fully with others, expert and lay, community and government, and to admit and negotiate uncertainties.

For all policies, issues of capacity to implement and commitment to the aims of the policy are relevant. Typically, these concerns are influenced by such factors as: the priority given to the policy by the implementing agency; the agency's relative power in the system of government given that the policy is likely to conflict with the mandate of other (probably much older and more powerful) agencies; finance agencies arguing that ecological protection is too expensive; the technical expertise in the agency for producing the detailed risk assessments; implementation plans – in particular, monitoring for compliance and taking enforcement action against violations; and the degree of popular support. For ecologically based policies, we should add three factors, which, although they would always be potentially important in most policy areas, are critical here:

- the high degree of uncertainty in almost every aspect of ecology;
- the highly contested nature of many important ecological concepts, with the result that ecologists are not an effective lobby group – although certain individuals have high public profiles; and
- the apparent threat of many ecologically derived policies to consumer lifestyles.

There is also the factor of large-scale or even global ecological bottom lines or carrying capacities, an assertion seen in some quarters as very controversial and inextricably linked with racism and first world imperialism. Many economists argue that human ingenuity and ability to substitute human for degraded natural capital mean that bottom lines can be avoided. However, to most ecologists the implication that the globe is infinite is not worth discussion. These differences of opinion and worldview highlight the need to confirm and strengthen the imperfect 'epistemic community' of ecology (see Walker, Chapter 12, this volume).

In this concluding chapter we draw out the main themes from the volume's diverse and interdisciplinary offerings.

Ignorance and uncertainty in science and all knowledge

Uncertainty (or, more broadly, ignorance (Smithson 1989)) is pervasive throughout society, and there are many different types of uncertainty and ways of thinking about the subject, just as there are types of knowledge. Problems are framed and decisions made according to the degree and type of uncertainty. The limitation of a fixation with probabilistic efforts – which afflicts much of the risk field – can be easily appreciated by examining the uncertainty stratagems used in fields such as politics, law or medicine, which are often intrinsically social in character, such as the 'burden of proof' concept in law and the verbal qualifiers used in medicine such as 'reasonable medical certainty' (Ricci 1995). Rayfuse and Wilder (Chapter 7, this volume) discuss legal uncertainty in the context of fisheries management. The essentially social nature of the policy process where risk is concerned, involving a discourse between technical experts, decision makers, the corporate world, and the wider public underlines how the omission of these broader dimensions of uncertainty undermines the viability of the process itself.

Uncertainty is a powerful tool of social control, both for society as a whole and for subgroups within it. In particular, control can be achieved:

- through creating taboos or treating knowledge or information as irrelevant;
- through the use of specialist discourse and ritual which excludes 'outsiders';
- through maintaining the status quo by burden of proof arguments (the status quo is kept until the evidence against it is overwhelming – the criteria for assessing evidence being decided by those interested in maintaining the status quo), and

■ through the deliberate creation of uncertainty, for example by hiding or distorting information, by being ambiguous or vague, or by undermining and thereby casting doubt on the arguments of others.

Our definition and framing of uncertainty therefore must be broad enough to encompass these issues in addition to the concerns of probability, chaos, vagueness and so on. It is important that assessments of uncertainty do not ignore illegal and informal activity simply because it is generally not well documented. Almost all groups seeking to influence policy exploit, encourage or hide uncertainty. It is used to undermine policy proposals or to discredit scientists and others with different views.

The implication is that knowledge is not simply an objective reality awaiting discovery. Anyone who doubts that science is constructed should consider the contested nature of almost all basic ecological concepts; or see Tickle (Chapter 11, this volume) on the acid rain issue and Rayfuse and Wilder (Chapter 7, this volume) on fisheries. A recent review by Matthews (2000) argues that subjectivity is an unavoidable and necessary part of the scientific research process. Inevitably, this leads to arguments over validity and over which knowledge should be privileged in terms of informing policy – a position held firmly by economics.

We need to confront the awkward question: if we had near-full knowledge (which is the most we can ever hope for), would it make any difference to policy? Approaches focused on eliminating uncertainty generally assume that existing scientific and other knowledge is fully incorporated into decision making. This is rarely, if ever, the case, as trade-offs are usually negotiated between the numerous competing interests, with the politically most powerful voice achieving its interest (as with the agricultural subsidies discussed in Chapter 8); and this is a legitimate part of the democratic process. However, we seek mechanisms to ensure that science is properly heard; so that trade-offs are made in full knowledge of the ecological implications.

The urgency of mitigation policies for ecological degradation

Full certainty is unlikely ever to arrive, and even if it did, we might not have the resources necessary to act on it. We have therefore to make decisions in the face of great uncertainty; in doing so we need to acknowledge our ignorance and recognize that the decision making must be framed so that, as new knowledge inevitably arises, it can be incorporated.

In some cases, more than this is needed, as when key ecosystems, refuges, biota or processes are under threat. Most attention in the past has been focused on protected areas and reserves – and these are very important. However, most biota is outside and will remain outside formal reserves. (Chapter 3 (among other chapters in this volume) draws attention to migration paths and ecological processes outside reserves.) It is important not to lose sight of this fact and to ensure that policies and implementation modes take this reality into account. Reserves need defending and protecting and there are established agencies and pressure groups for this purpose. However, off-reserve protection may be an even higher priority, because of the relatively weak protection such areas have; because most ecosystem functions and biota are outside reserves; and because it offers a hedge against uncertainty. Off-reserve protection poses much greater challenges for public commitment and support – and therefore for policy design and implementation as, among other things, uncertainty increases with scale. Working off-reserve in production landscapes, ecologists (and others) must engage with a wider peer community than the colleagues and ecosystem managers encountered in reserved landscapes. The managers of landscapes – farmers, foresters, community groups, indigenous peoples – become as relevant as any peer reviewer or organizational line manager. A current thrust in policy, and therefore an imperative for ecologists, is for such participation and inclusion in policy and in any research that informs policy. For natural scientists, this is unfamiliar territory, although much has been learned in recent years.

Two other broad policy thrusts are needed: a *precautionary approach* and *contingency planning* for the inevitable residual hazard. Harte and Gough (Chapter 8) suggest that New Zealand's Resource Management Act embodies a precautionary approach to development – although others argue that the practice may fall well short of this. A variation is the 'critical loads' approach as discussed by Tickle (Chapter 11, this volume). Contingency planning is less well represented. Ecologists need to be able to frame recommendations in a positive way, recognizing that it is often very difficult, if not impossible, to halt development for long, but that it is often possible to influence the details of that development. This advocates an approach that can often appear to be supportive of developments that are ecologically damaging – but is justified by the pragmatic need for damage control.

Implementing policy

Effective policy means effective implementation. Increasingly, this requires negotiated co-operative arrangements between a range of groups. It may mean expanding the concept of policy implementation away from sole reliance on government bodies, to encompass all stakeholders. It also means acknowledgement of other constraints to the policy process, such as the privileged position in government of certain ideologies such as economic 'rationalism' – and the need to develop processes which give non-economic sources of knowledge similar access to policy makers. Attempts should be made to move from an ad hoc project-based approach towards a strategic approach more in keeping with the long-term aims of sustainable development.

In many cases this will require building local capacity as well as local commitment. A monitoring programme may offer possibilities here, especially if locally run; for example, people may become aware through their own monitoring of the sources of problems in their communities. This may help build commitment to attempting to deal with them. It also helps capacity by encouraging the acquisition of basic technical knowledge and through networking with others confronting similar problems.

The prospects for implementation may be improved by other factors which may seem far from our immediate concerns. For example, attitudes and culture: satisfactory ecological outcomes may follow the abandonment of the 'frontier' mentality in Australia's north, and an appreciation that the land is far from limitless (Chapters 5 and 8) – a similar comment could be made for many other parts of the world, such as Alaska; and that environmental constraints are real and may overwhelm technology and economics. Woinarski and Dawson (Chapter 5, this volume) argue that extensive undisturbed environments are increasingly seen as 'an asset and not an affront'. Mason and Michaels (Chapter 4, this volume) put a similar view.

As well documented by all the chapters in this volume, our ecological knowledge is poor and we must expect (and hope) that it will be subject to constant improvement. Policies and implementation processes must take account of this and encourage the incorporation of new knowledge and, much more difficult to achieve, constant learning from mistakes (Chapter 8), and abandonment of long-held beliefs when shown to be deficient.

Monitoring: its function and dimensions

How can monitoring have an influence on policy? Or, more importantly, how can it influence ecologically important decision making? What is an appropriate allocation of resources compared with other aspects of ecological research and repair?

These questions pervade, at least implicitly, much of this volume. A more basic question may be whether monitoring results are likely to influence policy. In the absence of processes to ensure incorporation of results into policy would we perhaps be better off concentrating on monitoring a limited number of marker species and processes, and spending resources on institutionalizing linkages between science and policy?

Ecologists may need to be far more disciplined in devising and justifying monitoring programmes. Such programmes are – usually incorrectly – seen as unjustified or even a waste of resources by many of those who allocate funding. The OECD countries are richer than ever but, paradoxically, most of the public sector, including science, operates in an environment of severe scarcity, and we need to adapt, to articulate clear priorities, and to broaden the constituency for monitoring. For example, universal access to good quality monitoring information combined with viable legal standing could make all citizens potentially powerful protectors of the environment (this can also be achieved informally, see Chapter 11). A useful start would be to develop monitoring which, from the outset, is integrated with the policy process as a decision support tool (see Chapter 2).

Enormous resources are available for monitoring and modelling global climate change. Arguably, the resources consumed in this enterprise are second only to those used by economic monitors and modellers. Ecologists could examine the institutional arrangements embodied in the IPCC (Intergovernmental Panel on Climate Change) and GEF (Global Environment Facility) for lessons and opportunities. Whatever our criticisms of the IPCC and its 'consensus science' (see Boehmer-Christiansen, Chapter 6 this volume), it has been successful at capturing world attention and influencing policy – something normal fragmented science has singularly failed to achieve.

Less recognized than ecological monitoring, but just as crucial, is policy monitoring. All too often, the efficacy of policy and management interventions must be guessed at, in the absence of clear goals or the information streams necessary to inform evaluation. Cast in an 'adaptive' vein, such feedback loops are essential or learning cannot occur. There are two aspects to this. One is the design of ecological monitoring systems that will pick up the impact of policy interventions on the condition of ecosystems. The other is the monitoring of uptake, effectiveness, compliance or reaction to specific policy instruments (such as participatory programmes, or tradable rights markets in water or fish). The former is directly relevant to ecologists, the latter less so. The rise of the 'indicator industry' and of mechanisms such as state of environment reporting or green accounting have thus far failed to close the loop between ecological and policy monitoring, and are, therefore, of limited utility in terms of informing better policy (Dovers, forthcoming).

The changing context of all monitoring efforts needs to be recognized. The practical manifestations of a 'market' rationality – privatization, corporatization, contracting out, public sector shrinkage, etc – have, as yet, scarcely explored implications for very basic environmental information such as weather and water flows, let alone for new, difficult issues such as biodiversity.

Scale: space and time

Many writers argue that over the time spans that matter to the biosphere the issues that concern us today are of little relevance, and that we are overly preoccupied with a tiny part of history. Even when we consider the future we rarely go beyond a few generations. While acknowledging this view, there are a few perspectives worth raising.

Many of the contributors argue explicitly or implicitly that a static view is manifestly inadequate; and that ecological 'function' or process rather than simply the static 'state' should always be considered.

If we join the brief time period we are considering with space, a more complete picture emerges. Within a few generations, molecular contaminants have spread to every part of the globe. The local destruction of a small forest is unimportant regionally unless it was one of the last refuges for already stressed wildlife; and in any case the forest may have grown in the last couple of centuries. The same can be said about local problems with fisheries, wetlands, feral animals, streams, waste disposal, and so on. However, these minor local problems add up to degradation on a continental (such as Australia) or even global scale.

Australians are increasingly viewing their country from a very long perspective by attempting to incorporate the period of Aboriginal settlement of at least 60 000 years into ecological thinking which at most has generally considered the last 200 years of non-Aboriginal settlement. This is in sharp contrast to most scientists from Europe, and may set Australian ecology in an appropriate timescale. A growing field of study where ecologists have had a major impact on another discipline (at least, more than when they have encountered, for example, economists) is that of environmental history, especially where such inquiries seek to link past states to present choices (Dovers 2000). Interaction between natural scientists and the disciplines of the humanities may be more fruitful than with the social sciences, in terms of the influence of ecology on broader patterns of social thought.

Spatial and political realignments in ecology are observable, in reaction to dominance of the science and the literature by a few northern hemisphere countries. A richer and more realistic science of ecology – but of course a more complex and perhaps confusing one – is emerging from collaborations such as 'Southern Connections', which encourages interaction between southern hemisphere ecologists to explore commonalities and contrasts not explored sufficiently in a US-dominated global literature. In ecologically relevant areas of policy such groupings are emerging too, for instance, through the 'Valdivia Group' of southern countries working together on issues under the Convention on Biological Diversity. 'Ghettos' and separation are a risk in such developments, but, properly managed, they offer diversity and vigour.

Democracy: can it deliver ecological sustainability?

Traditionally, the only moral philosophy democracy has embraced is that of individual rights, often known as 'first-generation' human rights, such as freedom of expression. For all their advantages and success – especially for the industrialized west – it is difficult to reconcile such rights with the collective action and sense of responsibility needed for sustainable development. Recently, the image of democracy (and individual rights) has become almost inseparable from the ideas of economic rationalism, particularly in the English-speaking countries. The main concepts here – those of competition and market forces – may work strongly against sustainable development and ecological integrity. The strong trend towards a 'negotiated' state (May et al. 1996) with its implicit rejection of ecological bottom lines – because everything can be traded or bargained away – is also likely to work against ecological advice, given the contested nature of much of that advice and the very weak institutional arrangements for getting it into the policy process.

However, some other institutions central to modern democracy may help to counter some of these trends. Open access to information and an independent legal system, for example: the latter may assume increasing importance if 'second-generation' human rights, such as the right to a clean environment, assume greater prominence. The growing 'environmental justice' movement is driving these changes (Bosselmann and Richardson 1999).

In any case, appropriate institutions to facilitate movement towards sustainable development are required. These would include institutions to enable ecological science to inform policy. The concept of environmental security, in place of military security, may provide a way forward – although this concept is also highly contested and potentially dangerous when utilized as a metaphor and justification for conflict and national interest (Barnett 2000). Natural scientists need to be wary of the political meaning and potential of such apparently convenient and powerful terms such as environmental 'security'.

Moving ahead while coping with new challenges

This book explores science and public policy with a view to influencing policy to secure outcomes more in keeping with sustainable development. Our world is one where a relatively few transnational companies (predominantly concerned with energy, transport and finance) increasingly dominate global affairs and trillions of dollars are traded daily. The slogan 'free trade' is used to prevent action on environmental (and also human rights) matters, now with strong supranational largely unaccountable power through the World Trade Organization; and governments are selling their power to private companies, often in an international marketplace, so that key functions formerly exercised by government may now be carried out by a private company in a foreign country, often protected by the imperatives of commercial confidence. In such circumstances, perhaps a focus on state power and policy making is wishful thinking.

This is not to sound unduly negative, only to emphasize that the ambit of national public policy – always restricted – is now significantly more restricted than a decade ago. This may be arguable for the very large economies, but is certainly the case for smaller ones. The European Union provides some limited counterbalancing to this trend. The Union is concerned with promoting unrestricted movement of goods, services and people. But it has also paid attention to environmental standards and social policy, and has worked to raise the standards of many member countries in these areas.

Ecologists wishing to influence activity on the ground have four broad choices (although in reality the categories are not as clear as set out here):

1. They can continue to try to seek credibility through sound science and to influence national policy processes, especially the shape of institutions such as the land tenure system and legal liability for environmental damage; and they can work to remove policies which result in environmental damage. These latter policies include many agricultural subsidies (see Chapter 8 concerning New Zealand's high country, and worldwide debates on subsidized irrigation).

2. A second area of activity is to work more at the local level as discussed by Johnson *et al.* in Chapter 3 – a need also evident in many of the world's fisheries (Chapter 7, this volume), and epitomized by landcare (Chapter 2, this volume). There is no question that this is where many critical ecological decisions are taken, but we should not fool ourselves that local communities are necessarily interested in ecological issues. The general need is to build local commitment and capacity, and increasingly to find ways of insulating local activity from the broader and often capricious policy environment.

3. The third approach is to work more with developers, businesses and their associations and to attempt to negotiate and establish better processes directly with businesses – and through consumer pressure. Some good precedents exist – for instance, at the international sectoral scale, the Forest Stewardship Council and Marine Stewardship Council, and the Montreal process of defining criteria of sustainable forest management (UNEP 1999). The efficacy of such market-oriented approaches and international standards will depend upon, inter alia, the quality, timing and accessibility of scientific input.

4. Finally, ecologists can engage with international organizations and work with international law, to promote change in the processes of policy development and implementation, as well as in policy substance. Organizations and law here would include those which appear to be ecologically destructive, as well as those dedicated to promoting policy based on the best available science and practice. Many chapters contain examples of the different aspects and capacities of international organizations, networks and law, including Chapters 3, 6, 7 and 11.

We need precaution, and we do not need delay in adopting a cautious approach. Delay is a favourite tactic of those wishing to protect the status quo, to carry on as before: the 'We don't know so don't worry' syndrome. Change is delayed, and policy development and implementation are delayed, typically while further science is undertaken. Scientists may be willing partners in this strategy, believing that the science is currently inadequate for policy determination. However, experience suggests that development is still highly likely to proceed.

Commercial pressures, both domestic and international, as well as the priorities of governments, have made and will continue to make it difficult for ecological science to have the influence it seeks – even if it could overcome its internal conflicts. Whichever course is taken (ideally all four listed above should be pursued), attention should be devoted to contingency planning for the inevitable ecological crises. Also, the inevitable ecological crises, whether sharp or creeping, provide windows of opportunity for ecologists armed with relevant information to wield greater than usual influence outside their discipline, at a time when attention is focused. As windows of opportunity for policy change, crises can be mishandled through the influence of poor-quality, hastily gathered or simply convenient scientific information. To engage, ecologists must know the policy process, or at least deal through those who do. It is worth separating immediate contingency response and ongoing reform to increase resilience – post-crisis arrangements may be lasting (Busenberg 1999), and, if so, then the influence of the scientific information used at the time will be enduring as well.

In such crises, and indeed at all other times, ecology is not the sole preserve of ecologists, even if they wished it to be. Misused as it may be in policy debates, it has wide ownership and political potency. Ecologists need to accept that and work accordingly. The three 'rationalities' mentioned in this chapter – economic, ecological, communicative – may appear irrational to each other. Mixing the 'hard science' of an ecological rationality with the discursive, culturally sensitive style of a communicative rationality presents particular problems for the scientific ecologist. Human knowledge is uneven, poorly shared, and always deeply political, and the interaction of rationalities in policy debates is, of course, political too. Ecologists may have to accept that a certain degree of advocacy is unavoidable, and that a great degree of negotiation, respect and mutual learning is inevitable.

References

Barnett, J. (2000) Destabilizing the environment-conflict thesis. *Review of International Studies* 26: 271–88.

Bosselmann, K. and Richardson, B. (1999) (eds) *Environmental Justice and Market Mechanisms: Key Challenges for Environmental Law and Policy*. London: Kluwer Law International.

Brundtland, G.H. (2000) Our Common Future and ten years after Rio: how far have we come and where should we be going? In: Dodds, F. (ed.) *Earth Summit 2002: a New Deal*. London: Earthscan, 253–63.

Busenberg, G.J. (1999) The evolution of vigilance: disasters, sentinels and policy change. *Environmental Politics* 8: 90–109.

Dovers, S. (2000) On the contribution of environmental history to current debate and policy. *Environment and History* 6: 131–50.

Dovers, S. (forthcoming) Informing institutions and policies. In: Higgins, J. and Venning, J. (eds) *Towards Sustainability: Systems for Monitoring Sustainable Development*. Sydney: University of NSW Press.

Dovers, S. and Handmer, J.W. (1992) Uncertainty, sustainability and change. *Global Environmental Change*. December: 262–76.

Dryzek, J. (1987) *Rational Ecology: Environment and Political Economy*. Oxford: Basil Blackwell.

Matthews, A.J. (2000) Fact versus factions: the use and abuse of subjectivity in scientific research. In: Morris, J. (ed.) *Rethinking Risk and the Precautionary Principle*. Oxford: Butterworth-Heinemann, 247–82.

May, P., Burby, R., Ericksen, N., Handmer, J.W., Dixon, J., Michaels, S. and Smith, D.I. (1996) *Environmental Management and Governance: Intergovernmental Approaches to Hazards and Sustainability*. London: Routledge.

Ricci, P. (1995) Uncertainty in human health risk assessment: approaches, measures and methods. Unpublished paper. University of Wollongong.

Smithson, M. (1989) *Ignorance and Uncertainty: Emerging Paradigms*. New York: Springer-Verlag.

UNEP (United Nations Environment Program) (1999) *Global Environmental Outlook 2000*. London: Earthscan.

Index

Page numbers in *italic* refer to figures.

Abatement Strategy Assessment
    Model (ASAM) 253
Aboriginal people, northern
    Australia *see* northern
    Australia
accountability of policy makers 177
acid rain 236–7, 281
    critical loads *see* critical loads
    internationalization of science and
        policy community, UK 249
    policy issues 252–6
    research 239–45, 248–9
    science evolution, UK 238–51
    sustainability 252–6
    uncertainty issues 252–6
adaptive management 10, 17, 18, 31
    bioregional management 48, 60–1
    fire management 199, 202
Adirondack Park, NY, USA 66–80
Agenda 21: 147, 150
agribusiness 268
agriculture
    northern Australia 92–5
    subsidization, New Zealand 172–3
AIDS 264
air pollution, transboundary 241
algal blooms 18
Amboseli National Park, Kenya 50–1
Anadromous Stocks in the North
    Pacific Ocean, Convention for
    the Conservation of 149

analytical methods, environmental
    assessments 34
animals
    fire and 194, 202
    introduced, New Zealand 169
    mammals, northern Australia
        84–6
    post-fire recovery of species 194
Annapurna region, Nepal 55
Antarctic and Marine Living
    Resources Conservation
    Convention 148
anti-ecological rigidity 268
applied science 6, 37–8
aquatic systems 18
ASAM (Abatement Strategy
    Assessment Model) 253
auditing, ignorance 8
Australia 1–2, 8–9, 14
    biodiversity, fire impacts on, south
        eastern Australia 191–203
    bureaucracy 270–1
    coal exports 127
    democratic policy-making
        processes 282
    economists, influence of 279
    fire impacts on biodiversity, south
        eastern Australia 191–203
    fisheries 139–40, 152
    difficulties 156
    improving enforcement 162–3

lack of enforcement 159–60
legal uncertainty 158–9, 162
nature 153–5
policy uncertainty 157–8, 161–2
scientific uncertainty 156–7,
 160–1
sustainability 155–6, 160–2
Great Barrier Reef Marine Park
 57, 59
Kakadu National Park *see* Kakadu
 National Park
northern *see* northern Australia
policy communities 274
policy distortions 271–2
wetlands 210–14, 219, 220–3,
 224–5, 229–30
authoritative knowledge and public
 policy 285–6
authority, lack of, bioregional
 management 52–4

Before–After–Control–Impact (BACI)
 experimental design 32
billabongs 213
biodiversity
 definition of biological diversity 149
 ecosystem stability and 44
 fire and 191–203
 loss 27
 marine 139
 pastoralism, impacts of 96
 practical uncertainty, approach to,
  fisheries management 149
Biodiversity Convention 149, 150,
 151, 299
biophysical problems, sustainability 3
bioregional management 43–4
 authority, lack of 52–4
 financial capacity, lack of 54–6
 government roles 58
 human capacity, lack of 54–6
 incentives
  lack of 52–4
  use in motivation 58–9
 institutions
  arrangements 44–5
  co-operation, lack of 49–52
  lack of capacity 54–6

key characteristics 47–8
land tenure, lack of 52–4
local communities
 adaptive management 60–1
 capacity 44–8
 defining roles in management
  56–8
 incentives for motivation 58–9
 information, access to 59–60
 strategies for engagement
  56–63
obstacles to local participation
 48–56
research 47, 61–3
stakeholders
 getting to know 57–8
 information, access to 59–60
 involvement 46–8
uncertainty and 44–8
biosphere 27
 reserve planning 68–9
biota 26–7
 fire regimes, responses to 193–4
Bonaire Marine Park, Saba 53
boundaries, wetlands 213–15, 230
broad-area burning 197–8
Brundtland Commission 68
bureaucracies
 climate change 117–24
 convenience, serving 281
 dominance of bureaucrats 270
 economists 279
 greenhouse gas inventories 123
 power relationships 271
 public policy processes 270–1
burning *see* fire
bushfires 192

California 51, 58
Canada, wetlands 220, 225
capacity
 bioregional management 44–8,
  54–6
 building 123
carbon sinks 124
carbon taxes 126
catchment management 230
CBA (cost-benefit analysis) 9

Central Electricity Generating Board, UK (CEGB) 239–44, 248–9
centralized planning 45, 46
challenges for ecology 14–17
China 127
civic science 17
CJD (Creuzfeld-Jakob disease) 281
clean energy 125
climate
  change 116–17
    contested subject 122
    Framework Convention (FCCC) 8, 117–22
    Intergovernmental Panel see Intergovernmental Panel on Climate Change
    modelling and monitoring 297
    research 128
    scientific information on 131
    uncertainty 122
  policies 134–5
  protection regime 123
  strategies as side effects of energy policies 126
  see also global warming; greenhouse gases
Climate Change Convention (FCCC) 8, 117–22
closed-door approaches 281–2
closures 262–3
CLRTAP see Long-Range Transboundary Pollution Convention
co-operation, lack of, bioregional management 49–52
co-operative exchanges of information 60–1
co-operative leadership, bioregional management 48
co-operative skill development, bioregional management 48
coal
  alliance against unsustainable use 124–7
  electricity generation 126–7
  power stations fired by 126, 242
  restrictions on use 124

unsustainable use, alliance against 124–7
coastal wetlands 224–5
Code of Conduct for Responsible Fishing, FAO 147, 150–1
Colombia, Sierra Nevada de Santa Marta 60
commercial pressures 302
communication
  broader audiences 16–17
  ecosystem management 37–8
  ignorance 15
  importance 13
  risk 15
  uncertainty 15
communicative rationality 292, 302
competition 268
consensus science 297
conservation
  New Zealand pastoral high country 173
  planning, northern Australia 102–4
  wetlands 225–6, 229
Conservation Biologists, Society for 30
Conservation of Anadromous Stocks in the North Pacific Ocean Convention 149
Conservation of Antarctic and Marine Living Resources Convention 148
conservation reserve system, northern Australia 103
consultancy, professional 6, 37
contemporary approaches to ecosystem management 30–3
contingencies 17–19
contingency planning 39, 295
contract zones 273
convenience, bureaucracies serving 281
cores, bioregional management 47
corridors, bioregional management 47
cost-benefit analysis (CBA) 9
Costa Rica, La Amistad Biosphere Reserve see La Amistad Biosphere Reserve

cotton, northern Australia 94
Creuzfeld-Jakob disease (CJD) 281
crisis
  conceptions of 69
  management 17
critical loads 238
  policy issues 253–6
  precautionary principle 254
  science evolution, UK 245–51
  specific ecosystems, effects on 250
critical thought 266
cross-cutting issues 131
Cross River National Park, Nigeria
  59

dam proposals 271
data
  acquisition
    environmental assessments 34
    fisheries management 146–8
    monitoring as basis for 35
    scientific, governments controlling
      and interpreting 272
    use, environmental assessments
      34
  see also information; scientific
    information; time series data
decentralization 52
decision support techniques 31
Deer Creek community, California
  58
Deerlodge Forest Plan, Montana
  51
deforestation 45, 52, 271
degradation
  mitigation policies 294–5
  New Zealand pastoral high
    country 170, 171, 173
  northern Australia 96
  wetlands 224–5, 228
delay 301–2
democratic governments
  ecological sustainability and
    299–300
  policy-making 282
  weakness 273
desertification 52
developing countries 122–3

development
  northern Australia see northern
    Australia
  wetlands, affecting 220–2, 227–8
developmentalism 268
discursive rationality 292, 302
diseases, exotic 18
distant management 93, 95
distortions, public policy 271–2
driftnet fishing 149
droughts 18

earth systems research 127–9
EC see European Union
ecological considerations,
    sustainability 2
ecological economics 38
ecological information, acquisition
    13–14
ecological integrity 2, 11
ecological modernization 237–8
ecological monitoring see monitoring
ecological rationality 292, 302
ecological risk assessment (ERA) 10,
    15
Ecological Society of America
    (ESA-US) 30
ecological sustainability 8, 14, 69,
    299–300
ecological uncertainty, sources 178
ecologically safe fire windows 200
ecologically sustainable development
    (ESD) 8, 14
ecology
  challenges for 14–17
  core elements 29
  discipline of 28–9
  ecosystem management and 28–30
  knowledge base, incompleteness
    12
  policy and 10–14
  policy linkages enhancement 15
  political attractiveness of 134
  profession 29–30
economic considerations,
    sustainability 2
economic rationalism 268, 279, 296,
    299

economic rationality 302
economic sustainability 47, 69
economic uncertainty 44–5, 46–8,
    178
economics, ecological 38
economists 279
ecostewardship partnerships 73
ecosystem management 17, 18
    adaptive processes 61
    Adirondack Park 70
    communication 37–8
    contemporary approaches 30–3
    contingency planning 39
    ecology and 28–30
    monitoring, building capacity for
        38
    improving 33–6
    institutional arrangements,
        implications for 37
    problem definition 37–8
    for sustainability 291–302
ecosystems 26–7
    complexity 27–8
    human disturbance detection 36
    management see ecosystem
        management
    practical uncertainty, approach to,
        fisheries management 148–9
    relative importance 229–30
    restoration 74–5
    services 30
    stability, biodiversity and 44
Ecuador, Galapagos National Park
    53, 54
EEZ see exclusive economic zones
EIA (environmental impact
    assessments) 31, 32
electricity generation 124, 126–7
emergency planning 18
emissions of greenhouse gases see
    greenhouse gases
EMS (environmental management
    systems) 31
endangered species
    mining impacts, northern Australia
        100–1
    recovery plans 17
    USA 58

energy
    clean 125
    efficiency 122
    electricity generation 124, 126–7
    nuclear power 124–5
    policies, climate strategies as side
        effects of 126
    see also coal; natural gas
enforcement
    fisheries management 145–6,
        151–2
    Australia 159–60, 162–3
Enlightenment enterprise 285
Enlightenment fallacy 131
environment
    as 'just another issue' 268
    political attractiveness of 134
environmental assessments
    analytical methods 34
    'context for' 34–5
    data acquisition and use 34
    precautionary approach 33–4
environmental groups 275–6
environmental impact assessments
    (EIA) 31, 32
environmental impacts
    mining 98–101
    pastoralism 96
environmental justice 299–300
environmental management systems
    (EMS) 31
environmental policy see policy;
    public policy
environmental risk assessment
    10
environmental security 300
environmental standards 300
epistemic communities 237–8,
    245–51, 255, 277–9, 282,
    284, 285
epistemological uncertainty 6
ERA (ecological risk assessment) 10,
    15
erosion 52, 96
ESA-US (Ecological Society of
    America) 30
ESD (ecologically sustainable
    development) 8, 14

European Union (EU)
  acid rain policy 255–6
  environmental standards 300
exclusive economic zones (EEZ) 140,
      144, 150
exotic diseases 18

FAO (Food and Agriculture
      Organization) 147, 150–1
FATE model of vegetation dynamics
      196
FCCC (Framework Convention on
      Climate Change) 8, 117–22
field experimental studies 29
financial capacity, lack of,
      bioregional management 54–6
fire
  biodiversity and 191–203
  broad-area burning 197–8
  ecologically safe fire windows
      200
  fuel-reduction burning 198
  litter reduction burning 194
  management
    adaptive 199, 202
    determination of regimes 200
    monitoring 201–2
    objectives 200
    protection versus 197–8
  as management tool 191
  modelling approaches 195–6
  resource managers, implications
      for 198–202
  nutrients 194–5
  policy implications 198–202
  protection versus management
      197–8
  regimes
    ecological impacts 192–4
    historical, attempts to replicate
      199
    meaning 192
fire-dependent species 200
fire-sensitive species 200
FireNet 202
Fish and Wildlife Service, US see
      United States: Fish and
      Wildlife Service

fisheries 138–40
  Australia see Australia
  international management 141–2
    biodiversity approach to
      practical uncertainty 149
    data acquisition 146–8
    ecosystem approach to practical
      uncertainty 148–9
    enforcement 145–6, 151–2
    legal uncertainty 144–5, 150–1
    practical uncertainty 143–4,
      148–50
    precautionary principle 147
    responses to enforcement
      problems 151–2
    scientific uncertainty 142–3,
      147–8
    techniques 149–50
    uncertainty, responses to 146–51
  nature 140
  sustainability and 141
flag states
  enforcement 152
  responsibilities 145–6
flags of convenience 145
Food and Agriculture Organization
      (FAO) 147, 150–1
forestry
  New Zealand pastoral high
      country 173
  northern Australia 89–92
forests
  deforestation 45, 52, 271
  management 52
  northern Australia 84
  planting to fix carbon 124
fossil fuel interests 125–6
Framework Convention on Climate
      Change (FCCC) 8, 117–22
fuel-reduction burning 198
funding for bioregional management
      55

Galapagos National Park 53, 54
gas see natural gas
GEF (Global Environment Facility)
      121, 123
genetically modified seed 268

Geographic Information Systems
(GIS) 59–60
GHG see greenhouse gases
GIS (Geographic Information
Systems) 59–60
Global Environment Facility (GEF)
121, 123
global fisheries see fisheries
global warming
coal, alliance against unsustainable
use 124–7
policy advice from science 132–4
public policy 264
research effort 133
science as legitimator of politics
and trade 116–35
see also climate; greenhouse gases
globalization 273
governments
bioregional management roles 58
decision-making bodies 122
involvement, New Zealand pastoral
high country 173, 175
scientific data, controlling and
interpreting 272
weak 269–70, 273
see also democratic governments;
public policy
Great Barrier Reef Marine Park 57,
59
greenhouse gases (GHG)
'business as usual' 268
concentrations, stabilization 118,
123
emissions 117
coal, restrictions on use 124
reduction 119–20, 124, 126
stabilization 125–6
trading arrangements 120–1
inventories 123
research effort 133
see also climate; global warming
greenline parks 68–9
growth 123, 283

habitat destruction by wildfire 18
Hague conference of parties to
FCCC 121–2

harvesting, sustainable 101–2
Hazardous Substances and New
Organisms Act, New Zealand
183
hazards 32
highly migratory fish stocks (HMFS)
144
UN Fish Stocks Agreement 147,
149, 150
human activities 26–7
disturbance in ecosystems,
detection 36
unprecedented rates of change due
to 13
human capacity, lack of, bioregional
management 54–6
Human Environment, United
Nations Conference on 239
human rights 299–300

ICSU (International Council for
Scientific Unions) 128–9
IGBP (International Geosphere
Biosphere Programme) 128
ignorance
agriculture, northern Australia
92–5
auditing 8
communicating 15
forestry, northern Australia 89–92
in knowledge 293–4
northern Australia 83–107
mining, northern Australia
97–101
pastoralism, northern Australia
95–7
policy and 3–4
precaution and other approaches
7–10
risk and 3–4
in science 293–4
sustainable harvesting, northern
Australia 101–2
uncertainty and 3–4, 6, 176
implementing policy 296
impossibility results 285
in-discipline consensus on key policy
issues 15

incentives, bioregional management
   see bioregional management
indeterminacy 176, 182–6
India, bureaucracy 271
indicator industry 297–8
indicator species 36
Indonesia 53
information
   acquisition, time-consuming and
      expensive nature 13–14
   bioregional management 47, 48,
      59–60
   co-operative exchanges of 60–1
   disregard for, northern Australia
      104
   see also data; communication;
      scientific information; time
      series data
institutional arrangements, ecosystem
      management implications for
      37
institutions, bioregional management
      see bioregional management
integrity, ecological 2, 11
inter-generational equity 2
intergovernmental institutions 128
Intergovernmental Panel on Climate
      Change (IPCC) 117
   consensus science 297
   functioning 129–31
   politicization of, policy advice
      from science and 132–4
   science advice, earth systems
      research underpinning 127–9
intermittent wetlands 211
international co-operation,
      bioregional management 48
International Council for Scientific
      Unions (ICSU) 128–9
international fisheries management
      see fisheries
International Geosphere Biosphere
      Programme (IGBP) 128
international organizations, engaging
      with 301
internationalization 249, 268
intertidal wetlands 219–20
intra-generational equity 2

introduced species 18
   New Zealand pastoral high
      country 169
   plants, northern Australia 96
inventories, greenhouse gases 123
IPCC see Intergovernmental Panel on
      Climate Change
irreversibility 18
irrigation, northern Australia 93–4

Kakadu National Park 49–50, 55–6,
      59, 60, 99
Kasungu National Park, Malawi
      59
Kenya, Amboseli National Park
      50–1
keystone species 36
Khao Yai National Park 54
knowledge
   ignorance in 293–4
   incompleteness of base in ecology
      12
   sharing, bioregional management
      48
   use, bioregional management 47
Korea, Mt Sorak National Park 52

La Amistad Biosphere Reserve 54–5,
      56, 59, 60–1
land management
   New Zealand pastoral high
      country 172
   northern Australia 102–4
land ownership
   northern Australia 88–9
land tenure
   lack of, bioregional management
      52–4
   New Zealand pastoral high
      country 169–70, 171, 174,
      175
   northern Australia 88–9
land use, New Zealand pastoral high
      country 171–2
landscape approach to management
      230
large regions, bioregional
      management 47

Law of the Sea Convention *see* UN
    Convention on the Law of the
    Sea
leadership, bioregional management
    47, 48
least-effort basis 281
legal uncertainty
    fisheries management 144–5,
        150–1
    Australia 158–9, 162
legislation 269
licensing, fisheries 149–50
lifestyles, ecological science as threat
    to 291–3
litter reduction burning 194
local communities
    assertiveness 45
    bioregional management *see*
        bioregional management
local level action 301
logging 53
Long-Range Transboundary
    Pollution Convention
    (CLRTAP) 241, 245; 246–7,
    256
long-term monitoring 16, 38
long-term research 16
loss of species 26–7, 30
LRTAP Convention *see* Long-Range
    Transboundary Pollution
    Convention
lynx reintroduction 75

macro-problems 6, 11
mad cow disease 281
Malawi, Kasungu National Park 59
mammal fauna, northern Australia
    84–6
management
    adaptive *see* adaptive management
    bioregional *see* bioregional
        management
    decisions, uncertainty in 44–5
    distant 93, 95
    fisheries *see* fisheries
    wetlands 229, 230
mangroves 219
Maori 168–9, 175

marine biodiversity 139
market-oriented approaches 301
market rationality 298
Matobo National Park, Zimbabwe
    59
matrices, bioregional management
    47
maximax criteria 9
maximum sustainable yields (MSY)
    143
MCS (monitoring, control and
    surveillance) 146
meso-problems 4–6, 11
methodological uncertainty 6
micro-problems 4, 6
minimax criteria 9
minimax regret criteria 9
mining, northern Australia 97–101
mitigation
    ecological degradation 294–5
    wetlands development 227–8
MNC *see* multinational corporations
mobile links 36
modelling
    climate change 297
    fire 195–6
    techniques 31–2
monitoring 13
    bioregion changes 61
    bioregional management 47
    climate change 297
    control and surveillance (MCS)
        146
    dimensions 296–8
    data acquisition, basis for 35
    ecological 297
    ecosystem management, building
        capacity 38
    fire management 201–2
    function 296–8
    long-term 16, 38
    long-term time series data
        collection 36
    of policy 297–8
    selected organisms 35–6
Montana, Deerlodge Forest Plan
    51
moose reintroduction 75

motivation, use in bioregional
management 58–9
MSY (maximum sustainable yields)
143
Mt Sorak National Park, Korea 52
multinational corporations (MNC)
268, 273, 300

National Biodiversity Council (NBC)
30
National Coal Board, UK (NCB)
243–4
national parks 52, 53–4, 55–6, 57,
59, 60, 69
native title, northern Australia 88–9
natural disasters 61–2
natural gas 124
'dash to gas' 126
natural resources areas,
constitutional protection 71
natural systems, complexity 12–13
nature, concepts of 69
NBC (National Biodiversity Council)
30
NCB (National Coal Board), UK
243–4
Nepal, Annapurna region 55
networks, policy 276–7
New Zealand pastoral high country,
South Island 167–8
degradation 170, 171, 173
government involvement 173, 175
Hazardous Substances and New
Organisms Act 183
history of settlements 168–76
indeterminacy 182–6
land management 172
land tenure 169–70, 171, 174,
175
land use 171–2
Maori 168–9, 175
pastoral leases 169–70, 171,
174–5
policy issues 177–9, 180–6
precautionary principle 182–4
Resource Management Act 173–6,
183
runholders 169, 174–5

social context 185–6
soil conservation 172
stakeholder participation 185–6
subsidization of agriculture 172–3
sustainability as social concept
180–2
sustainable management 173–6,
180, 182, 187
uncertainty 177–9, 182–6
NGO see non-governmental
organizations
Niger 52, 58
Nigeria, Cross River National Park
59
'no regrets' options 10
non-fossil energy interests 125
non-governmental organizations
(NGO) 122, 247–9, 255, 275
North Pacific Ocean, Convention
for the Conservation of
Anadromous Stocks in the
149
North York Moors National Park,
UK 59
northern Australia
Aboriginal people 88
traditional ownership 98–100
conservation reserve system 103
development 83–4
agriculture case study 92–5
distant management 93, 95
forestry case study 89–92
land management and
conservation planning case
study 102–4
mining case study 97–101
pastoralism case study 95–7
philosophy 86–7
sustainable harvesting case study
101–2
environmental setting 84–6
forests 84
geographic setting 84–6
information, disregard for 104
land ownership and tenure 88–9
mammal fauna 84–6
native title 88–9
pastoral leases 88, 95, 103

research 105–6
savannas 84
wildlife 103–4
Norway, acid rain 239
nuclear power 124–5
nuclear safety 277
null hypotheses 34, 36
NUSAP (numeral-unit-spread-
assessment-pedigree) notation
9
nutrients
assimilation, wetlands 220
fire and 194–5

oceans, fisheries *see* fisheries
off-reserve protection 295
oil spills 18
optimum sustainable yield (OSY)
143
outwelling hypothesis 217–19

Panama 60
pastoral leases
New Zealand pastoral high
country 169–70, 171, 174–5
northern Australia 88, 95, 103
pastoralism
New Zealand pastoral high
country *see* New Zealand
northern Australia 95–7
performance assurance bonds 9, 31
plants
FATE model of vegetation
dynamics 196
fire and 192–4, 195–6, 201
introduced species, northern
Australia 96
vital attributes modelling 196
policy
acid rain, issues 252–6
advice from science on global
warming 132–4
critical loads, issues 253–6
democratized approach 186
ecology and 10–14
environmental 266–9
fire implications 198–202
ignorance and 3–4

implementing 296
in-discipline consensus on key
issues 15
learning 17
linkages within ecology 15
macro-problems 11
meso-problems 11
monitoring 297–8
New Zealand pastoral high
country, issues 177–9, 180–6
processes, active engagement
15–16
research related to 177
risk and 3–4
sustainability agenda 1
sustainability problems 3
uncertainty 3–4
fisheries management, Australia
157–8, 161–2
windows 19
see also public policy
political uncertainty, acid rain 252
politics
acid rain and critical loads, UK
236–57
closures 262–3
processes 262–3
science and, working together
122–4
see also public policy
pollution 18
assimilation, wetlands 220
control, Mediterranean 278
uranium mining, northern
Australia 97
Popperian philosophy 29
population viability analysis (PVA)
10, 15, 31, 196
positivism 265
post-normal science 6, 37
poster species 75
postmodernism 265
power generation 124, 126–7
power relationships, bureaucracies
271
power stations, coal-fired 126,
242
PP *see* precautionary principle

practical uncertainty, fisheries
    management 143–4, 148–50
precaution, ignorance and 7–10
precautionary principle (PP) 2, 31
  adaptive management 18
  climate protection 123
  critical loads 254
  environmental assessments 33–4
  fisheries management 147
  mitigation 295
  New Zealand pastoral high
      country 182–4
  uncertainty, dealing with 8–10
pressure groups 269–70
private land development 72
problems
  definition, ecosystem management
      37–8
  sustainability 3–7, 11
productivity, wetlands 217–19
professional consultancy 6, 37
public lands, local and regional
      involvement in management
      45
public participation 10
public policy
  authoritative knowledge and
      285–6
  communities 272–6
    epistemic 277–9, 282, 284, 285
  distortions 271–2
  making 262
    bureaucracy 270–1
    democracies 282
    terms of reference 272–3
  national
    influencing 301
    restricted ambit 300
  networks 276–7
  persistence
    bureaucratic power relationships
        271
    as source of uncertainty 264
  processes 266
    bureaucracy 270–1
    environmental policy 266–9
    sociology of public policy 269
    weak government 269–70

regimes 272–3
science and 265, 279–82
  the place of science 282–3
  scientists in public debate 283–5
sociology of 269
see also politics
PVA see population viability analysis

quantifiable risk 6
quantitative risk assessment 9, 31,
      36
quantitative risk-based approaches
      15

rabbits, New Zealand pastoral high
      country 172
radiation, uranium mining, northern
      Australia 97
RAINS (Regional Acidification
      Information and Simulation)
      model 250, 253, 255
Ramsar Convention 210–13, 226
rationalism 268, 279, 285, 296,
      299
rationalities 291–2, 298, 302
reclamation, wetlands 216
recreation, New Zealand pastoral
      high country 173
recreational amenities 71
Regional Acidification Information
      and Simulation model see
      RAINS
regions, bioregional management see
      bioregional management
regulatory science 280
reintroduction of species 74–9
relativism 285–6
renewables 124–5
research
  acid rain 239–45, 248–9
  bioregional management 47, 61–3
  climate change 128
  communicating risk, uncertainty
      and ignorance in 15
  earth systems 127–9
  global warming 133
  northern Australia 105–6
  long-term 16

policy, related to 177
sustainability agenda 1
wetlands 228
reserves 295
resilience 27
resource centres 62
Resource Management Act, New
Zealand 173–6, 183
resource managers, fire, implications
for 198–202
resource use optimization 18
Responsible Fishing Code of
Conduct 147, 150–1
restoration, bioregional management
48
rice, northern Australia 92
Rio Declaration on Environment and
Development 8, 150
risk 176
communicating 15
definition 32
quantifiable 6
uncertainty, ignorance and policy
3–4
risk assessment
application 31
difficulties 32–3
ecological (ERA) 10, 15
environmental 10
fire management options 196
lacking 94
quantitative 9, 31, 36
World Trade Organization
requirements 292
see also environmental
assessments; environmental
impact assessments
runholders, New Zealand pastoral
high country 169, 174–5

Saba, Bonaire Marine Park 53
safe minimum standards 9, 31
saltmarshes 217–19, 221, 227–8
satellite tracking 150
savannas, northern Australia 84
scale, space and time 298–9
Scandinavia, acid rain 239–41,
244

science
acid rain, policy interactions and
236
advice from IPCC, earth systems
research underpinning 127–9
authority in public debate 280
ecological modernization and
237–8
global warming, policy advice on
132–4
ignorance in 293–4
as legitimator of politics and
trade, global warming 116–35
politics and, working together
122–4
post-normal 6, 37
and public policy see public policy
regulatory 280
social construction 265, 280
standards 281
uncertainty in 293–4
scientific data, governments
controlling and interpreting
272
scientific information
on climate change 131
inadequacy of existing mechanisms
for using 13
scientific uncertainty, 176
acid rain 252
bioregional management 44–5
fisheries management 142–3,
147–8
Australia 156–7, 160–1
greenhouse effect 133
sources 178
scientists in public debate 283–5
Scotland, acidity 239
seas, fisheries see fisheries
security 300
seed varieties, patenting 268
sewage disposal 272
SFS see straddling fish stocks
Sierra Nevada de Santa Marta,
Colombia 60
simulation techniques 31–2
social acceptance, bioregional
management 47

social considerations, sustainability 2
social construction of science 265, 280
social context, New Zealand pastoral high country 185–6
social learning 186
social science 181
social uncertainty 44–5, 46–8
Society for Conservation Biologists 30
socio-economic issues 131
sociology of public policy 269
soil conservation, New Zealand pastoral high country 172
sorghum, northern Australia 93
South Carolina, ACE basin project 55
Southern Connections 299
space and time 298–9
species loss 26–7, 30
species reintroduction 74–9
stakeholders
  bioregional management see bioregional management
  New Zealand pastoral high country 185–6
START (System for Analysis, Research and Training) 128
straddling fish stocks (SFS) 140, 144
  UN Fish Stocks Agreement 147, 149, 150
subjectivity 294
subsidization of agriculture, New Zealand pastoral high country 172–3
Surface Waters Acidification Programme, UK (SWAP) 243–5
surprise 17–19
sustainability
  acid rain 252–6
  biophysical problems 3
  clean energy 125
  critical loads significance 254
  definition 68
  different meanings 66–7
  ecological 8, 14, 69, 299–300

ecological considerations 2
economic 47, 69
economic considerations 2
ecosystem management for 291–302
fisheries management 141
  Australia 155–6, 160–2
institutional arrangements 37
macro-problems 6, 11
meso-problems 4–6, 11
micro-problems 4, 6
as overriding goal 267
policy agenda 1
policy problems 3
problems 3–7, 11
research agenda 1
as social concept 180–2
social considerations 2
uncertainty 2–3
sustainable growth 123
sustainable harvesting 101–2
sustainable management
  fisheries 140, 141
  New Zealand pastoral high country 173–6, 180, 182, 187
SWAP (Surface Waters Acidification Programme), UK 243–5
Sweden, acid rain 239
system dynamics, time series data on 31–2
System for Training, Analysis, Research and Training (START) 128

TAC (total allowable catches) 143
Task Force on Integrated Assessment Modelling (TFIAM) 249
technical uncertainty 6
technology
  sharing, bioregional management 48
  transfer 123
temperature exposure, fires 192
tenure see land tenure
TFIAM (Task Force on Integrated Assessment Modelling) 249
Thailand 54

timber, salvage or preservation after natural disasters 73–4
time series data
  long-term collection, monitoring for 36
  on system dynamics 31–2
time spans 298–9
total allowable catches (TAC) 143
tourism 54, 173
trading arrangements, greenhouse gases 120–1
transboundary air pollution 241
transboundary fish stocks 140, 144
transnational companies see multinational corporations
tussock grass 169, 178–9

UK see United Kingdom
UN Agreement on Straddling Fish Stocks and Highly Migratory Fish Stocks see UN Fish Stocks Agreement
UN Conference on Environment and Development (UNCED) 1, 150
UN Conference on the Human Environment 239
UN Convention on the Law of the Sea (UNCLOS) 143–5, 149
UN Economic Commission for Europe (UN ECE) 238
  Convention on Long-Range Transboundary Pollution see Long-Range Transboundary Pollution Convention
  critical loads 246–50, 253, 256
  Task Force on Integrated Assessment Modelling (TFIAM) 249
  Working Group on Effects (WGE) 249
UN Environment Programme (UNEP) 128
UN Fish Stocks Agreement 147, 149, 150
UN Framework Convention on Climate Change (FCCC) 8, 117–22

UNCED (UN Conference on Environment and Development) 1, 150
uncertainty
  acid rain, issues 252–6
  bioregional management and 44–8
  characterization of 176–7
  climate change 122
  communicating 15
  conservation planning, northern Australia 102–3
  economic see economic uncertainty
  epistemological 6
  fisheries, international management see fisheries
  high 37
  ignorance and 3–4, 176
  legal see legal uncertainty
  in management decisions 44–5
  methodological 6
  mining, northern Australia 97–101
  New Zealand pastoral high country 177–9, 182–6
  northern Australia 83–107
  policy and 3–4
  practical, fisheries management 143–4, 148–50
  precautionary principle 8–10
  risk and 3–4, 176
  scientific see scientific uncertainty
  social 44–5, 46–8
  social control, tool of 293–4
  sources 178, 264
  sustainability 2–3
  taxonomy of 176
  technical 6
UNCLOS see UN Convention on the Law of the Sea
UNEP (United Nations Environment Programme) 128
United Kingdom
  acid rain and critical loads 236–57
  North York Moors National Park 59
United Nations see entries beginning with UN

United States
   Adirondack Park 66–80
   bioregional management 51, 55, 58
   centralized planning 45
   Endangered Species Act 58
   energy policies 127
   federal agency funding 55
   Fish and Wildlife Service (USFWS)
      210–11, 214, 215
   national parks 52, 53, 57
   wetlands 210–11, 228
uranium mining, northern Australia
   97–8
USA see United States
USFWS see United States: Fish and
   Wildlife Service

Valdivia Group 299
value, wetlands see wetlands
vessel monitoring systems (VMS)
   150
vested interests 269–70, 275
vital attributes modelling, plants 196
VMS (vessel monitoring systems)
   150

WCED (World Commission on
   Environment and
   Development) 1
WCRP (World Climate Research
   Programme) 128
weak governments 269–70, 273
wetlands 209
   boundaries 213–15, 230
   coastal 224–5
   conflicting agendas 225–6
   conservation 225–6, 229
   definitions 209–13, 215, 230
   degradation 224–5, 228
   development affecting 220–2,
      227–8
   functions, assessing 220–4

intermittent 211
intertidal 219–20
losses 224–5, 228
management 229, 230
mining impacts, northern Australia
   98
mitigation 227–8
nutrient and pollution assimilation
   220
productivity 217–19
protection 229, 230
Ramsar Convention 210–13, 226
regional inventories 226
research 228
urgency of action 224–5
value 215–20, 221–4, 227, 228–9
WGE (Working Group on Effects)
   249
wildfires 18, 197
wildlife, northern Australia 103–4
wise use doctrine 275
WMO (World Meteorological
   Organisation) 128
wolf reintroduction 75–9
Working Group on Effects (WGE)
   249
World Bank 119–21, 122
World Climate Research Programme
   (WCRP) 128
World Commission on Environment
   and Development (WCED) 1
World Meteorological Organisation
   (WMO) 128
World Trade Organization (WTO)
   292, 300

Yellowstone National Park 53, 57
Yosemite National Park 53

zero emission power generation 126
Zimbabwe, Matobo National Park
   59